Gross Anatomy for Physician Assistants

A Clinical Application Approach
Third Edition

Shireen Rahman, Ed.D., ATC

Please note, it is assumed and expected that the student enter into this course with a thorough base of knowledge of Anatomy and Physiology. The student is responsible for understanding adjacent anatomical structures that may impact function of the clinically relevant topics.

This Lab Manual will provide the student with information that will help connect all structures however it is the student's responsibility to use this to their advantage.

Kendall Hunt
publishing company

www.kendallhunt.com
Send all inquiries to:
4050 Westmark Drive
Dubuque, IA 52004-1840

ISBN 978-1-7924-6280-1

Published in the United States of America

Contents

Getting Started

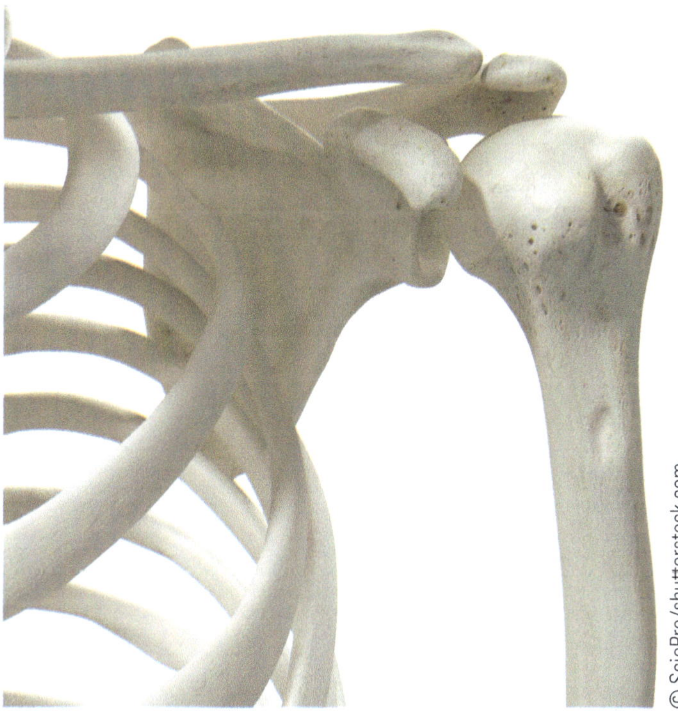

Anatomical Directional Cues
Skeletal Landmarks
Joint Types

Anatomical Directional Cues

Terminology Glossary

Cue	Definition	Example
Anterior/Ventral	Towards the front	The quadriceps are anterior to the hamstrings Anterior tibialis refers to a muscle that is on the anterior surface of the tibia
Posterior/Dorsal	Towards the back	The hamstrings are posterior to the quadriceps Posterior tibialis refers to a muscle that is on the posterior aspect of the tibia
Inferior/Infra	Below (not under)	The knee is inferior to the hip. Infraglenoid tubercle refers to a bump below the glenoid fossa

Cue	Definition	Example
Superior/Supra	Above	The hip is superior to the knee Supraspinatus refers to a muscle that sits above the spine of the scapula
Medial	Towards midline of body	The ulna is more medial than the radius
Lateral	Away from midline of the body	The radius is more lateral than the ulna
Greater	Bigger relative to something smaller	Greater tuberosity is a bigger bump than the lesser tuberosity
Lesser	Smaller relative to something bigger	Lesser tuberosity is smaller than the greater tuberosity
Proximal	Closer to the center of the body	The proximal radioulnar joint is closer to the elbow
Distal	Farther from the center of the body	The distal radioulnar joint is farther away from the elbow (near the wrist)
Deep/Profundus	Under other layers	The vastus intermedius muscle is deep to the rectus femoris; The flexor digitorum profundus is deep to the other flexor muscles
Superficial	On the surface	The rectus femoris is superficial to the vastus intermedius The flexor digitorum superficialis is found above the profundus
Cranial	Towards the head	
Caudal	Towards the tail (coccyx)	

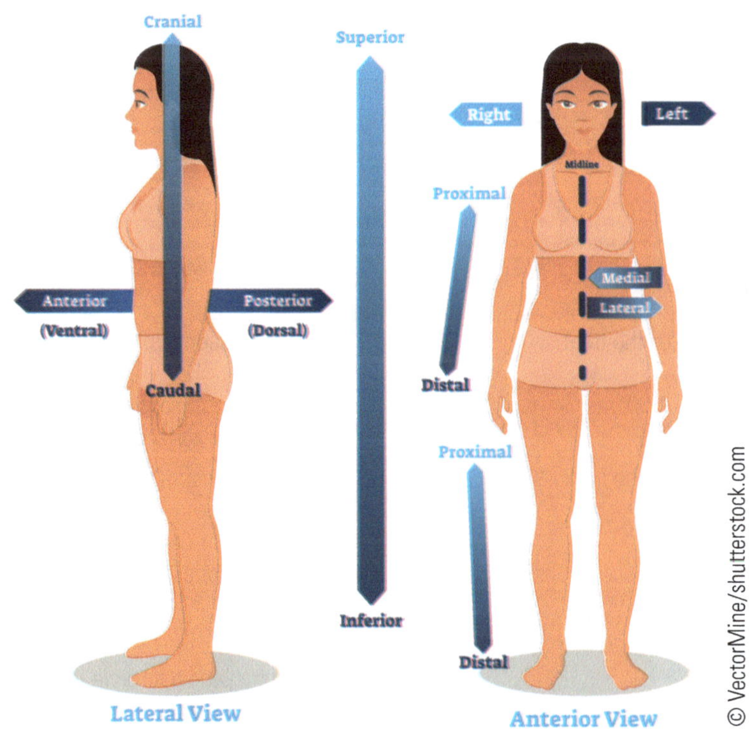

Anatomical Directional Cues

Plane and Axis

Understanding the planes will help you when referring to views on a radiograph/scan etc; also a movement moves ALONG a plane.

Understanding the axes will help you when referring to joint movements and how many joint movements are found within a joint; a movement moves AROUND an axis.

An axis is perpendicular to the plane

Plane	Description	Axis	Resulting Movements
Transverse/Horizantal or Axial (view)	Divides body into superior and inferior	Vertical	Rotation
Frontal/Coronal	Divides the body into anterior and posterior	Sagittal	Abduction/Adduction
Sagittal	Divides the body into left and right	Frontal *horizontal* axis	Flexion/Extension

From left to right: Axial view, Coronal view, Sagittal view of the knee

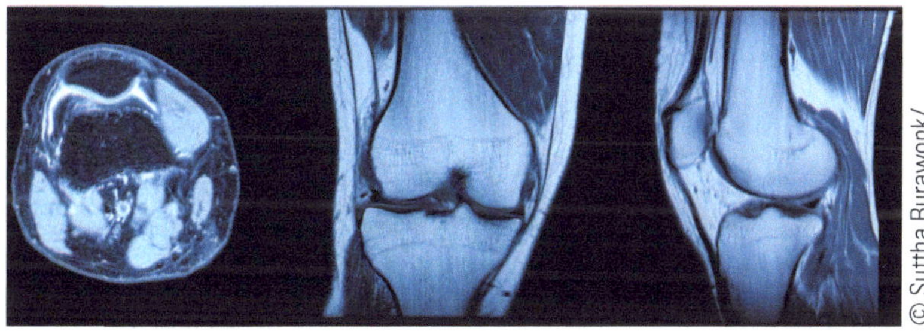

© Suttha Burawonk/ shutterstock.com

From left to right: Axial view, Coronal view, Sagittal view of the brain

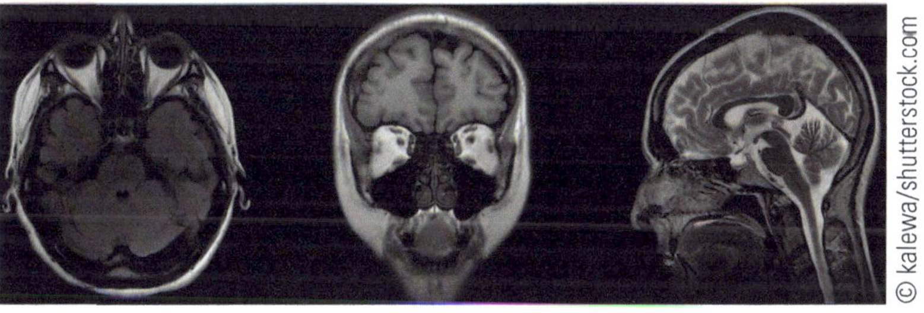

© kalewa/shutterstock.com

Skeletal Landmarks

Border	Edge; typically defines the outline of a bone	**Scapula** Medial, Lateral, Superior borders
Angle	Angled border	**Scapula** Inferior, superior
Notch	Typically u shaped; nerve or artery passageway	**Scapula** Scapular notch **Femur** Intercondylar Notch
Process	A protrusion that usually has a muscle attaching to it; typically jagged or sharp	**Scapula** Coracoid Process, Acromion Process
Tubercle	a small bump used for muscle or ligament attachment	**Scapula** Infraglenoid Tubercle

Tuberosity	larger bump used for muscle or ligament attachment	**Humerus** Greater and Lesser Tuberosity
Fossa	A shallow surface	**Scapula** Subscapular Fossa
Groove	Hollowed out surface usually holding a tendon or nerve	**Humerus** Intertubercular Groove
Condyle	Rounded articulating surface; usually comes in pairs	**Femur** Medial/Lateral Femoral Condyle
Crest	Ridge line of a bone	**Pelvis** Illiac Crest
Epicondyle	To the side of larger condyles; smaller rounded surfaces for muscle attachment	**Humerus/Femur** Medial Epicondyle
Facet	Flat, smooth surface; articulation area between bones	**Vertebrae** Superior Articular Facet
Foramen	Hole or passageway through bone	**Vertebrae** Vertebral Foramen
Line	A raised line on a bone usually serving as muscle attachment	**Tibia** Soleal Line

Joint Types and Movements

1. **<u>Fibrous Joint Type</u>**
   a. Bones held together by dense fibrous connective tissue
   b. Generally <u>immoveable</u> and fixed
   c. Three types include **Sutures, Gomphosis, Syndesmosis**

<u>Suture Fibrous Joint:</u> Found in the skull connecting a flat bone to a flat bone to create an immoveable joint; the **coronal suture** is found tying the frontal and parietal bones of the skull together.
<u>Gomphosis Fibrous Joint:</u> Found where **teeth fit into the mandible and maxilla.**
<u>Syndesmosis Fibrous Joint:</u> Found between long bones, held together by fibrous tissue/ligaments; limited mobility; the **distal tibiofibular joint.**

2. **<u>Cartilaginous Joint Type</u>**
   a. Bones held together by fibrocartilage and/or hyaline cartilage
   b. Added but limited mobility when compared to the fibrous joints
   c. Two types include **Synchondrosis and Symphysis**

<u>Synchondrosis Cartilaginous Joint:</u> Lined by hyaline cartilage and typically found in developing skeletons; **epiphyseal growth plates.**
<u>Symphysis Cartilaginous Joint:</u> Comprised of fibrocartilage between bones; **pubic symphysis, intervertebral discs.**

3. **<u>Synovial Joint Type</u>**
   a. Bones held together via an articular capsule- articular cartilage- and synovial cavity
   b. Typically reinforced by ligaments
   c. Mobility is based on its structure (examples on next page)

Did you know that Kelly Clarkson won the first American Idol?

Synovial Joint Types

Synovial Joint Type	Description	Number of Axes	Typical Movements	Examples	Analogy
Plane Joint (Gliding)	Flat with Flat	Non-axial	Gliding	Intercarpal Jt Articulating facets of vertebraes	Sliding book on a table Shuffling puzzle pieces
Hinge Joint	Convex with a concave *Collateral ligament restrict motions*	Uni-axial	Flexion/Extension	Interphalangeal Jt	Door hinge
Pivot Joint	A bony cylinder fitting into a ring (bone or ligament ring)	Uni-axial	Internal Rotation External Rotation	Atlantoaxial Jt Radioulnar Jts	Screw into nut Turning door knob
Condyloid Joint	Oval convex to opposite concave cup	Bi-axial	Flexion/Extension Abduction/Adduction	Radiocarpal Jt	
Saddle Joint	A "u" with a "n" surface Modified ball and socket	Bi-axial *May be considered a multi-axial*	Flexion/Extension Abduction/Adduction Minimal Rotation	1st Carpometacarpal Jt	Horse rider on saddle
Ball and Socket	Round convex with opposite concave cup	Multi-axial	Flexion/Extension Abduction/Adduction Internal/External Rotation	Glenohumeral Jt Femoral-Acetabular Jt	

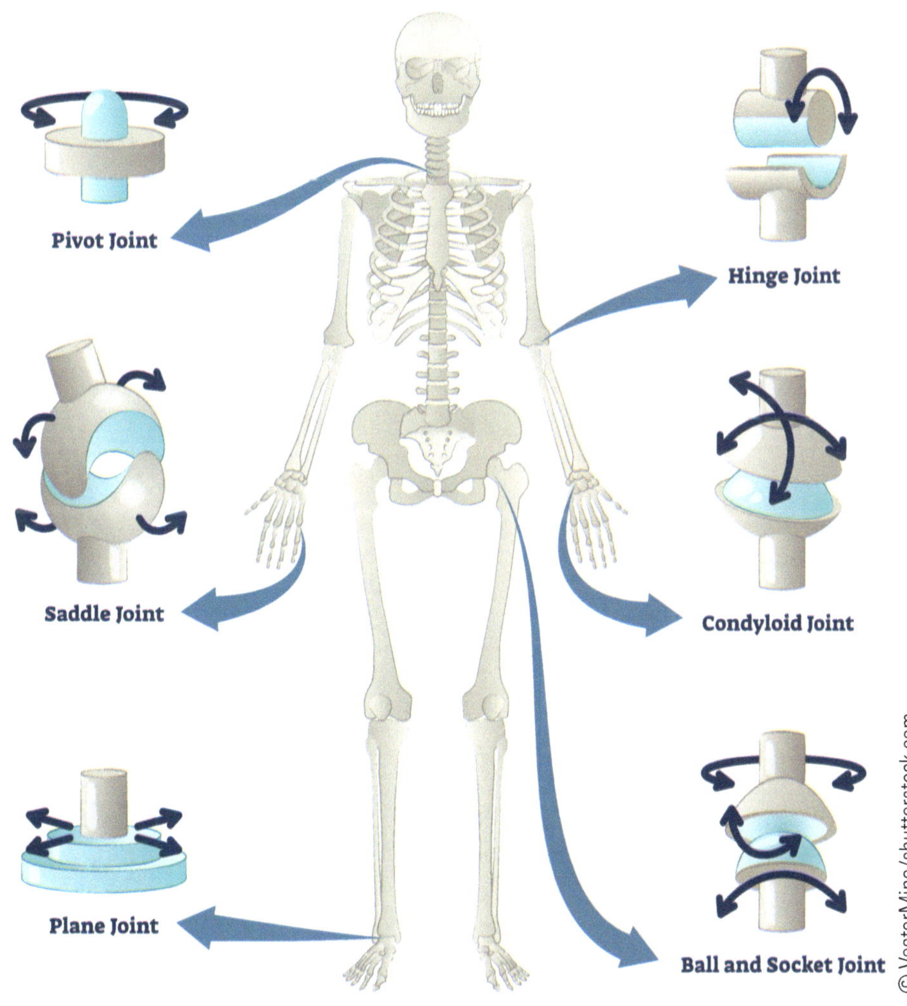

Upper Extremity

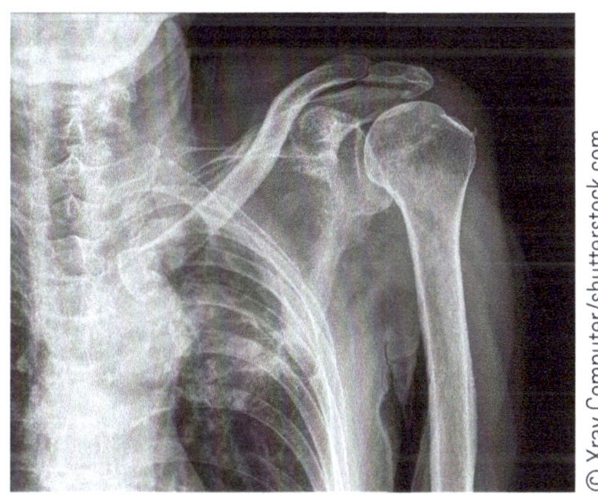

CHAPTER 2

© Xray Computer/shutterstock.com

Bony Landmarks
Joints and Ligaments
Muscles
Blood Supply
Innervations

Bony Landmarks of Upper Extremity

All bony landmarks are listed in a logistical assessment order- <u>*trace/outline, label and/or color accordingly*</u>. *Knowing the bony landmarks will help you soon understand muscle location, muscle compartments, and muscle function.*

Posterior Scapula

*Moving from the **acromion process (a)**, move medially along the **spine of the scapula (b)**. As the spine ends, you will find the **medial border(c)**. Moving superiorly, enter into the **superior angle(d)**, across the **superior border(e)**. Return to the spine and now move inferiorly along the medial border until it turns via the **inferior angle(f)**. As you leave the angle, you reach the **lateral border(g)**. Follow the lateral border superiorly until you reach a small bump, the **infraglenoid tubercle (h)**. Moving above the tubercle, find the **glenoid fossa(i)** and just superior to the fossa is the **supraglenoid tubercle (not shown)**. Return to the spine. The surface above the spine is the **surpaspinous fossa (j)** and the surface below the spine is the **infraspinous fossa (k)**.*

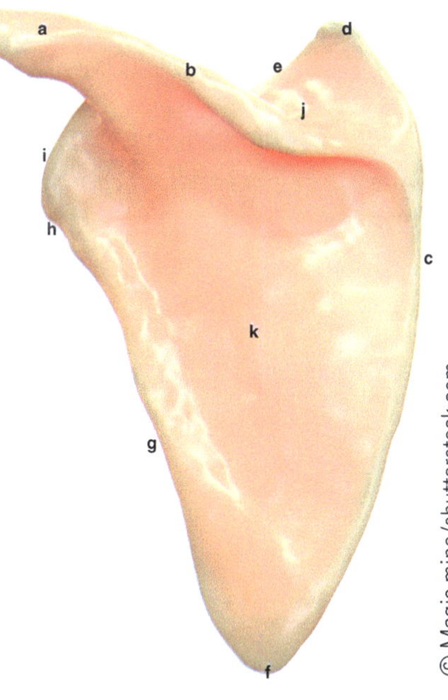

© Magic mine/shutterstock.com

Muscle insertion points

Acromion process: *Trapezius*
Spine of scapula: *Trapezius*
Medial border: *Rhomboids, Serratus Anterior*
Superior angle: *Levator Scapulae*

Anterior Scapula

*Start at the **superior angle** and move laterally across the **superior border.** You will soon dip into the **scapular notch** (houses suprascapular nerve). Upon leaving the notch, you will find the outward facing **coracoid process.** The large surface between the medial and lateral borders is the **subscapular fossa.***

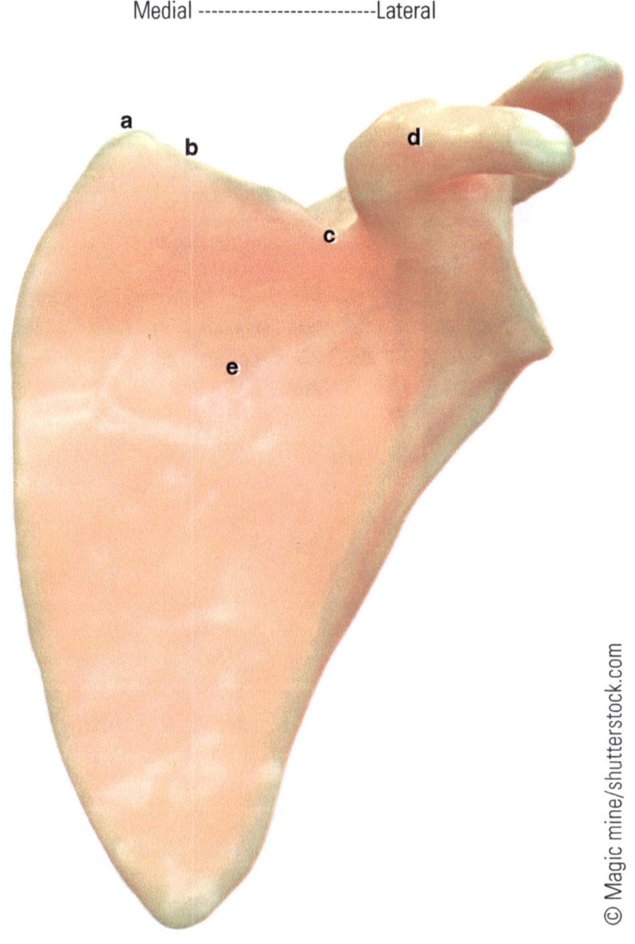

<u>*Muscle insertion points:*</u>
Superior angle: *Levator Scapulae*
Coracoid process: *Pectoralis Minor*

Lateral Scapula

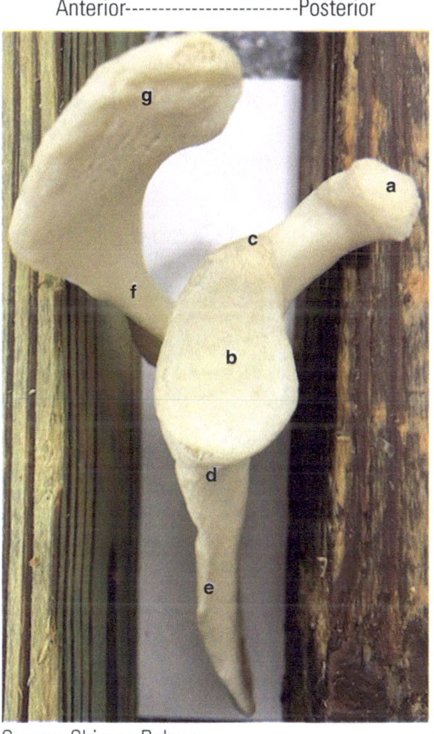

Source: Shireen Rahman

Start at the most anterior projection, the **coracoid process (a).** *Move toward the* **glenoid fossa (b)** *, just above is the* **supraglenoid tubercle (c)** *and below is the* **infraglenoid tubercle (d).** *Moving inferiorly you will be on the* **lateral border (e).** *Posteriorly you will find the* **spine of scapula (f)** *and* **acromion process (g).**

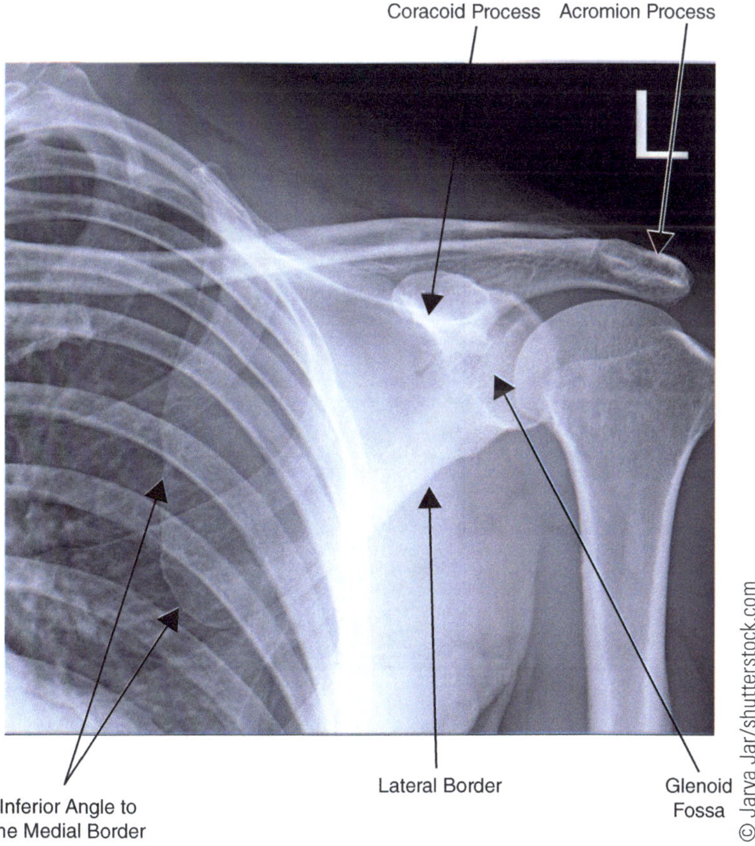

Clavicle

S-shaped bone located between the sternum and the acromion process of the scapula comprising the shoulder girdle.

Right Clavicle
(Collarbone)

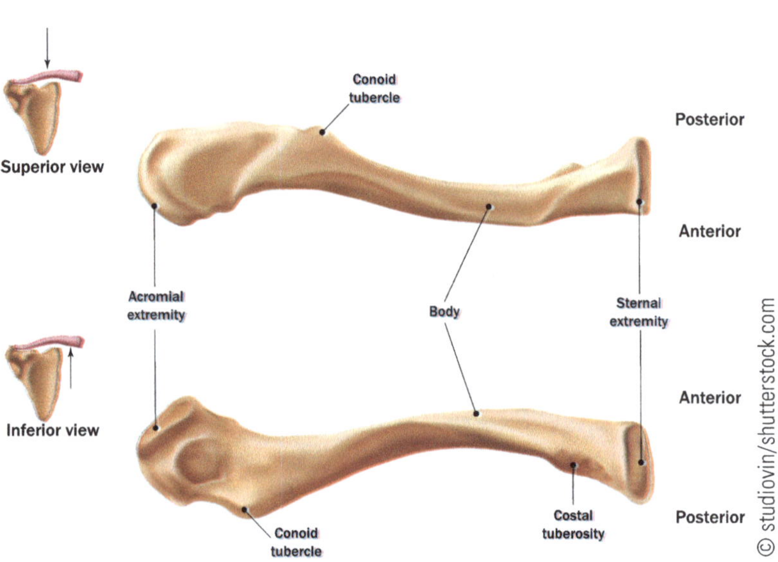

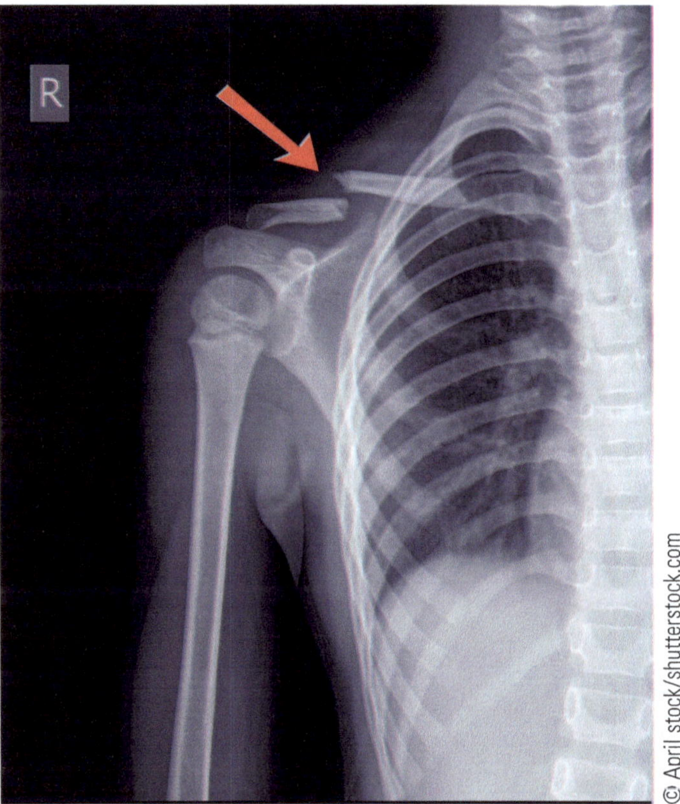

The majority of clavicular fractures are a result of a fall directly on the lateral shoulder. Direct trauma and falling on an outstretched hand are also causes. The most common fracture site is at the middle shaft of the clavicle. Important structures that must be considered are the subclavian artery and the brachial plexus.

Anterior View of the Proximal Humerus

*Start with the **head of the humerus (a)**, which articulates with the glenoid fossa of the scapula. The head empties into the **anatomical neck (b)**. Two bumps will appear- the **lesser tuberosity(c)** is the more anterior (and smaller), and the **greater tuberosity(d)** is more lateral (and larger). In between these tuberosities, you find the **intertubecular groove (e)**, which will house the tendon of the biceps brachii.*

*Follow the groove distally, and it empties onto the **surgical neck (f)**, named due to the high prevalence of fractures that occur here resulting in the need for surgery.*

*Distal to the surgical neck, the **deltoid tuberosity(g)**, which serves as an insertion point for the deltoid muscle, will be found on the lateral aspect of the **shaft of the humerus (h)**.*

<u>*Muscle insertion points:*</u>
Lesser Tuberosity:
Subscapularis, Latissimus Dorsi, Teres Major
Greater Tuberosity:
Supraspinatus, Infraspinatus, Teres Minor, (Pectoralis Major)
Deltoid Tuberosity: *Deltoid*

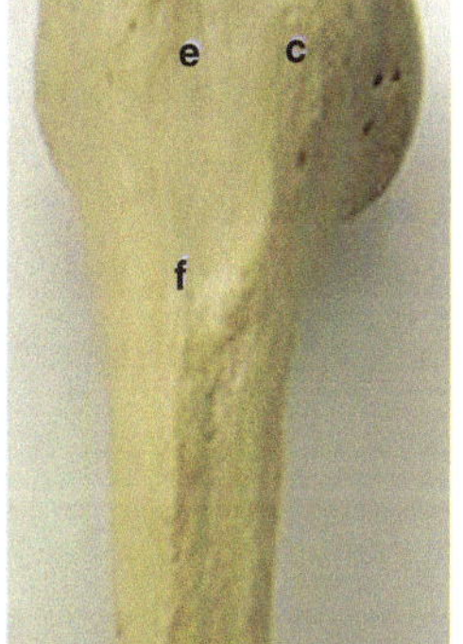

Source: Shireen Rahman

Anterior Distal Humerus

*Following the shaft (a) distally, the **lateral (b)and medial (c) supra-condylar ridges** will empty into the **lateral (d) and medial (e) epicon-dyles** (bumps to the side).*

Now focus on the landmarks between the epicondyles.

*The **capitulum (f)** is the round bump on the lateral side of the humerus (put a "cap" on the head of the radius); it will line up with the radius. On the medial side, the **trochlea (g)** will look like a spool of thread (note the 2 sides of "g").*

*Above the capitulum is the **radial fossa(h)**, where the radial head will move upon flexion of the elbow. Above the trochlea is the **coronoid fossa (i)**, where the coronoid process of the ulna will move upon elbow flexion. Also of note, the anterior start of the **ulnar groove (j)**, where the ulnar nerve will move from posterior to neutral forearm.*

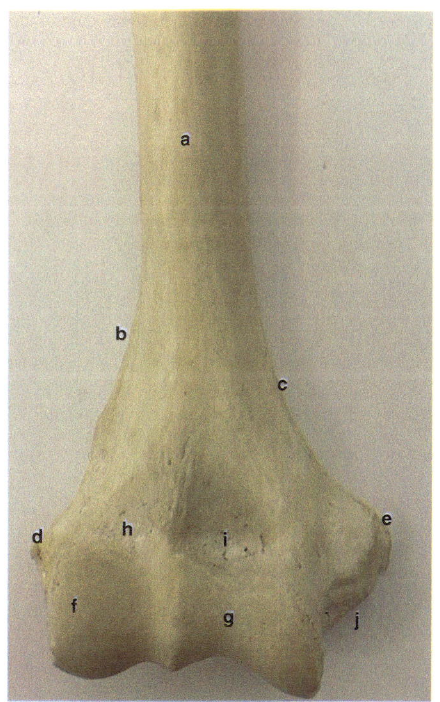

Source: Shireen Rahman

Posterior Distal Humerus

*Again, between the two epicondyles you will find the **olecranon fossa (a),** where the olecranon process of ulna will fit. Please note the **trochlea (b)** from the posterior view.*

Lateral ---------------------------Medial

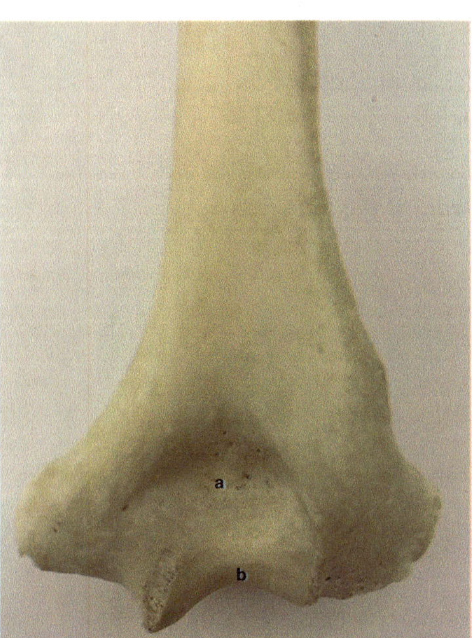

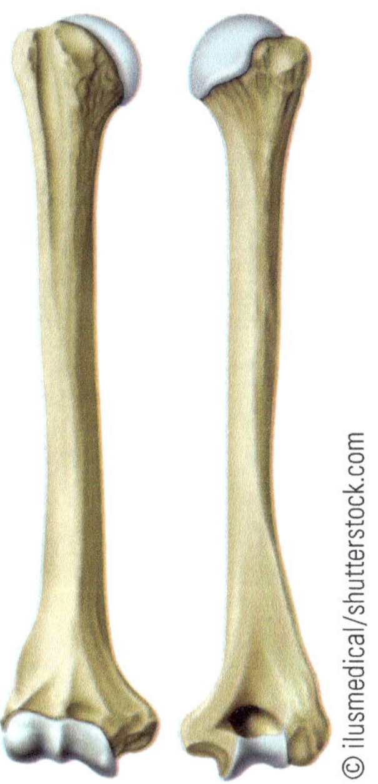

Anterior Proximal Ulna

Lateral(Radial) View

*Starting at the "u-shaped" structure, the **trochlea notch (a).** This notch falls between the **olecranon process (b)** on the posterior aspect and the **coronoid process (c)** forms the anterior aspect (the ski jump). Just inferior to the trochlear notch, you will find the **radial notch of the ulna(d),** for the radial head. Just inferior to the coronoid process is the **ulnar tuberosity (e).***

Posterior--------------Anterior

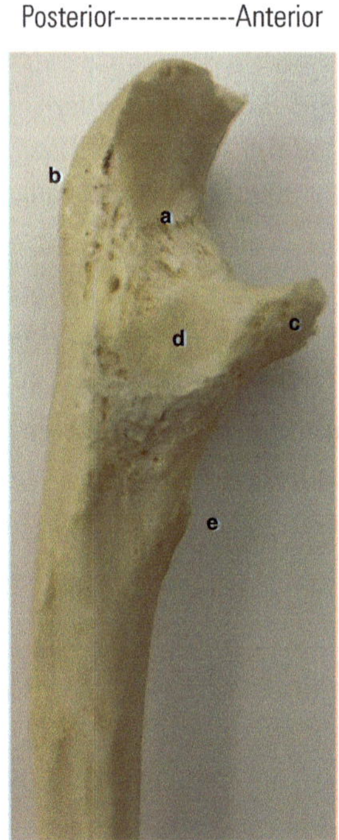

Source: Shireen Rahman

Distal Ulna

*Distally, we see the **head of the ulna (a)** and the **styloid process of the ulna (b)**, found on the medial aspect.*

Muscle insertion points:
Ulnar tuberosity: *Brachialis*
Olecronon Process: *Triceps Brachii*
Coronoid Process: *Pronator Teres*

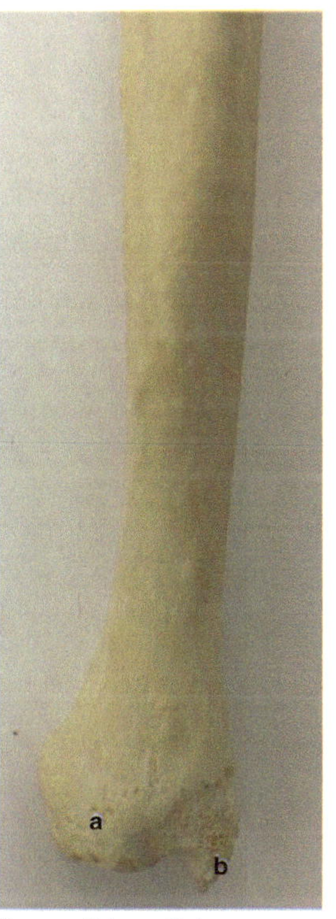

Source: Shireen Rahman

Proximal Radius

*Start with the **Head of the Radius (a)**, which articulates with the capitulum of the humerus and the radial notch of the radius. Just inferior and medial is the **radial tuberosity (b)**.*

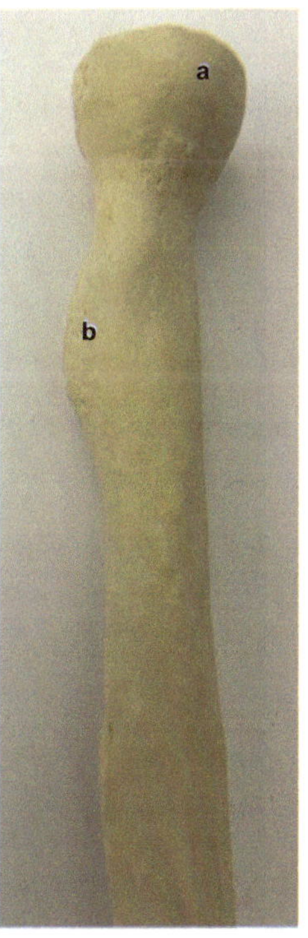

Source: Shireen Rahman

Distal Radius

Find the **styloid process of the radius (a)** and move medially to fund the **ulnar notch of the radius (b)**, which articulates with the ulna. On the most inferior surface of the radius you will find the **articulating surface for the carpals (c).**

<u>Muscle insertion points:</u>
Radial Tuberosity: *Biceps Brachii*
Radial Styloid Process:
Brachioradialis

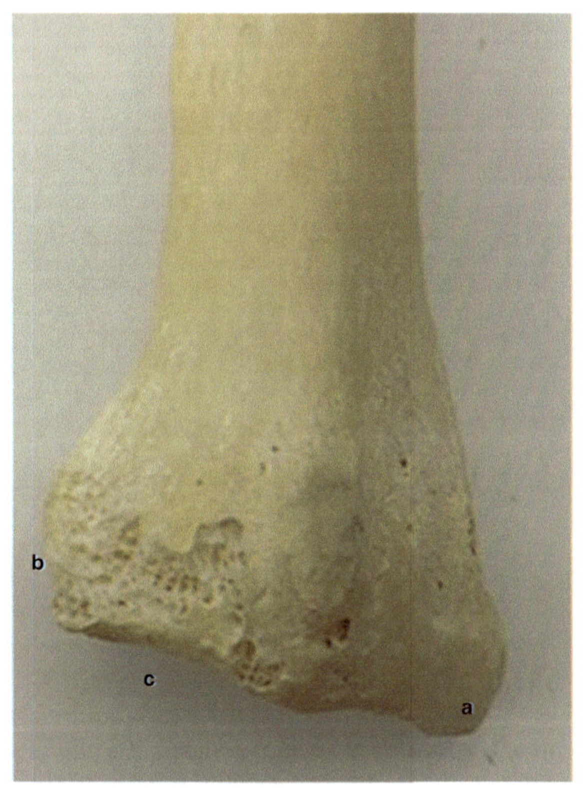

Source: Shireen Rahman

The Hand

The carpal bones will articulate with the radius. They will be arranged in a <u>proximal row</u> and a <u>distal row,</u> starting at the thumb side always.

<u>Proximal Row:</u> Sally lowers Tim's pants

Scaphoid (a) – Lunate (b) – Triquetrium (c) – Pisiform (d)

<u>Distal Row:</u> Tiny Tim cries home

Trapezium (e), Trapezoid (f) , (because I become before O), Capitate (g), Hamate (h)

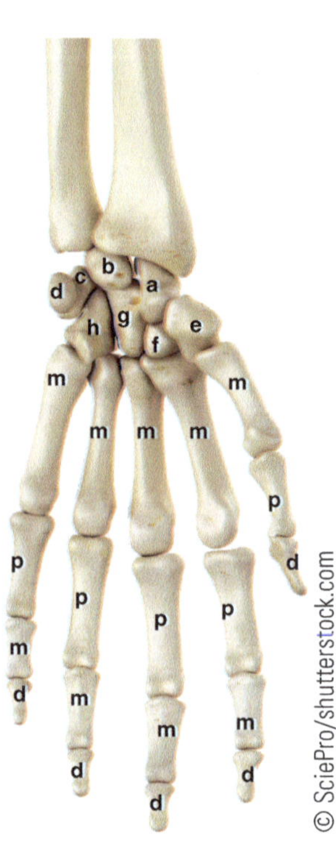

The **Metacarpals (m)** will be just distal to the distal row of carpals, followed by the **proximal (p), middle (m), and distal (d) phalanx.** The 1st will only have a proximal and distal phalanges, not a middle.

<u>Muscle insertions:</u>
The wrist flexors will insert on several on the palmar surface.
The wrist extensors will insert on several on the dorsal surface.

Joints and Ligaments of Upper Extremity

*Joints will provided information about the movement. Ligaments will **restrict** a movement as it lengthens*

<u>JOINTS</u>

Use the bony landmarks to help you with joints.

Joint	Bone to	Bone	Joint Type	Movement Type	Function
Glenohumeral	Glenoid Fossa of Scapula	Head of Humerus by ligaments	Ball and Socket	Multi-axial	F/E Abd/Add L/M Rotation Circumduction
Sternoclavicular	Manubrium of Sternum	Sternal end of Clavicle by ligaments	Saddle	Biaxial in structure but multi-axial in function	Allows movement of the clavicle to enable movement of other joints
Acromioclavicular	Acromion Process of the Scapula	Acromial end of Clavicle by ligaments	Gliding	Non-axial	Gliding
Scapulothoracic	Subscapularis Fossa of Scapula	Thorax by connective tissue	Gliding	Non-axial	Gliding
Humeroradial	Capitulum of Humerus	Head of Radius by ligaments	Hinge	Uniaxial	F/E
Humeroulnar	Trochlea of Humerus	Trochlea Notch of Ulna by ligaments	Hinge	Uniaxial	F/E
Proximal Radioulnar	Head of the Radius	Radial Notch of the Ulna by ligaments	Pivot	Uniaxial	Pronation/ Supination
Distal Radioulnar	Head of Ulna	Ulnar Notch of Radius by ligaments	Pivot	Uniaxial	Pronation/ Supination
Radiocarpal	Articulating surface of Radius	Proximal row of Carpals by ligaments	Condyloid	Biaxial	F/F Abd/Add
Midcarpal	Proximal row of Carpals	Distal row of Carpals	Gliding	Non-axial	Gliding
Intercarpal	Carpal	Carpal	Gliding	Non-axial	Gliding
Carpometacarpal	Distal row of Carpals	Base of Metacarpals 2-5	Gliding	Non-axial	Gliding
1st Carpometacarpal	Trapezium	1st Metacarpal	Saddle	Multi-axial	F/E Abd/Add Circumduction
Metacarpophalangeal	Metacarpals	Proximal Phalange	Condyloid	Biaxial	F/E Abd/Add
Proximal Interphalangeal	Proximal Phalange	Middle Phalange	Hinge	Uniaxial	F/E
Distal Interphalangeal	Middle Phalange	Distal Phalange	Hinge	Uniaxial	F/E

Take a breather—take in a walk—a moment for yourself

The Glenohumeral Joint

The head of the humerus fits into the shallow glenoid fossa of the scapula forming a ball and socket, multi-axial joint.

The GH joint is the most mobile of joint which leaves it susceptible to injury.

Ligaments, rotator cuff muscles, and the labrum will all contribute to the integrity of the joint.

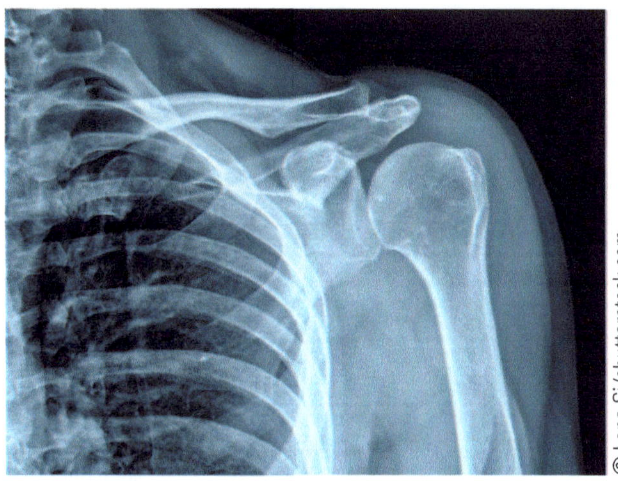

Integrity of the Glenohumeral Joint:

The **Glenoid Labrum** is a fibrocartilage rim around the glenoid fossa will deepen the socket. It is continuous with the Biceps Brachii tendon, superiorly.

The joint is also surrounded by a **joint capsule,** extending from the anatomic neck of humerus to the glenoid fossa.

Several ligaments and rotator cuff muscles reinforce the joint, while several bursa lend support to tendons at the joint.

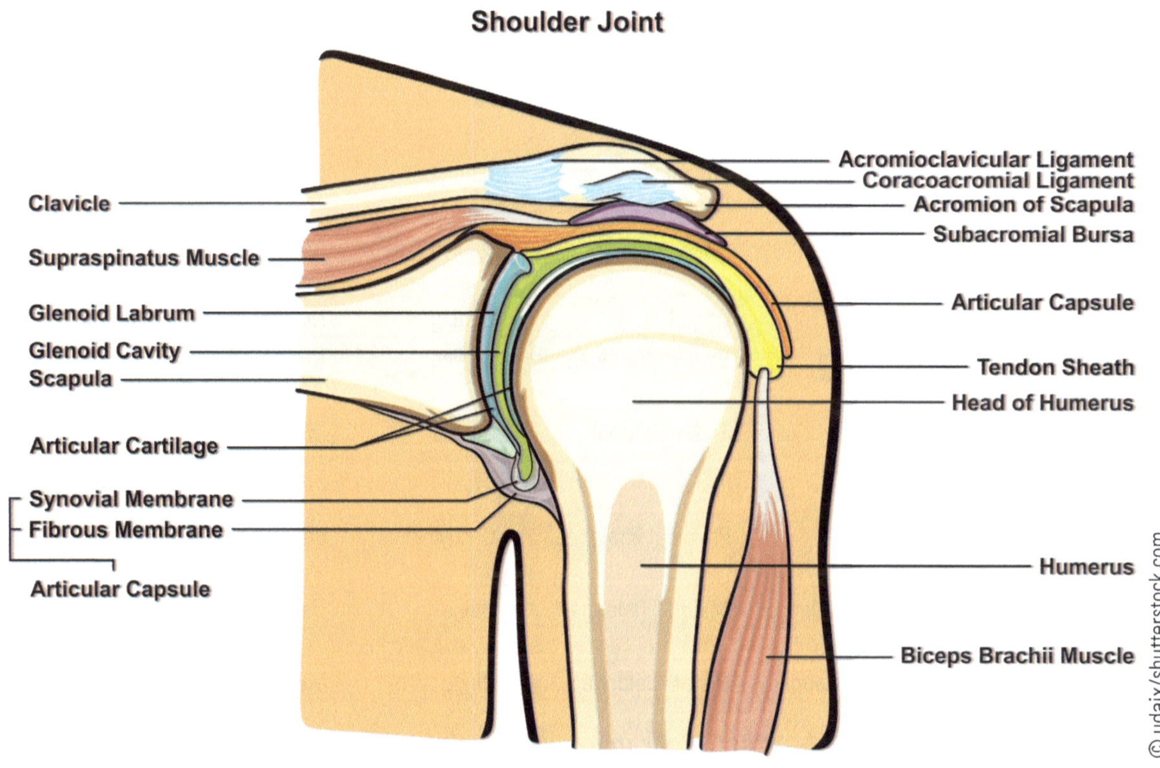

Glenoid Laburm:

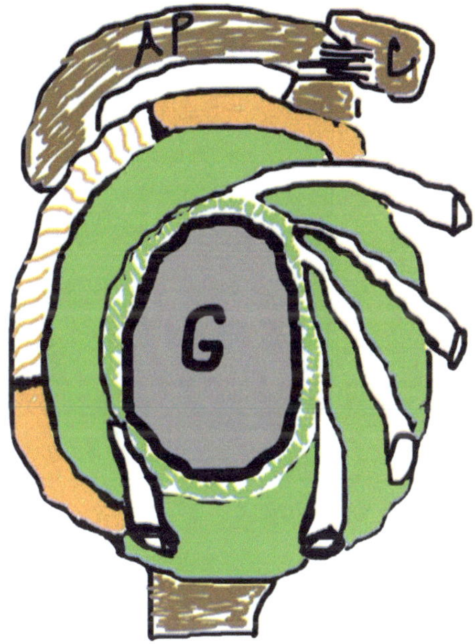

SLAP Tear: **Superior Labrum** *from* **Anterior to Posterior**

Bankart LesionL **Anterior-Inferior Labrum**

Glenohumeral Joint Ligaments

Draw and color accordingly

Coracohumeral Ligament:
From the base of Coracoid Process of the Scapula
Greater Tuberosity of the Humerus
Function(s)

Coracoacomial Ligament:
From the posterior Coracoid Process of Scapula
Acromion Process of Scapular
Function(s):

Transverse Humeral Ligament:
Moves across the Intertubecular Groove of Humerus
Function(s):

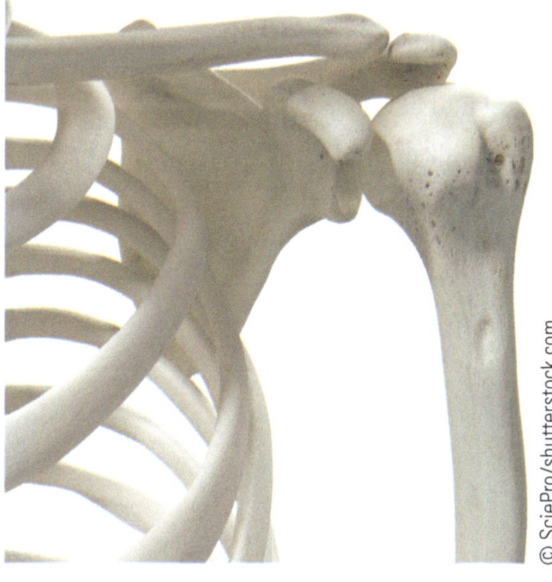

Glenohumeral Joint Ligaments

Superior Glenohumeral Ligament:
From superior/anterior Glenoid Fossa of Scapula →
Lesser Tuberosity of the Humerus
Function(s):

Middle Glenohumeral Ligament:
From anterior –mid Glenoid Fossa of Scapula →
Mid-anterior Lesser Tuberosity of Humerus
Function(s):

Inferior Glenohumeral Ligament:
From inferior Glenoid Fossa of Scapula →
Inferior Lesser Tuberosity/Anatomical Neck of Humerus
Function(s):

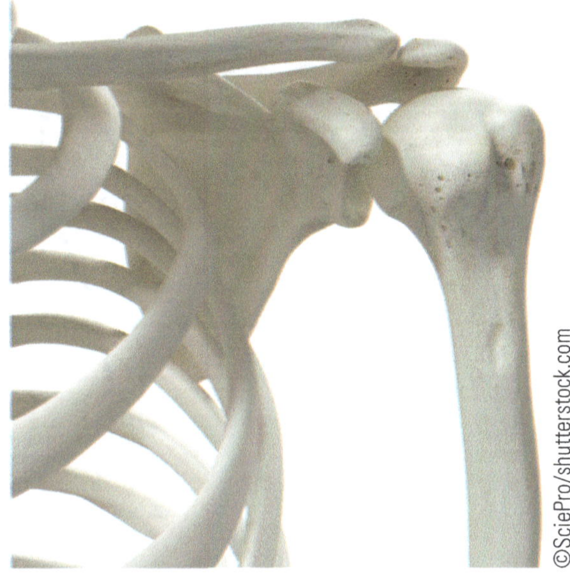

Sternoclavicular Joint Ligaments
<u>Color or trace/outline accordingly</u>

Anterior Sternoclavicular Ligament: (a)
Sternal End of Clavicle →
Manubrium of Sternum
Function(s)

Posterior Sternoclavicular Ligament:
(Not able to draw)
Sternal End of Clavicle →
Manubrium of Sternum
Function(s):

Interclavicular Ligament: (b)
Sternal End of Clavicle →
Sternal End of Clavicle
Function(s):

Costaclavicular Ligament: (c)
Clavicle to First Rib
Function(s)

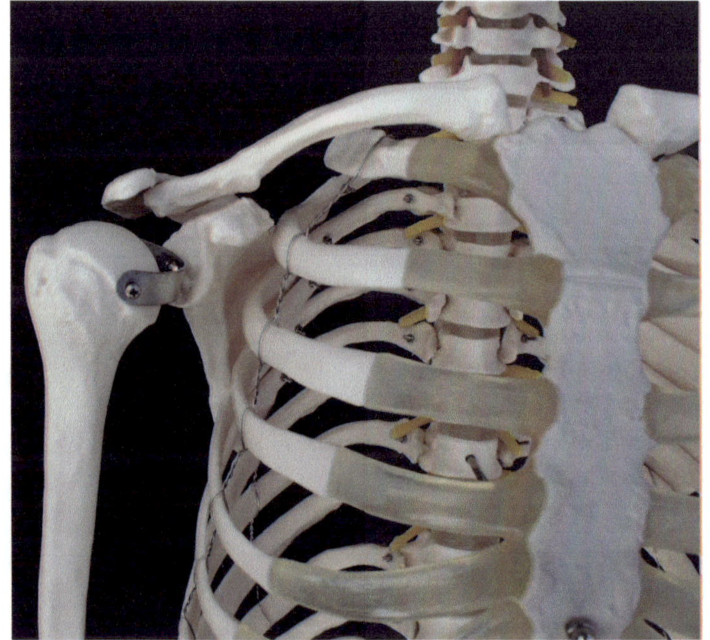

Acromioclavicular Joint Ligaments

Draw and Color Accordingly

Acromioclavicular Ligament:
Acromion Process of Scapula →
Clavicle
Function(s):

Coracoclavicular Ligaments:
Trapezoid Portion (Lateral, "LT"
Conoid Portion (Medial, "MC"- *Hammer Time*"
Function(s):

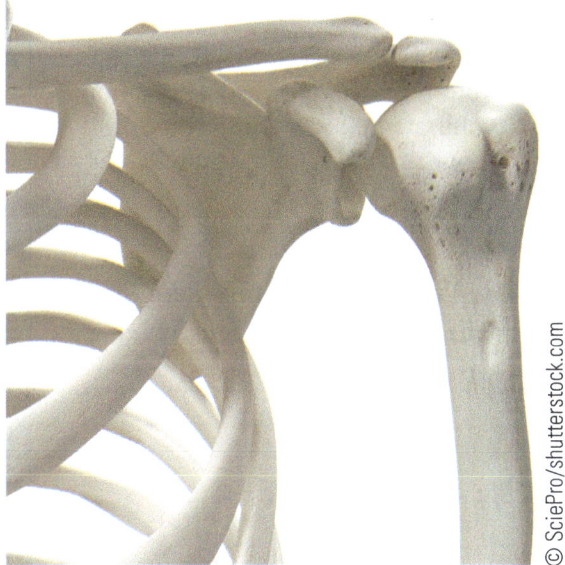

Elbow Joint

Carrying Angles of the Elbow

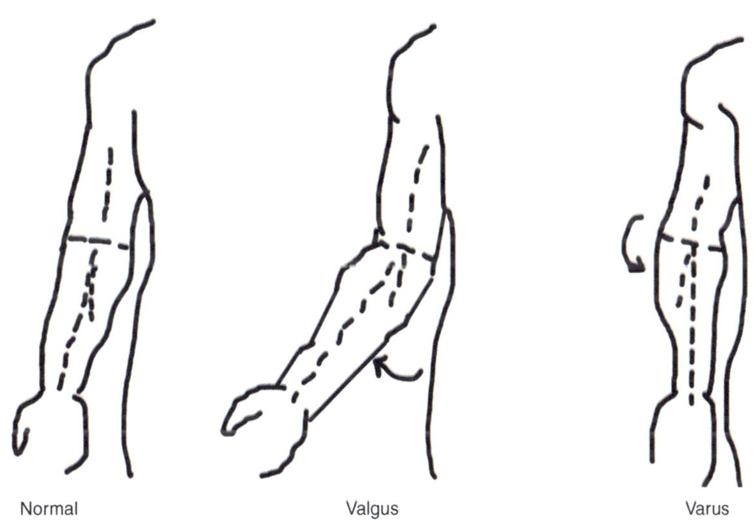

Cubitus Valgus

Cubitus Varus:

The Elbow Joint/Proximal Radioulnar Joint included

1. Contains articular cartilage/joint capsule/synovial membrane
2. Protected posteriorly by 3 **Olecronon Bursa**
   a. Intratendinous (Triceps Brachii)
   b. Subtendinous forming under tendon but on top of olecranon
   c. Subcutaneous
3. Strengthened by the collateral ligaments

Anular Ligament: (a; color accordingly)

Surrounds the Head of the Radius holding the head into the Radial Notch of the Ulna

Function: Stabilize Head of Radius and Proximal Radioulnar Joint

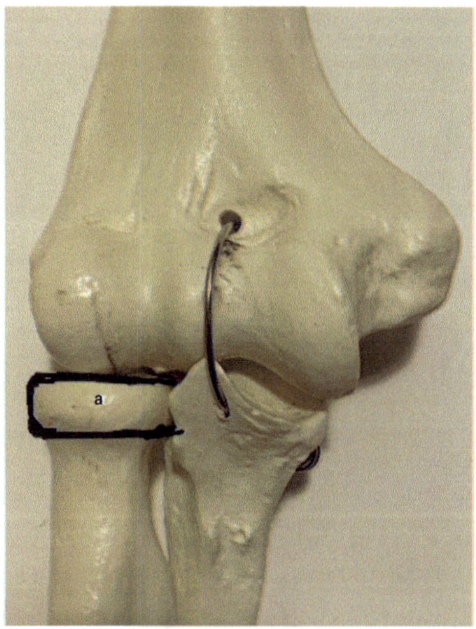

Source: Shireen Rahman

Radial Collateral Ligament (red)

Lateral Epicondyle of the Humerus → Anular Ligament of Radius (black)

Function(s):

Lateral Ulnar Collateral Ligament (blue)

Lateral Epidcondyle of Humerus → Lateral Ulna

Function(s):

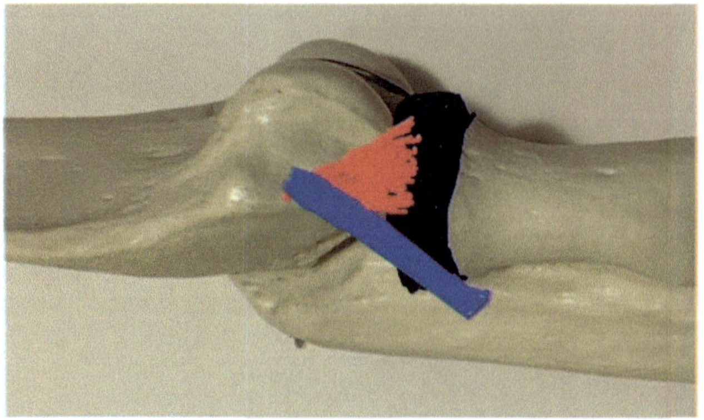

Source: Shireen Rahman

Ulnar Collateral Ligament:

3 aspects: (**Draw accordingly**)

1. Medial Epicondyle of Humerus (a) to Coronoid Process of Ulna (b) (draw)
2. Coronoid Process to Olecronon Process of Ulna (c) (draw)
3. Olecronon Process to Medial Epicondyle of Humerus (draw)

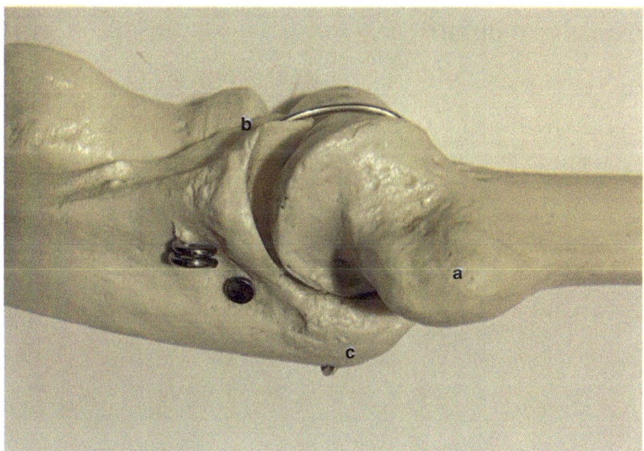

Source: Shireen Rahman

Time to Stretch or Scream, you choose!!!

Radioulnar Interosseous Membrane:

1. *Fibrous sheet which falls in between the radius and ulna*
2. *Holds the radius and ulna together*
3. *Serves as an attachment site for antebrachium muscles*
4. *Serves as a passageway for specific arteries*

Distal Radioulnar Joint

1. Contains a fibrocartilage articular surface and joint capsule/synovial membrane
2. The ulna will remain flexed while the radius moves to result in supination/pronation (pivot)
3. Reinforced by **Palmar and Dorsal Radioulnar Ligaments (shown below-*red*)**

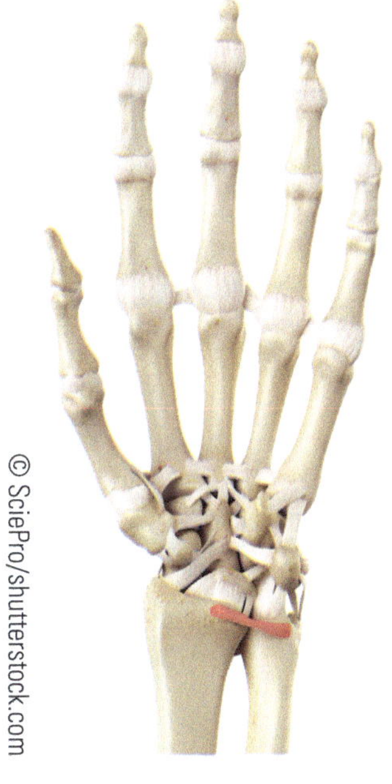

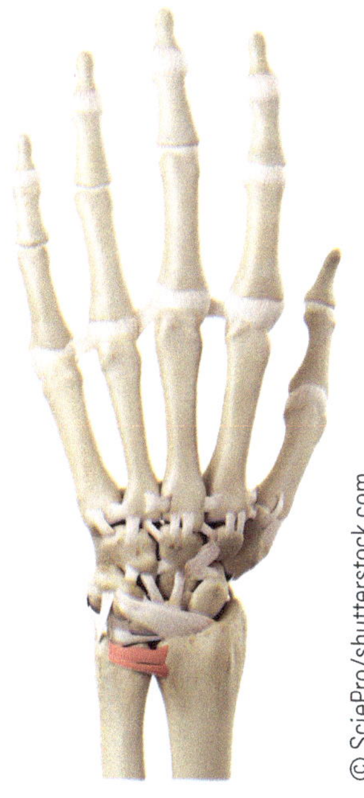

Hand

Hand Ligaments
<u>Palmar Surface Ligaments</u>
Draw/color accordingly

Palmar Radiocarpal Ligament:
Distal Radius   Carpal Bones (proximal and distal)

Radiocarpal Collateral Ligament:
Radial Styloid Process Scaphoid
Function: restrict excess   ulnar deviation (adduction)

Palmar Ulnocarpal Ligament:
Ulnar Styloid Process Lunate/Triquetrium

Palmar Intercarpal Ligaments:
Horizantal fibers between carpal bones

Palmar Carpometacarpal Ligaments
Distal row of Carpals to Metacarpals (vertical)

Palmar Metacarpal Ligaments
Base of metacarpal to Base of metacarpal (horizontal)

Palmar Ligaments
U shaped grooves at Head of metacarpal to Proximal Phalange
Reinforced by **Anular Ligaments**

Deep Transverse Metacarpal Ligament
Bridge between each Metacarpophalangeal Joint (horizontal)

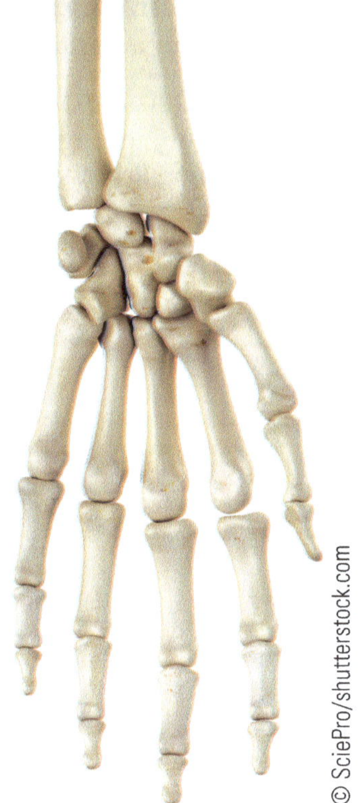

Dorsal Surface Ligaments
Label/color accordingly

Dorsal Radiocarpal Ligament:
Radius → Lunate/Triquetrium (v-shaped)

Ulnar Carpal Collateral Ligament:
Ulna Styloid Process → Triquetrium/Pisiform
Function: Prevent excess radial deviation (abduction)

Dorsal Intercarpal Ligaments:
Interconnects proximal and distal rows (horizontal)

Dorsal Carpometacarpal Ligaments:
Distal row of carpals to Metacarpals (vertical)

Collateral Ligaments: (to both sides)

Metacarpophalangeal Collateral Ligaments:
Found on both sides of the MCP joint (1-5)

Proximal Interphalangeal Collateral Ligaments:
Found on boths sides of the PIP and DIP joints

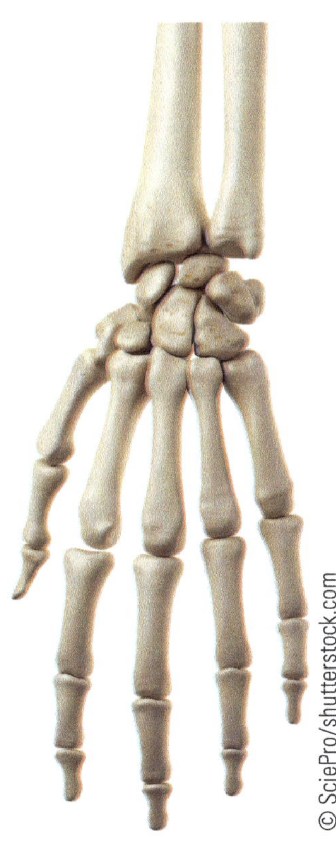

Flexor Retinaculum (Transverse Carpal Ligament)

1. A fibrous bridge running from the Scaphoid/Trapezium to the Pisiform/Hamate
2. Forms the **Carpal Tunnel** housing flexor tendons and the Median Nerve

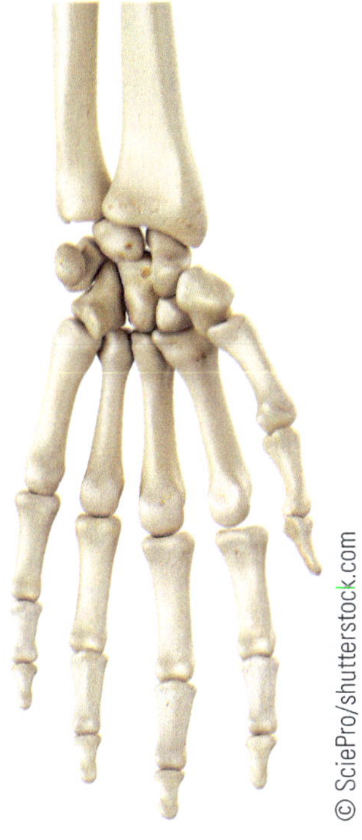

Extensor Retinaculum

1. A fibrous band running from the dorsal side of the radius to the Ulnar Styloid Process, Triquetrium, and Pisiform
2. Supports the extensor tendons

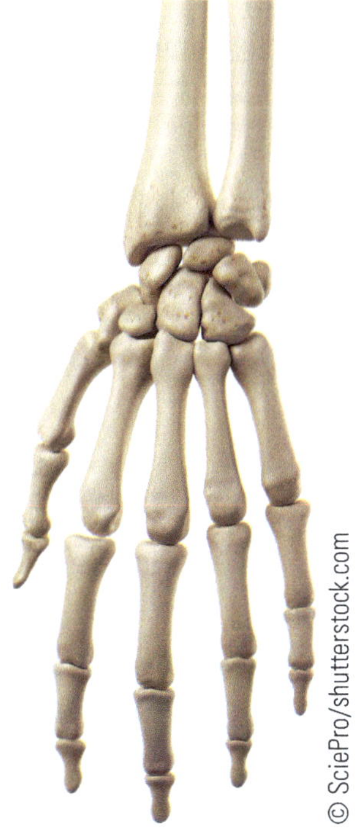

Carpal Tunnel:

1. Formed by the Flexor Retinaculum, which forms the superior roof of the tunnel
2. Contains the flexor tendons and the Median Nerve

Tunnel of Guyon:

1. Formed by an extension of the Flexor Retinaculum to the Pisiform/Hook of Hamate
2. The Flexor Retinaculum serves as the floor of the tunnel
3. Houses the Ulnar Nerve, Ulnar Artery/Vein

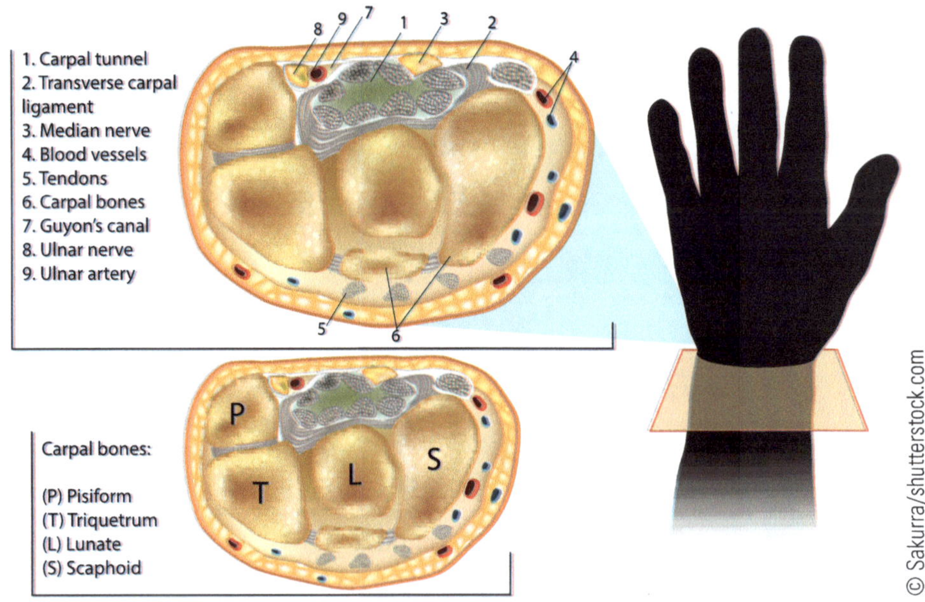

Ready for some muscles????
Maybe you should bake something for yourself first?!!?

Muscles of the Upper Extremity

A Brief Review of the Basics

What is an ORIGIN of a muscle?
1. **Immoveable** bony attachment of a muscle
2. "HOMEBASE"
3. Usually the more proximal aspect of the muscle (with exceptions)

What is an INSERTION of a muscle?
1. **Moveable** bony attachment of a muscle
2. "COLLEGE"
3. Usually the more distal aspect of the muscle (with exceptions)

(Continued)

How to determine FUNCTION of a muscle
1. The **insertion** always moves towards the **origin** as a muscle shortens (contracts)
2. "We all go home again"
3. Function move insertion to origin
4. Muscle fiber direction can help with function
   a. Pectoralis Major has some horizontal fibers = adduction
   b. Rectus Femoris has vertical fibers= flexion
5. A muscle moves the bone it will **insert** on as insertion point moves to the origin (immoveable)
   a. Deltoid inserts on the humerus, so it will be a humeral mover
   b. **Humeral movers INSERT on the humerus**
   c. **Scapular movers INSERT on the scapula**
6. A muscle will move as many joints as it crosses so know your joints!!! However, some of these movements are not necessarily the primary function of the muscle (we will only provide primary function not synergistic, secondary function)
7. **COMPARTMENTALIZATION** of muscles for location is the **SIMPLEST** way to study and avoid memorization of each and every individual muscle.

Trust Me Moment

First and foremost-before you start studying function, make sure you know how to **identify** each muscle. This will help you immensely when calculating function—
if you can see it in your mind—you can figure out function.

Identify muscles below:

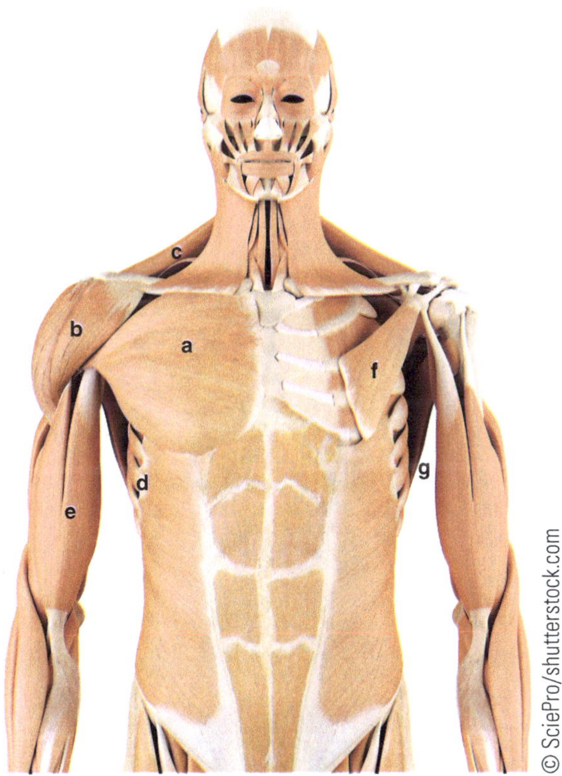

a) pectoralis major, b) anterior deltoid, c) trapezius, d) serratus anterior, e) biceps brachii, f) pectoralis minor, g) latissimus dorsi

Identify muscles below:

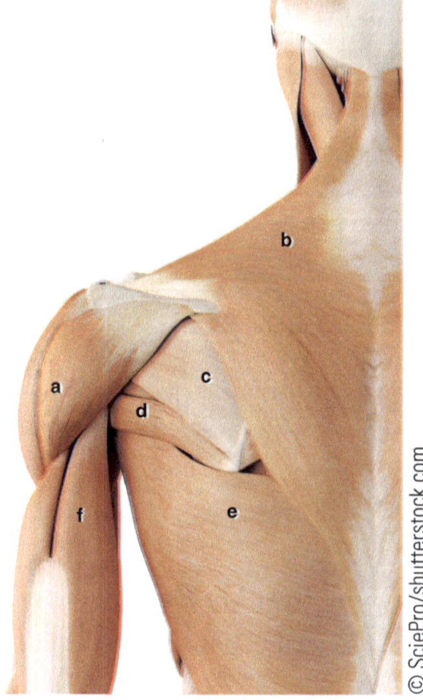

a) posterior deltoid, b) trapezius, c) infraspinatus, d) teres major, e) latissimus dorsi, f) triceps brachii

Identify the muscles below:

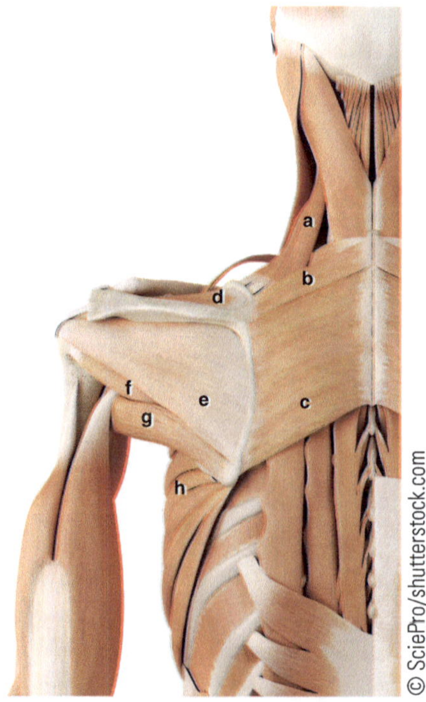

a) levator scapulae, b) rhomboid minor, c) rhomboid major, d) supraspinatus, e) infraspinatus, f) teres minor, g) teres major, h) serratus anterior

FUNCTION COMPARTMENT GOLDEN RULES

Know these first and foremost

Step #1: Be able to identify the muscles- see them in your brain.
Step #2: Compartmentalize do not memorize.
Step #3: Remember some muscles can fit into multiple compartments

Glenohumeral Joint Compartments

Compartment	Function	Example
Caps **over** the GH joint to get to humerus		Supraspinatus
Crosses **under** the GH joint to get to humerus		Pectoralis Major
Crosses GH anteriorly to get to humerus		Anterior Deltoid
Crosses GH posterior to get to humerus		Posterior Deltoid
Inserts close to the anterior head of the humerus		Subscapularis
Inserts close to the posterior head of the humerus		Infraspinatus

Scapulathoracic Joint Compartments

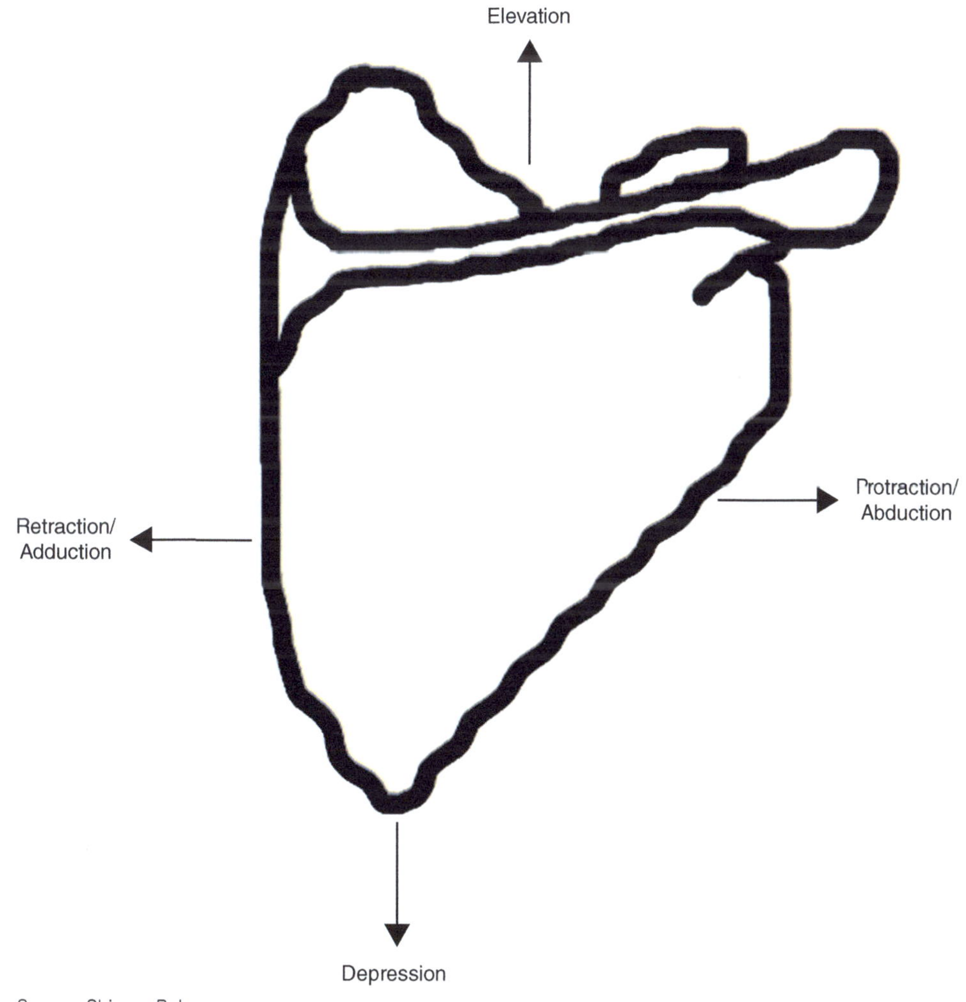

Source: Shireen Rahman

Compartment	Function	Example
Inserts on the superior aspect of scapula		Levator Scapulae
Inserts on the inferior aspect of the scapula		Trapezius
Inserts on the /medial posterior aspect of scapula		Rhomboids
Inserts on anterior aspect of the scapula		Pectoralis Minor

Elbow Joint Compartments

Crosses the joint anteriorly to reach radius/ulna, etc.   Flexion

Crosses the joint posterior to reach radius/ulna, etc.   Extension

Crossing between ulna and radius   Pronation/Supination

Radiocarpal Joint Compartments

Crosses the joint anteriorly to reach carpals, etc.   Flexion

Crosses the joint posteriorly to reach carpals, etc.   Extension

Crosses the joint laterally   Abduction

Crosses the joint medially   Adduction

Metacarpophalangeal/Interphalangeal Compartments

Crosses the joint anteriorly   Flexion

Crosses the joint posteriorly   Extension

Kelly wants you to remember . . ."What doesn't kill you makes you stronger!!!"

© Leonard Zhukovsky/shutterstock.com

As you learn the muscles to functions, remember your patient will mostly complain of pain or weakness during a movement versus naming a specific muscle. It is important to be able to relate the movement with the muscle.

Glenohumeral Muscles- Anterior Compartment   GH FLEXORS

If flexion is weak or elicits pain, think of which muscles fall within the anterior compartment.

All the following muscles are humeral movers (insert on the humerus)

Note how the following muscles cross the glenohumeral joint anteriorly:

Label Insertion/Origin; Color/Outline from Insertion to Origin

1. Anterior Deltoid

Insertion **moves to**. . .	Origin	Anterior Function
Deltoid tuberosity of humerus	Lateral clavicle	GH Flexion

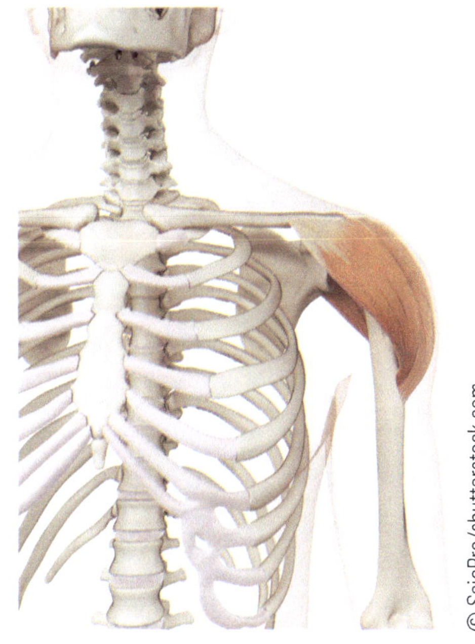

2. Pectoralis Major

Insertion **moves towards the**	Origin	Anterior Function
Crest of greater tuberosity of humerus	Clavicle Sternum Rectus Sheath	GH Flexion

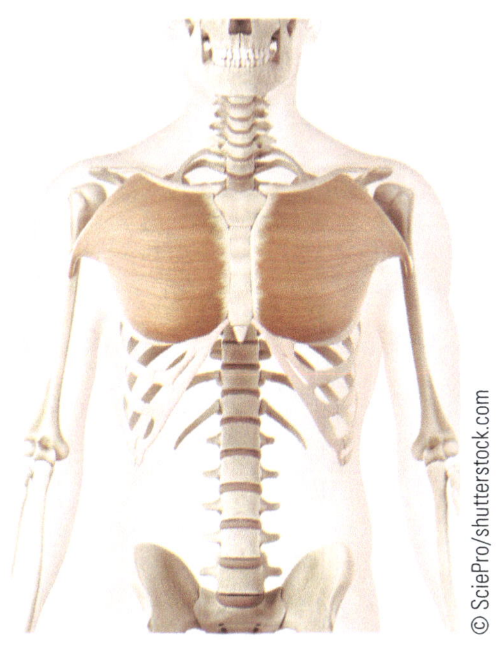

3. Coracobrachialis

Insertion **moves toward the**	Origin	Anterior Function
Crest of lesser tuberosity	Coracoid process of scapula	GH Flexion

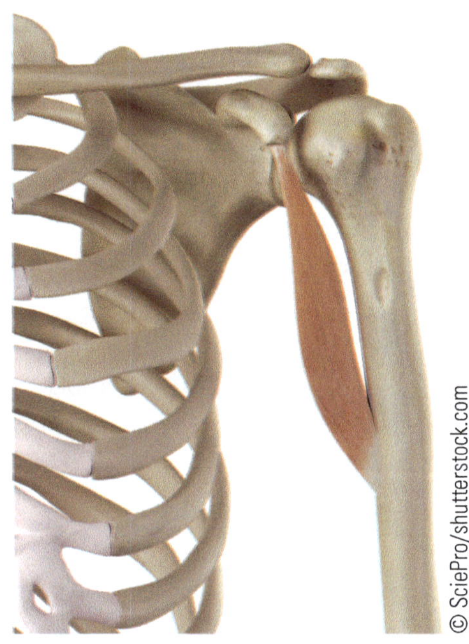

4. Biceps Brachii

Insertion moves towards the	Origin	Anterior Function
Radial tuberosity of radius	Long head: Supraglenoid tubercle of scapula Short head: Coracoid process of scapula	GH Flexion

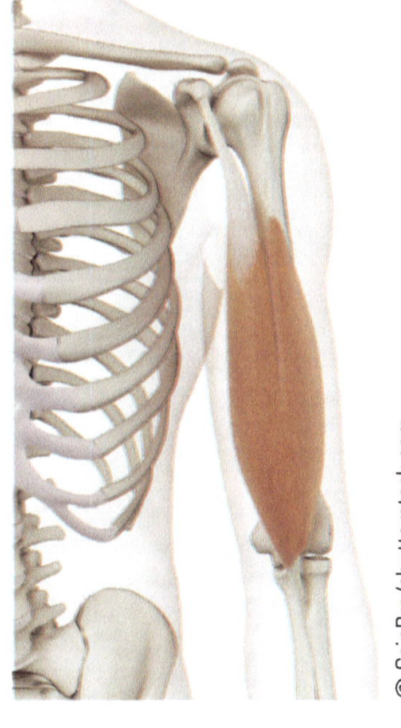

<u>Glenohumeral Muscles- Posterior Compartment</u>   <u>GH EXTENSORS</u>
If extension is weak or elicits pain, think of which muscles fall within the posterior compartment.

All the following muscles are humeral movers (insert on the humerus)

Note how the following muscles cross the glenohumeral joint posteriorly:

Label Insertion/Origin; **Color/Outline from Insertion to Origin**

1. **Posterior Deltoid**

Insertion moves towards	Origin	Posterior Function
Deltoid tuberosity of humerus	Spine of scapula	GH extension

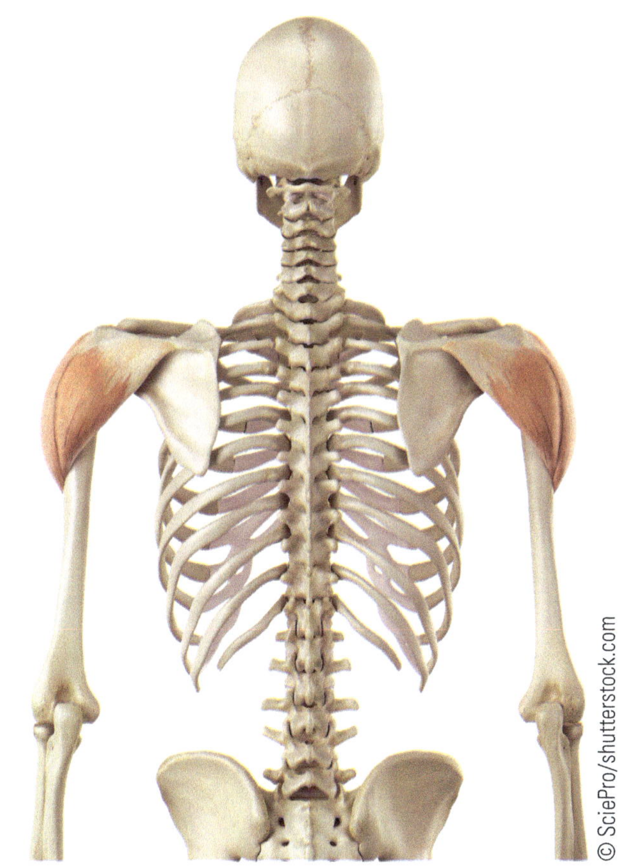

© SciePro/shutterstock.com

2. **Teres Major**

Insertion moves toward	Origin	Posterior Function
Crest of lesser tuberosity of humerus	Inferior angle of scapula	GH extension

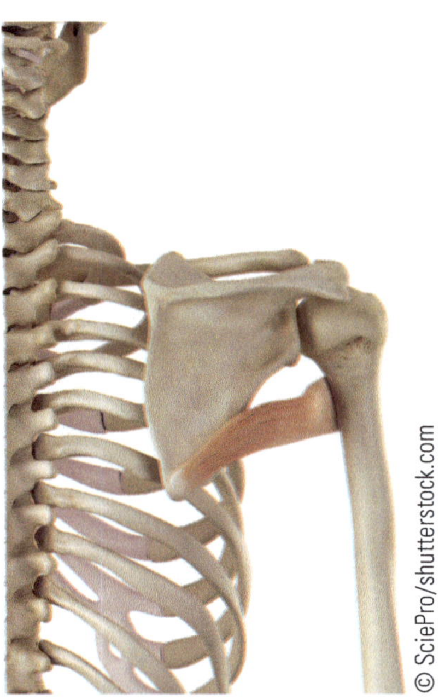

3. **Latissimus Dorsi**

Insertion moves towards the	Origin	Posterior Function
Crest of lesser tuberosity of humerus	Iliac crest (posterior) Thoracolumbar fascia Spinous process of lower thoracic vertebrae Lower ribs Inferior angle of scapula	GH extension

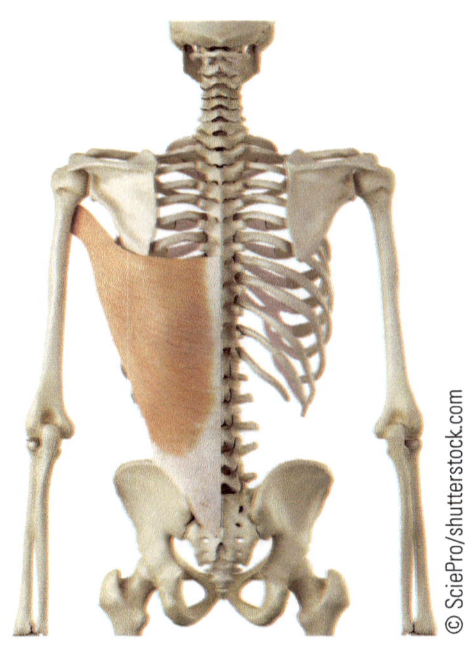

4. <u>**Triceps Brachii**</u>

Insertion moves towards the	Origin	Posterior Function
Olecronon of ulna	Long: Infraglenoid tubercle	GH extension

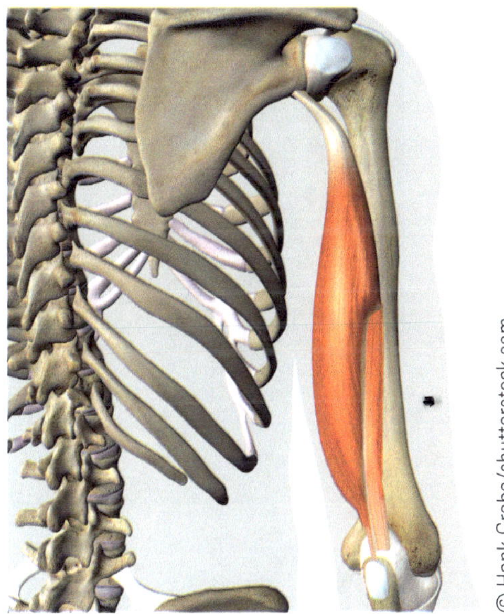

© Hank Grebe/shutterstock.com

Glenohumeral Muscles-Cap Over/Lateral Compartment   GH Abductors

If abduction is weak or elicits pain, think of which muscles fall within the lateral compartment.

All the following muscles are humeral movers (insert on the humerus)

Note how the following muscles cap over the GH joint to move to lateral humerus:

Label Insertion/Origin; Color from insertion to Origin

1. <u>**Supraspinatus**</u> (Rotator Cuff muscle)

Insertion moves towards the	Origin	Function
Greater tuberosity of humerus	Supraspinous fossa of scapula	GH Abduction

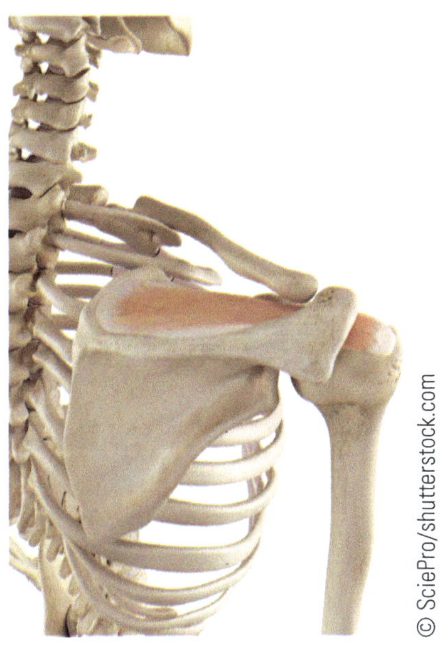

© SciePro/shutterstock.com

2. <u>Lateral Deltoid</u>

Insertion moves towards the	Origin	Function
Deltoid tuberosity of humerus	Acromion process of scapula	GH Abduction

<u>Glenohumeral Muscles-Under Arm/Medial Compartment</u>   <u>GH Adductors</u>

If adduction is weak or elicits pain, think of which muscles fall within the medial compartment.

All the following muscles are humeral movers (insert on the humerus)

Note how the following muscles cross under the GH joint to move to medial humerus:

Label Insertion/Origin; Color/Outline from Insertion to Origin

1. <u>Pectoralis Major</u>

Insertion moves towards the	Origin	Under Arm Function
Crest of greater tuberosity of humerus	Clavicle, Sternum, Rectus Sheath	GH adduction

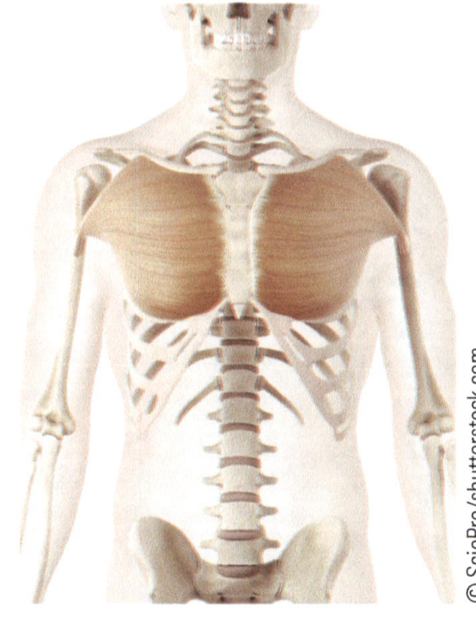

2. Coracobrachialis

Insertion moves toward the	Origin	Under Arm Function
Crest of lesser tuberosity	Coracoid process of scapula	GH adduction

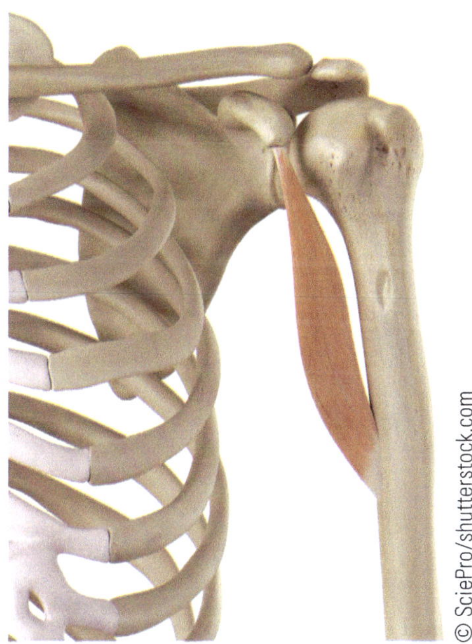

3. Latissimus Dorsi

Insertion moves towards the	Origin	Under Arm Function
Crest of lesser tuberosity of humerus	Iliac crest (posterior) Thoracolumbar fascia Spinous process of lower thoracic vertebrae Lower ribs Inferior angle of scapula	GH adduction

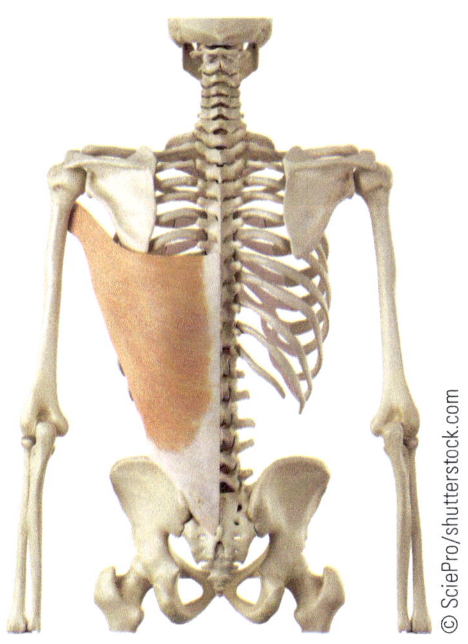

4. <u>Teres Major</u>

Insertion moves toward	Origin	Under Arm Function
Crest of lesser tuberosity of humerus	Inferior angle of scapula	GH adduction

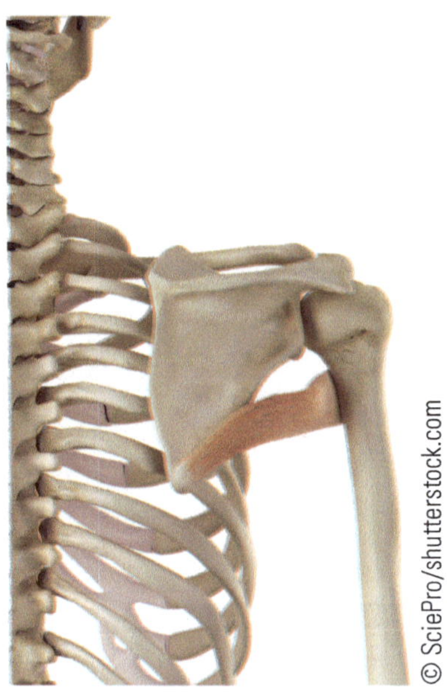

© SciePro/shutterstock.com

<u>Glenohumeral Muscles-Anterior Head of Humerus Compartment</u>   <u>GH Medial Rotation</u>

If medial rotation is weak or elicits pain, think of which muscles fall within the anterior head compartment.

All the following muscles are humeral movers (insert on the humerus)

Note how the following muscles attach to the anterior head region of humerus:

<u>Label Insertion/Origin; Color/Outline from Insertion to Origin</u>

1. <u>Subscapularis (Rotator Cuff)</u>

Insertion moves toward the	Origin	Anterior head Function
Lesser tuberosity of humerus	Subscapularis fossa	GH medial rotation

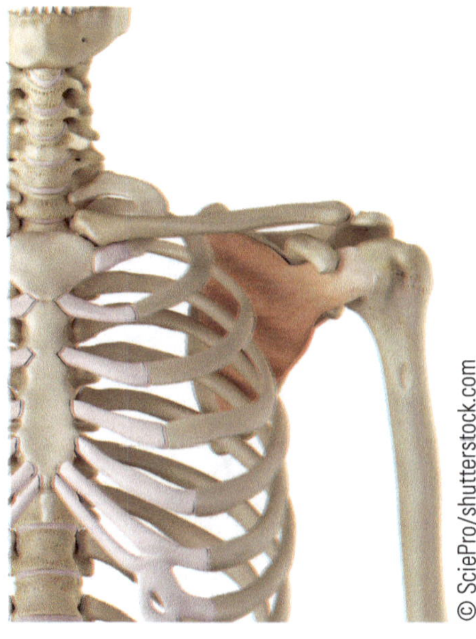

© SciePro/shutterstock.com

2. **Pectoralis Major:** note insertion at anterior crest of greater tuberosity

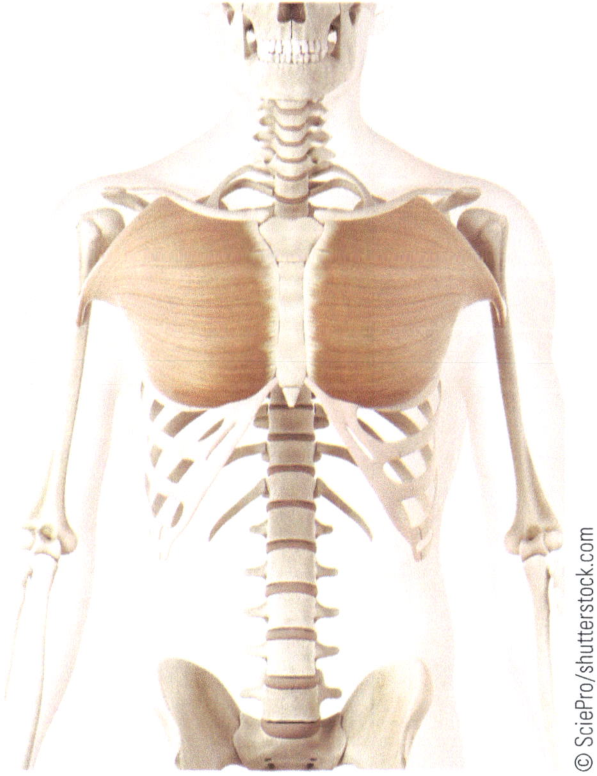

© SciePro/shutterstock.com

3. **Latissimus Dorsi:** Note insertion point on crest of lesser tuberosity

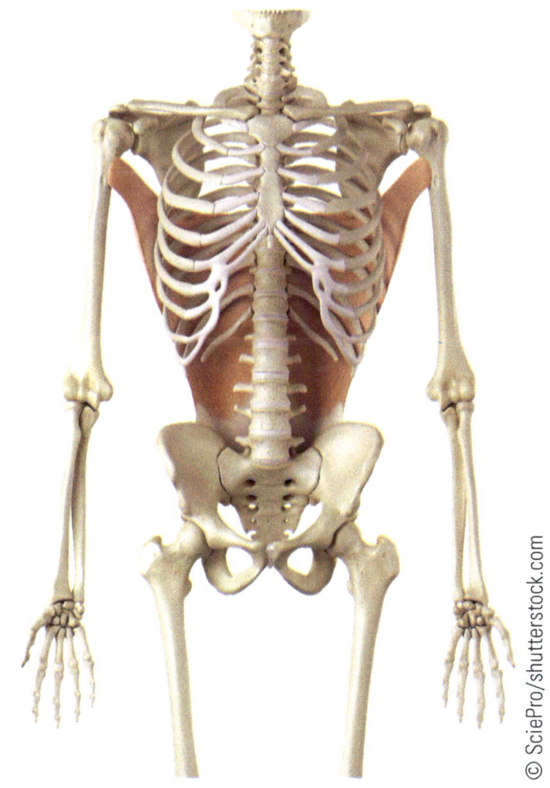

© SciePro/shutterstock.com

4. **Teres Major:** Note insertion point on crest of lesser tuberosity

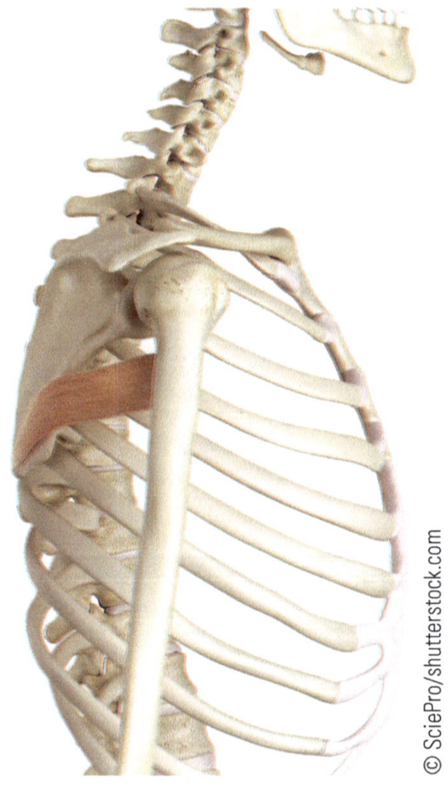

5. **Coracobrachialis-** note the insertion at the mid-medial aspect of humerus

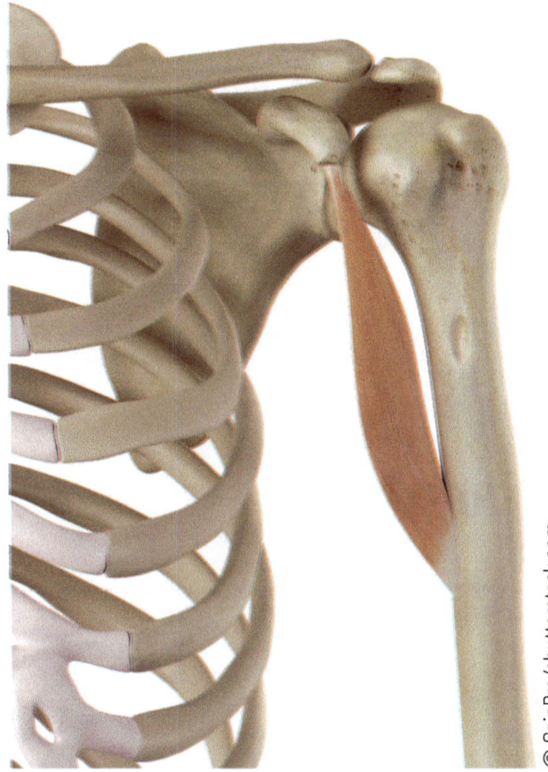

Glenohumeral Muscles-Posterior Head of Humerus Compartment   GH Lateral Rotation

If lateral rotation is weak or elicits pain, think of which muscles fall within the posterior head compartment.

All the following muscles are humeral movers (insert on the humerus)

Note how the following muscles attach to the posterior head region of humerus:

Label Insertion/Origin; Color/Outline from Insertion to Origin

1. Infraspinatus (Rotator Cuff)

Insertion moves toward the	Origin	Posterior Head Function
Greater tuberosity of humerus	Infraspinatus fossa	GH lateral rotation

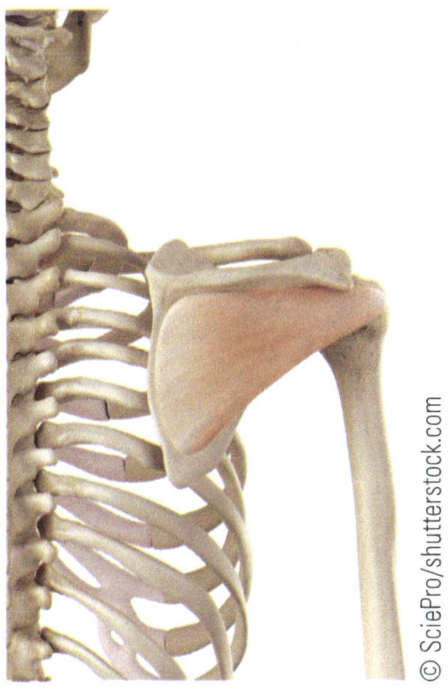

2. Teres Minor (Rotator Cuff)

Insertion moves toward the	Origin	Posterior Head Function
Greater tuberosity of humerus	Lateral border of scapula	GH lateral rotation

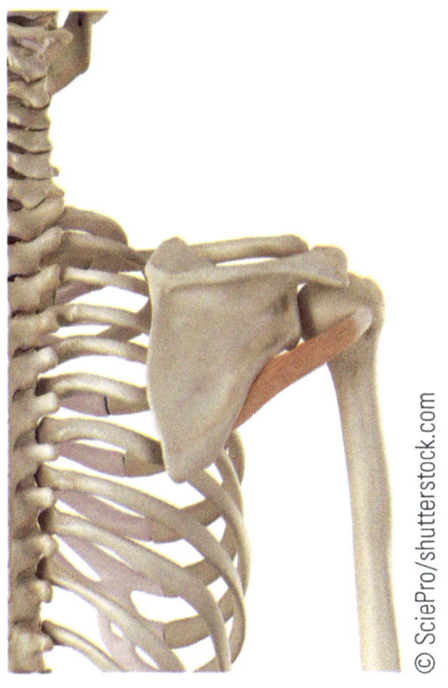

Now try some on your own . . .

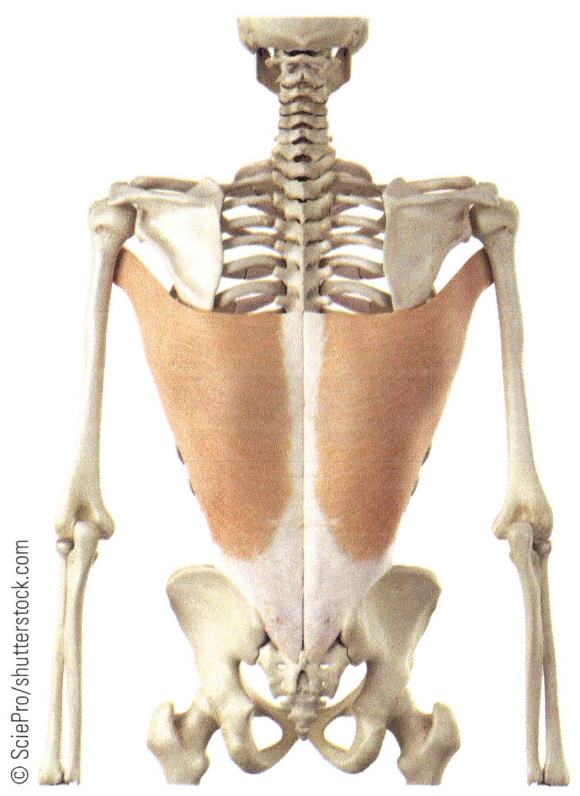

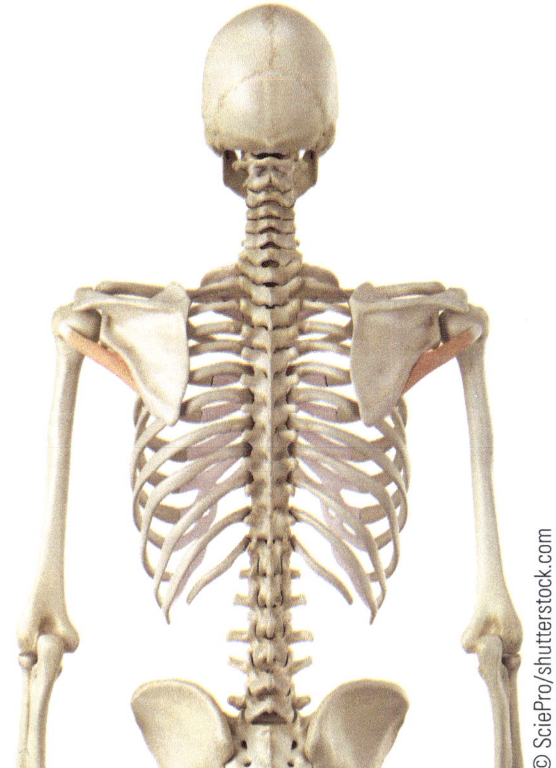

Name Muscle:

Name Muscle:

What compartment(s) do you see?

What Compartment(s)

What function(s) result?

What function(s) result?

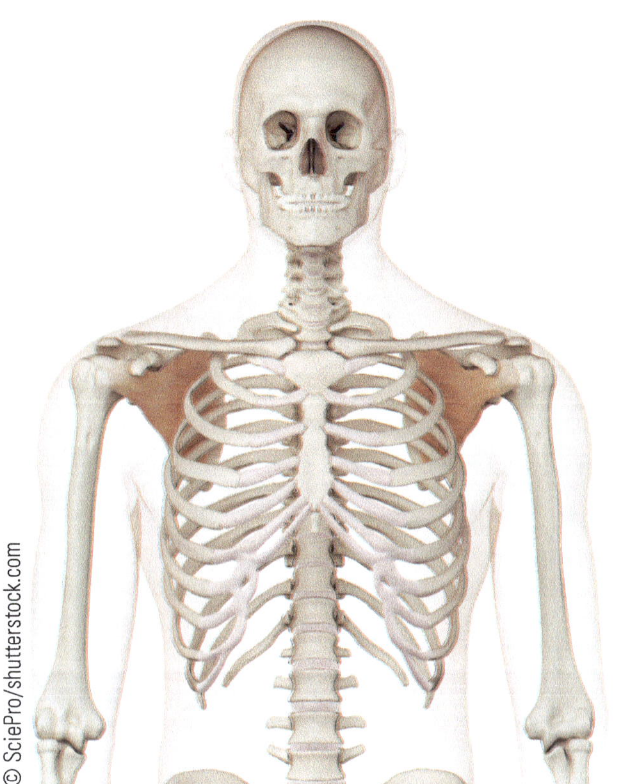

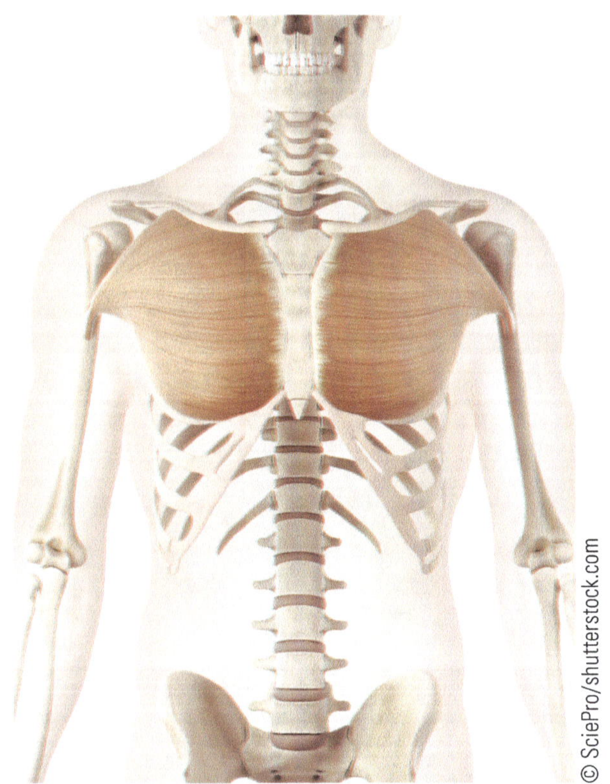

Name Muscle:

What compartment(s) do you see?

What function(s) result?

Name muscle:

What compartment(s) do you see?

What function(s) result?

Summary of Glenohumeral Movers:

What are the muscles of the anterior compartment resulting in GH Flexion?

What are the muscles of the posterior compartment resulting in GH Extension?

What are the muscles of the "cap over" or lateral compartment resulting in Abduction?

What are the muscles that cross under the arm or medial compartment resulting in Adduction?

What are the muscles of the anterior head of humerus compartment resulting in medial rotation?

What are the muscles of the posterior head of humerus compartment resulting in lateral rotation?

You are Brilliant!!!

Summary of Glenohumeral Movers:

What are the muscles of the anterior compartment resulting in GH Flexion?

Anterior Deltoid—Pectoralis Major—Biceps Brachii--Coracobrachialis

What are the muscles of the posterior compartment resulting in GH Extension?

Posterior Deltoid- Latissimus Dorsi- Teres Major- Triceps Brachii-

What are the muscles of the "cap over" or lateral compartment resulting in Abduction?

Middle Deltoid- Supraspinatus

What are the muscles that cross under the arm or medial compartment resulting in Adduction?

Coracobrachialis- Latissimus Dorsi-Teres Major-Pectoralis Major

What are the muscles of the anterior head of humerus compartment resulting in medial rotation?

Subscapularis- Pectoralis Major- Coracobrachialis- Latissimus Dorsi- Teres Major

What are the muscles of the posterior head of humerus compartment resulting in lateral rotation?

Infraspinatus- Teres Minor

You are So Brilliant!!!

Scapulothoracic Joint Muscles

**Remember, first know how to identify muscles-picture them in your mind-it will help with determining function.

Scapulothoracic Muscles- Superior Compartment   **ST Elevation**

If elevation is weak or elicits pain, think of which muscles fall within the superior compartment.

All the following muscles are scapular movers (insert on the scapula)

Note how the following muscles attach to the superior scapula

Label Insertion/Origin; Color from Insertion to Origin

1. **Levator Scapulae- DRAW ON THE PHOTO BELOW**

Insertion moves toward the	Origin	Superior scapula Function
Superior angle of scapula	C1-C4 transverse process	Elevation

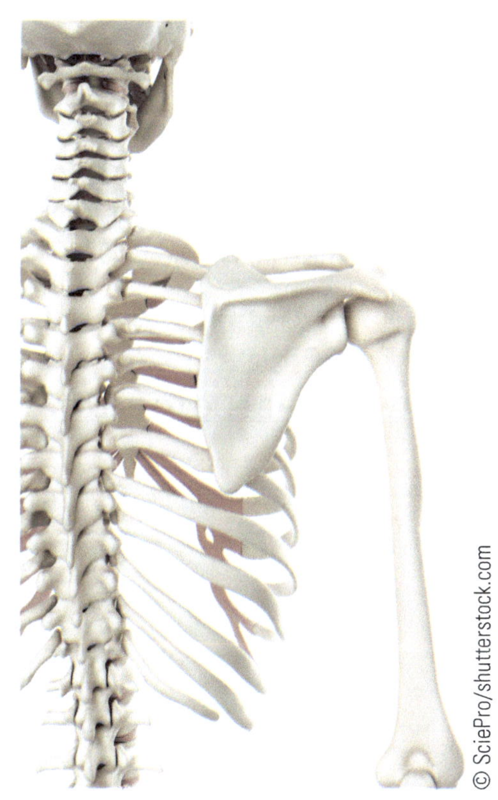

2. Descending Part of Trapezius- DRAW ON THE PHOTO BELOW

Insertion moves toward the	Origin	Superior scapula Function
Lateral clavicle	Occipital bone; spinous process C1-C7	Elevation *any many other secondary functions

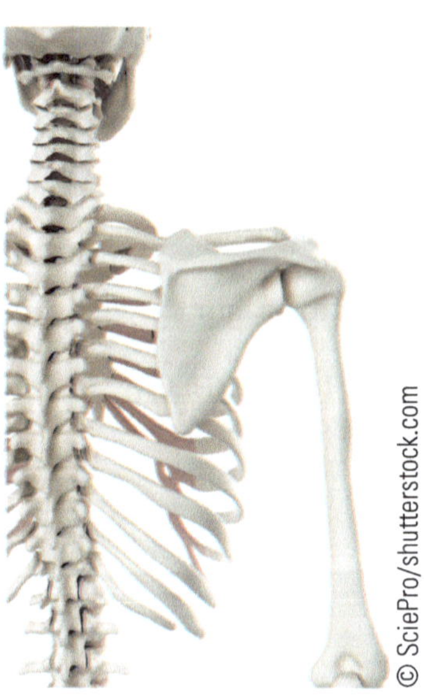

Keep working hard—you can do this!!!

Scapulothoracic Muscles- Inferior Compartment   ST Depression

If depression is weak or elicits pain, think of muscles within the inferior compartment.

All the following muscles are scapular movers (insert on the scapula)

Note how the following muscles attach to the inferior scapula

Label Insertion/Origin; Color from Insertion to Origin

3. Ascending Part of Trapezius- DRAW ON THE PHOTO BELOW

Insertion moves toward the	Origin	Superior scapula Function
Spine of scapula	Spinous process lower thoracic vertebrae (T5-T12)	Depression

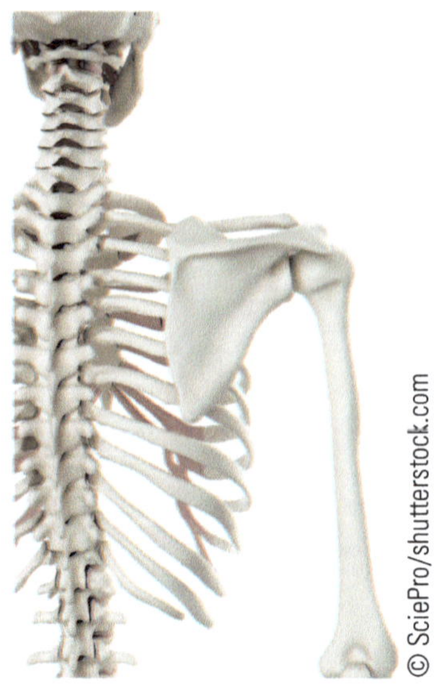

Scapulothoracic Muscles- Posterior/Medial Compartment   ST Retraction

If retraction is weak or elicits pain, think of muscles within the posterior/*medial* compartment.

All the following muscles are scapular movers (insert on the scapula)

Note how the following muscles attach to the posterior/*medial* scapula

4. <u>Rhomboid Major and Minor - DRAW ON THE PHOTO BELOW</u>

Insertion moves toward the	Origin	Superior scapula Function
Medial border of scapula	Spinous process C6- T4	Retraction

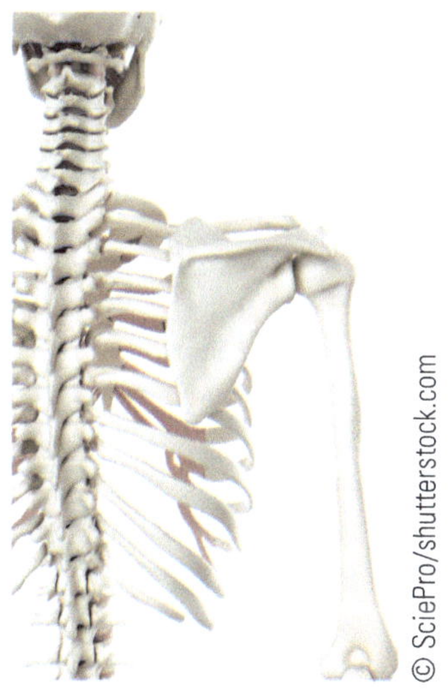

You're Doing It!!!

5. <u>Transverse Part of Trapezius - DRAW ON THE PHOTO BELOW</u>

Insertion moves toward the	Origin	Superior scapula Function
Acromion process of scapula	Spinous process T1-T4	Retraction

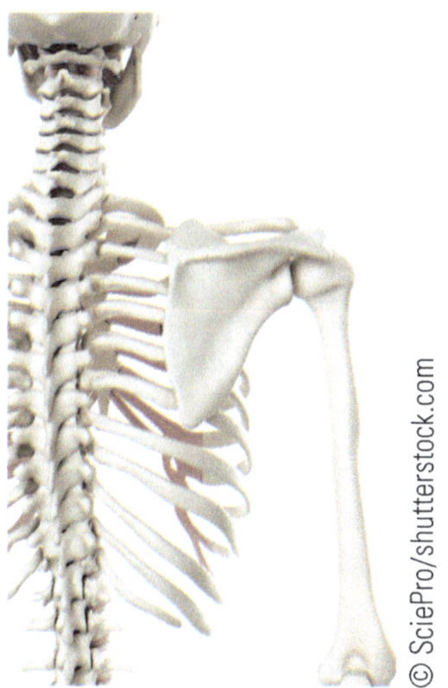

Scapulothoracic Muscles- Anterior Compartment   ST Protraction

If protraction is weak or elicits pain, think of muscles within the anterior compartment.

All the following muscles are scapular movers (insert on the scapula)

Note how the following muscles attach to the anterior scapula

1. Pectoralis Minor - DRAW ON THE PHOTO BELOW

Insertion moves toward the	Origin	Superior scapula Function
Coracoid process of scapular	Anterior aspect of ribs 3-5	Protraction

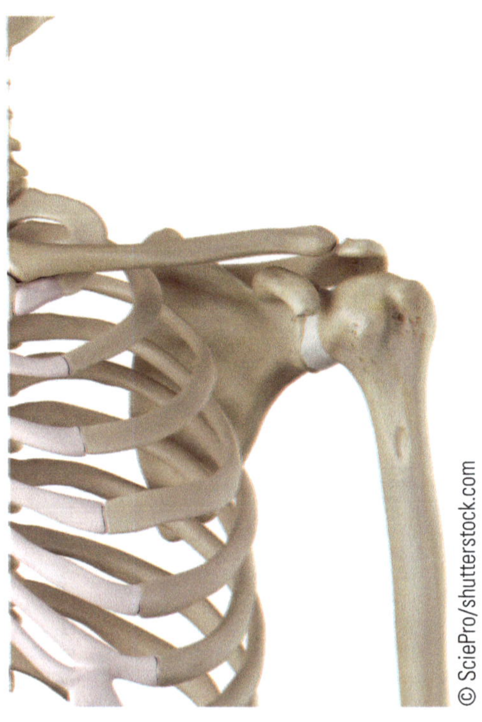

2. Serratus Anterior

Insertion moves toward the	Origin	Superior scapula Function
Medial border of scapula via subscapularis fossa	Lateral ribs 1-9	Protraction

So you know. . . .

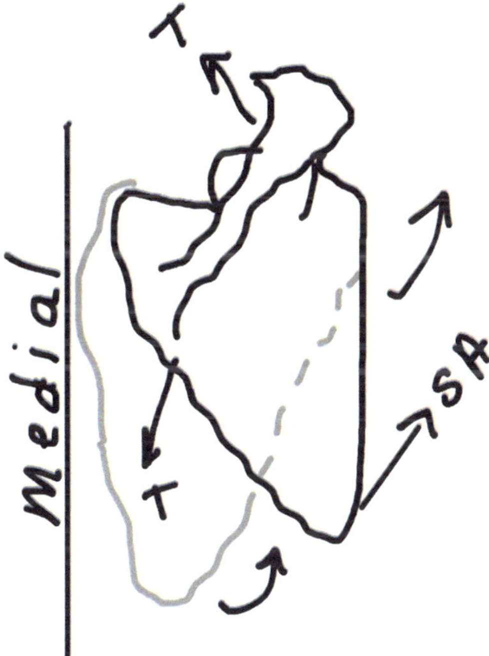

Source: Shireen Rahman

<u>Upward rotation</u> (final aspect of GH abduction) of the scapula is a result of a combination of muscles: the serratus anterior (SA), ascending (lower) and descending (upper) trapezius

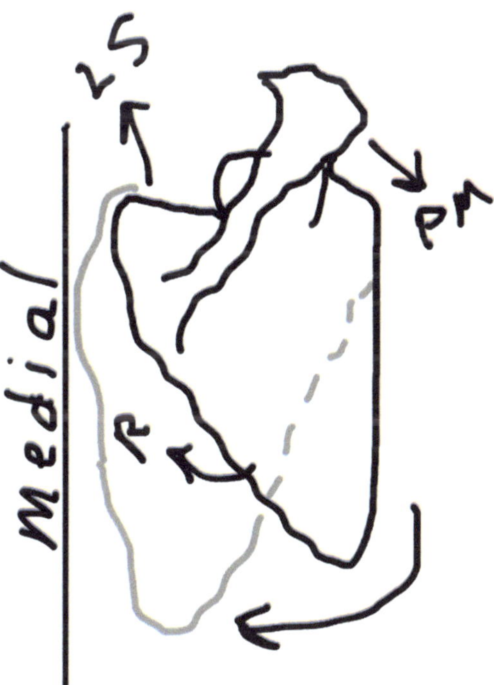

Source: Shireen Rahman

<u>Downward rotation of scapula</u> (returning from upward rotation) is a result of a combination of muscles: the pectoralis minor (PM), rhomboids (R), and levator scapulae (LS)

<u>**On to the Brachium**</u>

Elbow Muscles-Anterior Compartment Elbow Flexion

If elbow flexion is weak or elicits pain, think muscles in the anterior compartment.

All the following muscles are elbow movers (insert on the radius/ulnar)

Note how the following muscles attach to the anterior radius/ulnar.

Label Insertion/Origin; Color/Outline from Insertion to Origin

Anterior Compartment Muscles: Biceps Brachii, Brachialis, Brachioradialis

1. Biceps Brachii and Brachialis

	Insertion moves toward the	Origin	Anterior radius/ulna Function
Biceps Brachii	Radial tuberosity of radius	Long head: Supraglenoid tuberosity of scapula Short head: Coracoid process of scapula	Flexion *when elbow flexed it is also a supinator (proximal RU joint)
Brachialis	Ulnar tuberosity of ulna	Mid shaft of humerus	Flexion

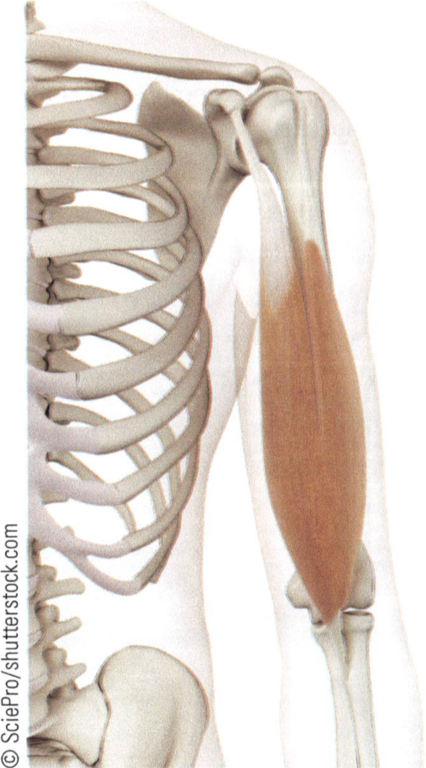

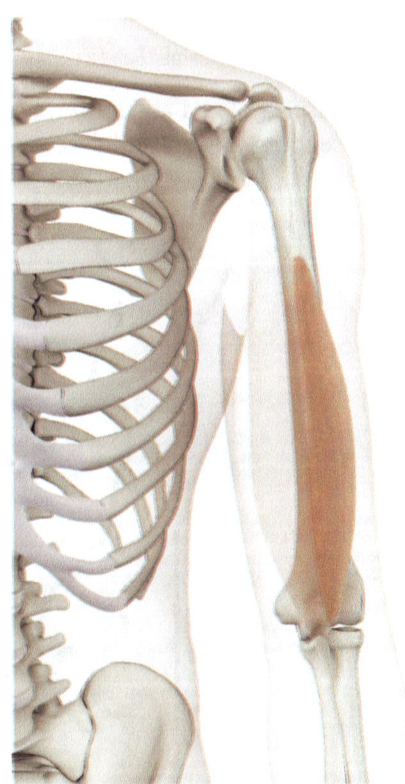

2. <u>Brachioradialis</u>

Insertion moves toward the	Origin	Anterior radius Function
Styloid process of radius	Lateral supracondylar ridge of humerus	Flexion of elbow *esp in pronation

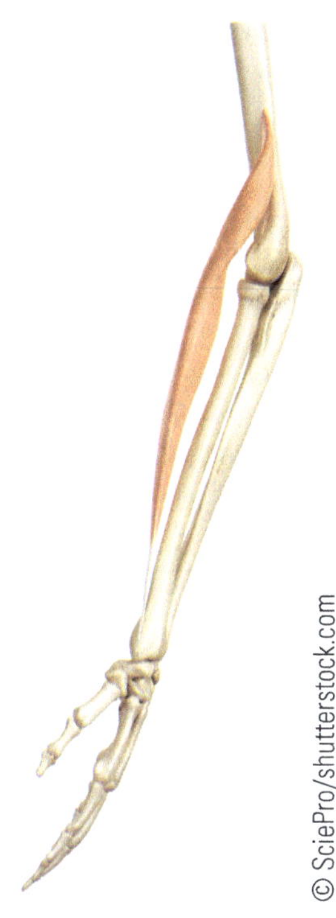

© SciePro/shutterstock.com

**Additional muscles will cross the elbow anteriorly but they will act as secondary flexors and will be discussed in the "antebrachium section"

© murbansky/shutterstock.com

Elbow Muscles-Posterior Compartment Elbow Extension

If elbow extension is weak or elicits pain, think muscles in the posterior compartment.

All the following muscles are elbow movers (insert on the radius/ulnar)

Note how the following muscles attach to the posterior radius/ulnar.
Label Insertion/Origin; Color/Outline from Insertion to Origin

Anterior Compartment Muscles: Triceps Brachii, Anconeus

3. Triceps Brachii and Anconeus

	Insertion moves toward the	Origin	Posterior radius/ulna Function
Triceps Brachii	Olecronon of ulna	Long head: Infraglenoid tubercle of scapula Medial head: Posterior humerus (medial) Short head: Posterior humerus (lateral)	Extension *medial head not show below as it is between and deep to long and short*
Anconeus	Olecronon of ulna on radial side	Lateral epicondyle of humerus	Extension *draw below*

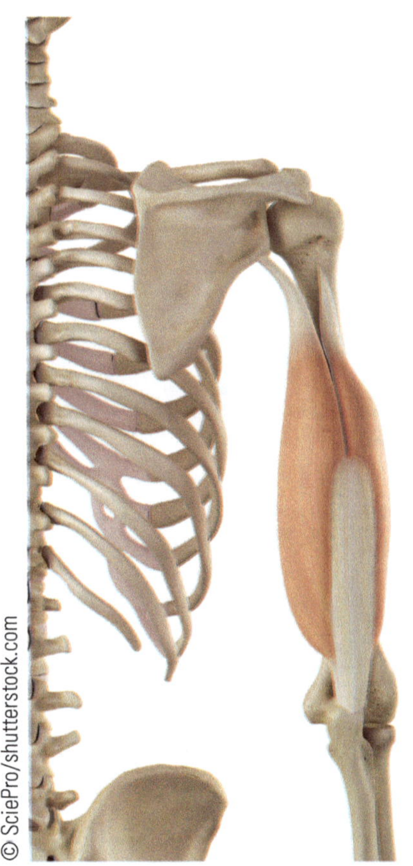

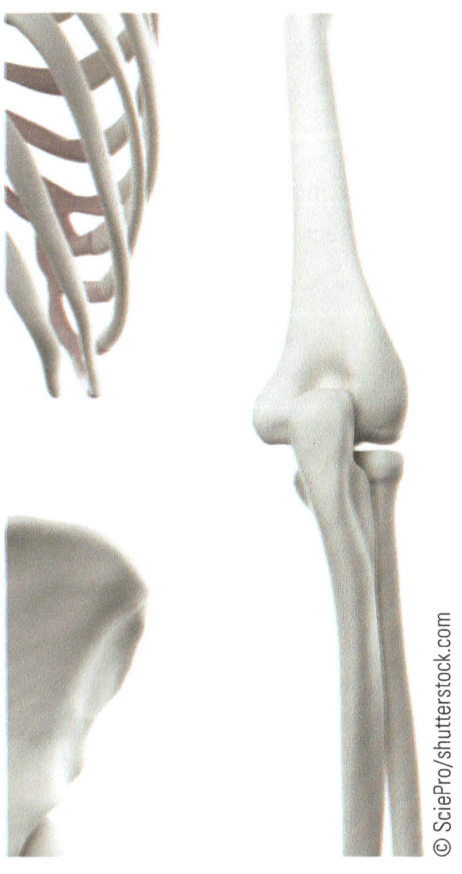

Quick Hit:

The **<u>Proximal Radioulnar</u>** joint is a pivot joint resulting in pronation and supination. You are supinating in anatomical position or <u>palm up</u> (carrying soup); you are pronating when palm is turned down (drop the soup). Both occur as the radial head rotates in the radial notch of the ulna.

The primary pronator at the proximal radioulnar joint is the **<u>Pronator Teres (a)</u>**
The primary supinator at the distal radioulnar joint is the **<u>Supinator</u>**
Note that both muscles will cross the RU joint

Don't you wish they were all this simple!!!

<u>Pronator Teres</u> inserts on radius and works its way back to the medial epicondyle of the humerus to radius; The <u>Supinator</u> inserts on the radius and works its way back to the olecranon of ulna.

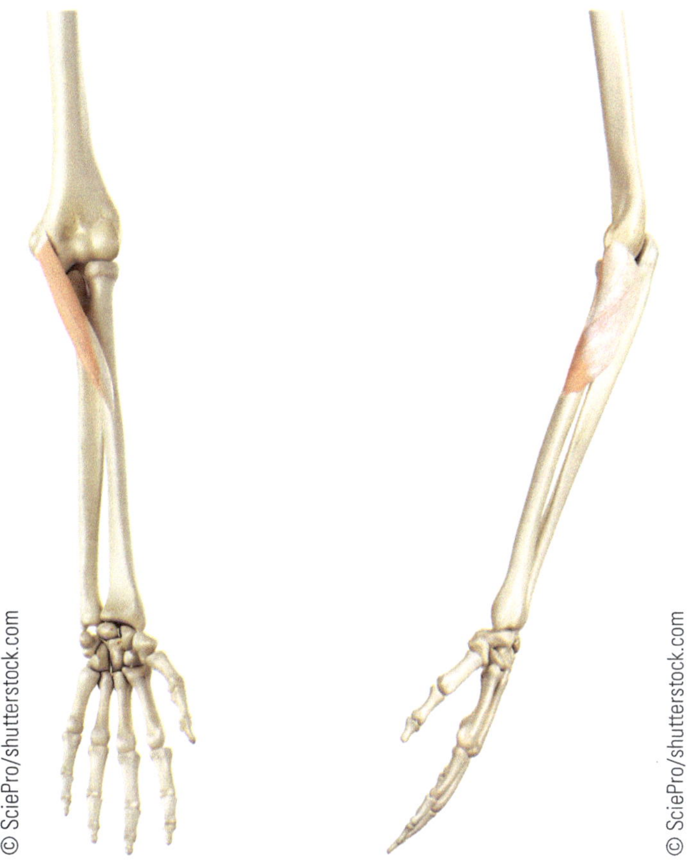

At the **<u>Distal Radioulnar</u>** joint, the **Pronator Quadratus** will enable pronation; it inserts on the radius and moves towards its origin at the ulna.

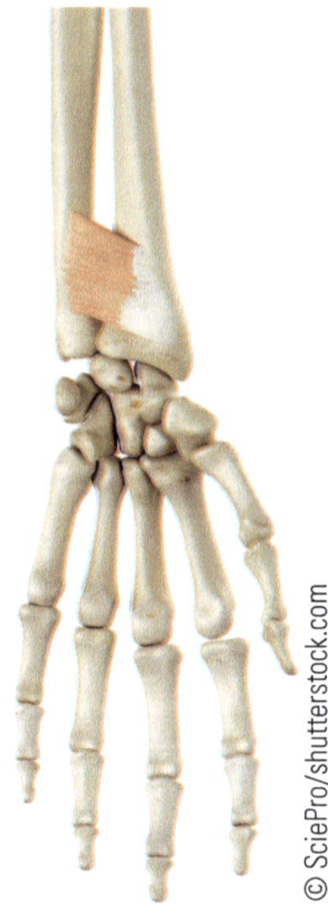

Muscles of Antebrachium

1. The key to the antebrachium muscles is to first know **how many joints** the muscle crosses; this will determing function(s); primary or secondary.
2. May originate at the humerus making them flexors or extensors of the elbow (primary or secondary)
3. Crosses <u>radiocarpal joint</u> (condyloid) anteriorly, posteriorly AND laterally or medially; think location!!
4. Some may cross over the <u>metacarpophalangeal joints</u>
5. Some may go all the way to the <u>distal interphalngeal joints</u>
6. <u>Where they cross-they have a function</u>
7. <u>Use the names of the muscles to help guide function and location</u>

Remember

Anterior compartment:	Flexion
Posterior compartment:	Extension
Radial side:	Abduction
Ulnar side:	Adduction

Anterior Compartment of Radiocarpal, MCP, and PIP/DIP Joints   Flexion

Medial epicondyle of the humerus will be a common origin
<u>PFPFF!!!!!</u> *(you can do better than that!!!)*

	Insertion moves toward the	Origin	Anterior Function
Pronator Teres	Radius	Medial epicondyle of humerus	Pronation of prox R/U joint
Flexor Carpi Radialis *Flexor= anterior* *Carpi= stops at carpals* *Radialis= radial side*	Base of 2nd metacarpal (palmar side)	Medial epicondyle of humerus	Elbow: weak flexion Radiocarpal: flexion abduction

(Continued)

	Insertion moves toward the	Origin	Anterior Function
Palmaris Longus	Palmar aponeurosis	Medial epicondyle of humerus	Elbow: weak flexion Radiocarpal: flexion
Flexor Digitorum Superficialis *Flexor=anterior* *Digitorum= digits* *Superficialis= superficial*	Middle phalanges 2-5 (palmar)	Medial epicondyle of humerus	Elbow: weak flexion Radiocarpal: flexion MCP: flexion DIP: flexion
Flexor Carpi Ulnaris *Flexor: anterior* *Carpi: ends at carpals* *Ulnaris: ulnar side*	Pisiform, hook of hamate, base of 5th metacarpal	Medial epicondyle of humerus	Elbow: weak flexion Radiocarpal:flexion adduction

Color/Outline the **PFPFF** muscles accordingly- minus the Pron Teres

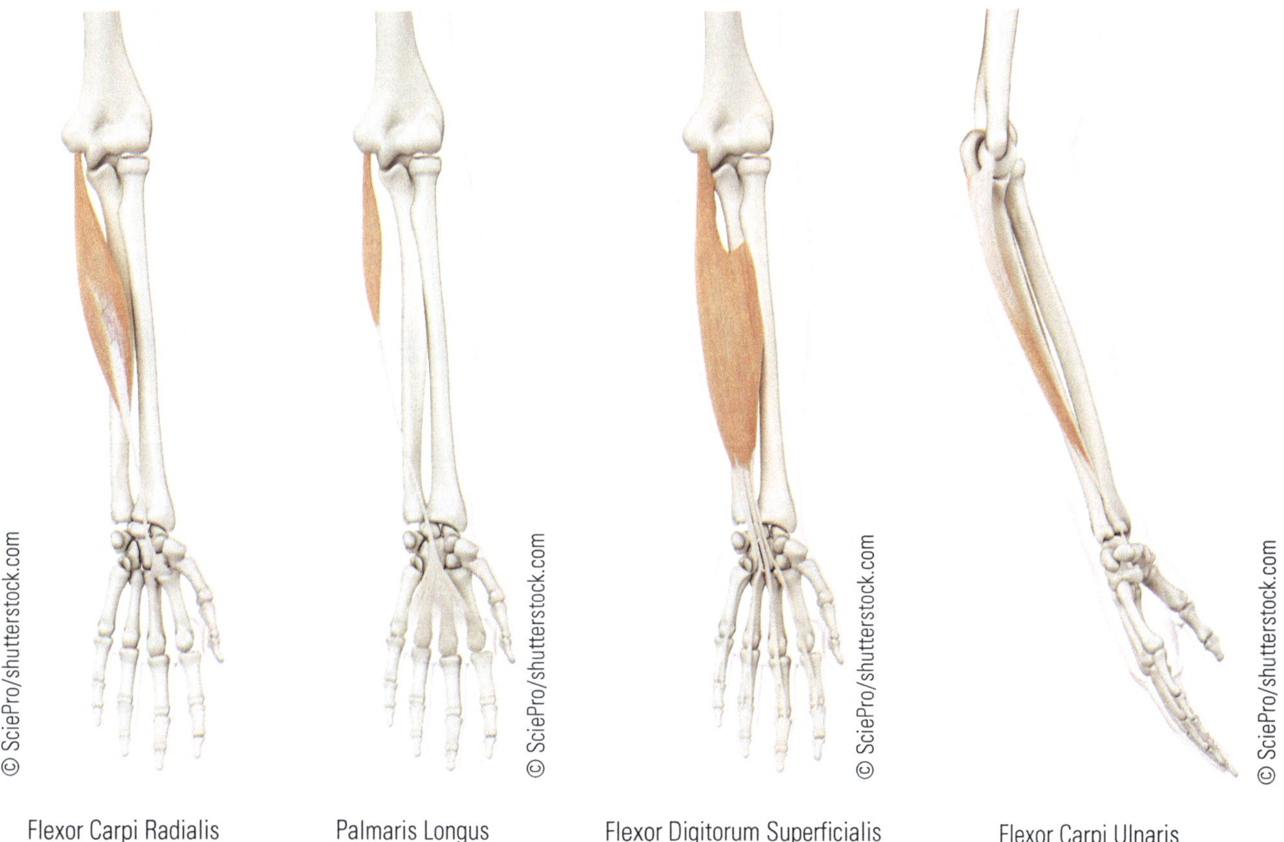

Flexor Carpi Radialis   Palmaris Longus   Flexor Digitorum Superficialis   Flexor Carpi Ulnaris

*Additional Anterior Antebrachial Muscles (Deep_*

1. **Flexor Pollicis Longus**

Insertion moves toward the	Origin	Anterior Function
Palmar distal phalange of thumb	Anterior radius	Radiocarpal; flexion 1st CMC: flexion 1st MCP: flexion 1st IP: flexion

2. **Flexor Digitorum Profundus**

Insertion moves toward the	Origin	Anterior Function
Palmar distal phalange 2-5	Ulna	Radiocarpal; flexion MCP: flexion PIP/DIP: flexion

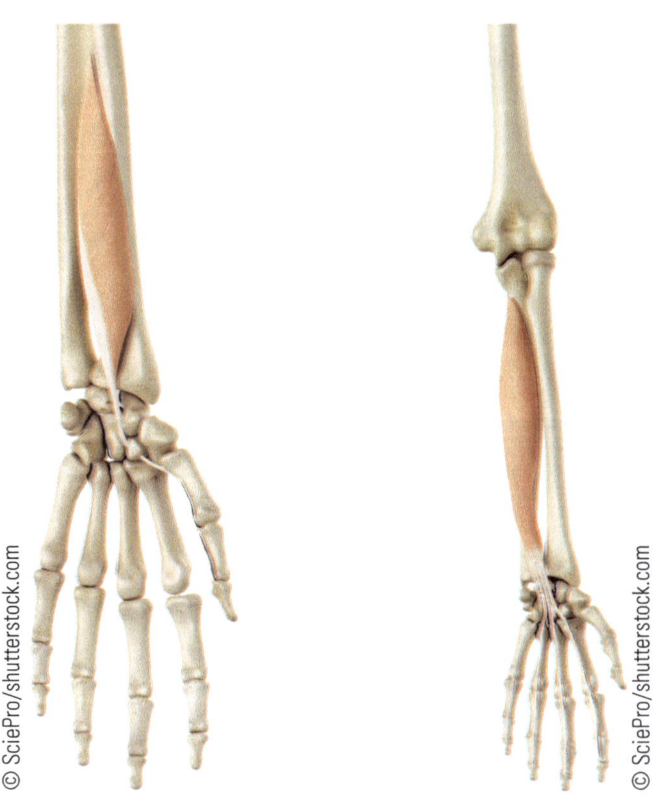

Remember
Posterior compartment: Extension
Radial side: Abduction
Ulnar side: Adduction

Posterior Compartment of Radiocarpal, MCP, and PIP/DIP Joints   Extensors

Lateral epicondyle of the humerus will be a common origin

LBDMU !!!!! *(Shout it out!!!)*

	Insertion moves toward the	Origin	Posterior Function
Extensor Carpi Radialis **L**ongus	Dorsal base of 2nd metacarpal	Lateral supracondylar ridge of humerus	Elbow: weak extension Radiocarpal: extension abduction
Extensor Carpi Radialis **B**revis	Dorsal base of 3rd metacarpal	Lateral epicondyle of humerus	Elbow: weak extension Radiocarpal: extension abduction
Extensor **D**igitorum	Dorsal distal phalange 2-5	Lateral epicondyle of humerus	Elbow: weak extension Radiocarpal: extension MCP/PIP/DIP: extension *Abduction of digits*

(Continued)

	Insertion moves toward the	Origin	Posterior Function
Extensor Digiti **M**inimi	Dorsal distal phalange 5	Lateral epicondyle of humerus	Elbow: weak extension Radiocarpal: extension MCP/PIP/DIP: extension *Abduction of 5th*
Extensor Carpi **U**lnaris	Dorsal base of 5th metacarpal	Lateral epicondyle of humerus	Elbow: weak extension Radiocarpal: extension adduction
Plus (deeper)			
Extensor Indicis	Dorsal 2nd distal phalange	Distal ulna	Radiocarpal: extension 2nd MCP/PIP/DIP: extension

Additional thumb muscles discussed later

<u>Color/Outline **LBDMU**</u> muscles- note difference in insertion

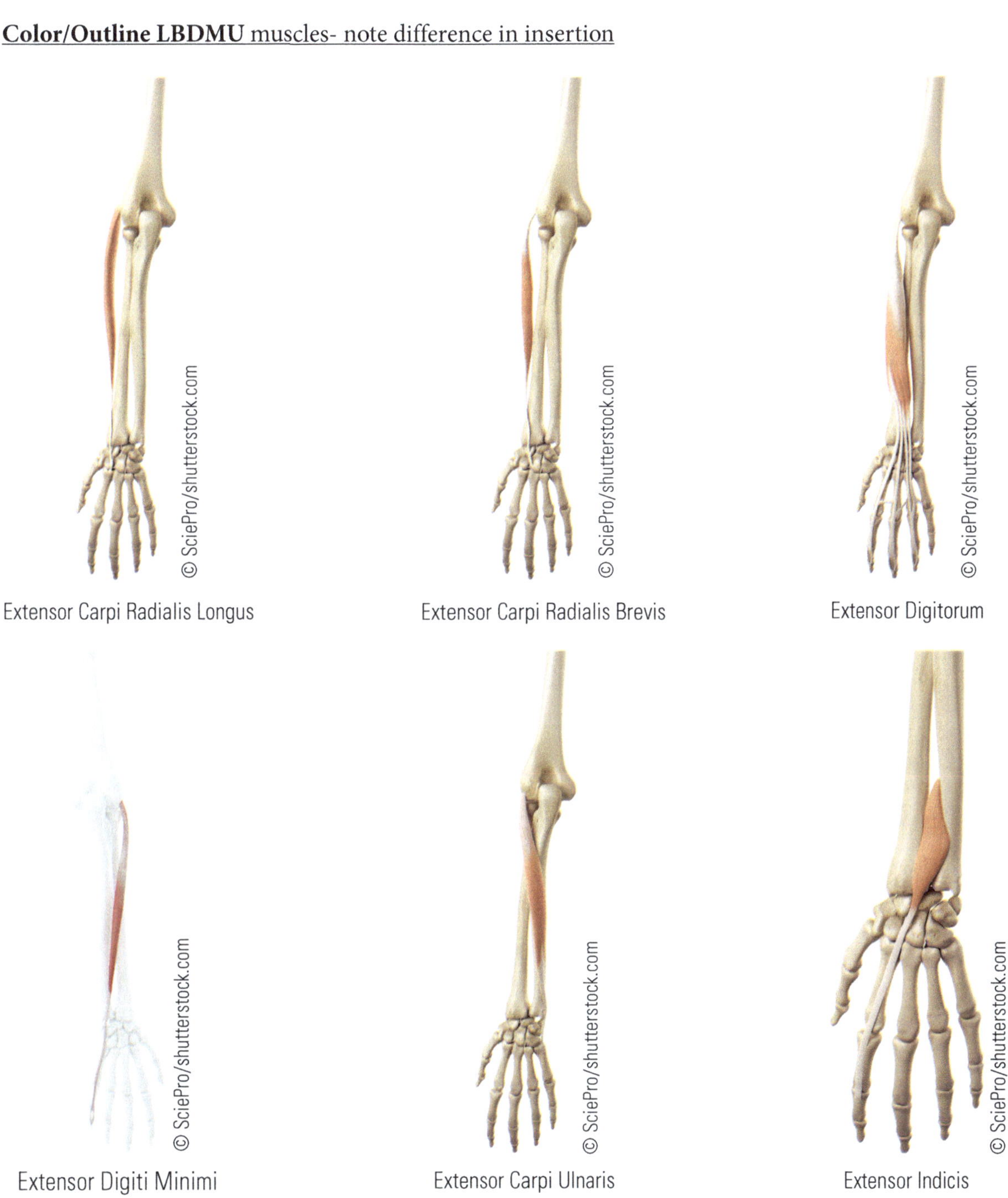

Extensor Carpi Radialis Longus   Extensor Carpi Radialis Brevis   Extensor Digitorum

Extensor Digiti Minimi   Extensor Carpi Ulnaris   Extensor Indicis

Anatomical Snuff Box : Deep extensor compartment

The "Brevis" Sandwich: With a **Scaphoid** basement (between EPL and EPB)

Abductor Pollicis Longus (bread)   **Extensor Pollicis Brevis** (meat)   **Extensor Pollicis Longus** (bread)

(Red, most lateral)   (Green, middle to Radius)   (Blue, most posterior to ulna)

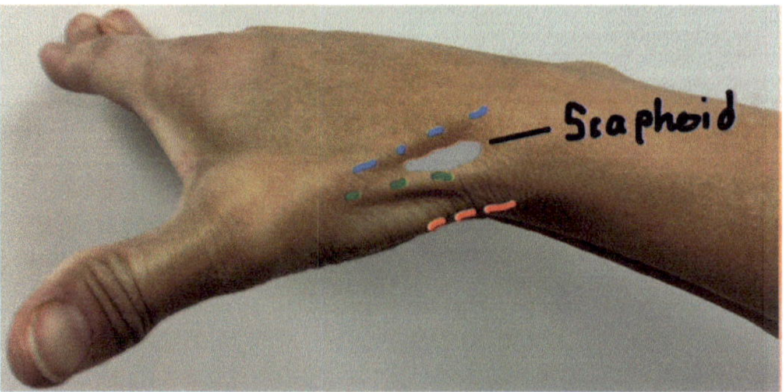

Source: Shireen Rahman

Hand Muscles (in brief)

Thenar Aspect: (thumb side)

Flexor Pollicis Brevis
Abductor Pollicis Brevis
Opponens Pollicis (deep to flexor and abductor)

Note that the Thenar and Hypothenar are connected via the **Flexor Retinaculum (Transverse Carpal Ligament)**

Hypothenar Aspect: *(pinky side)*

Flexor Digiti Minimi
Abductor Digiti Minimi
Opponens Digiti Minimi (deep)

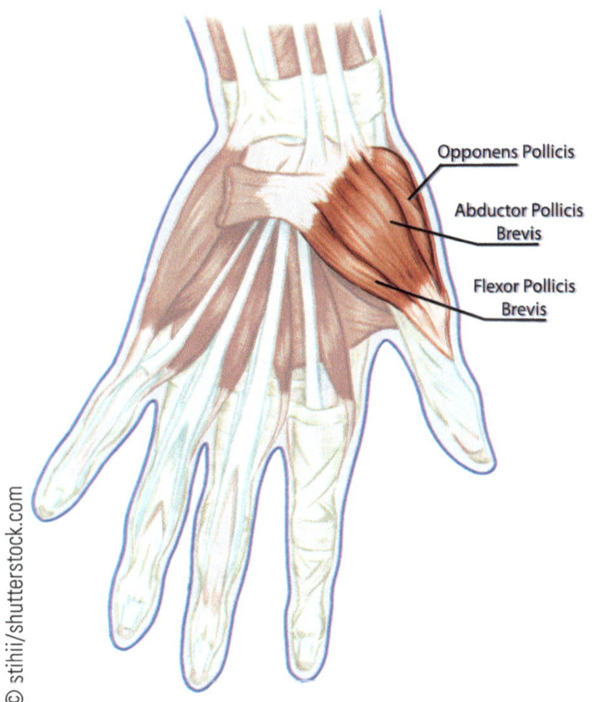

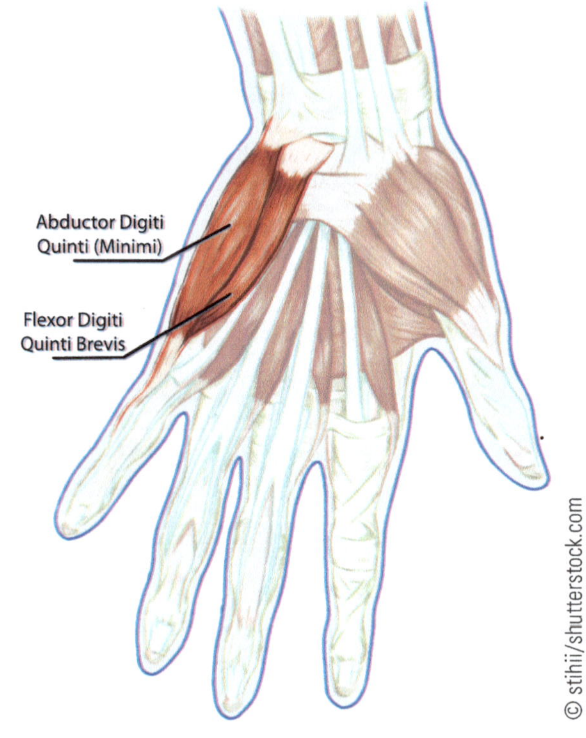

Opponens Pollicis and Opponens Digiti Minim

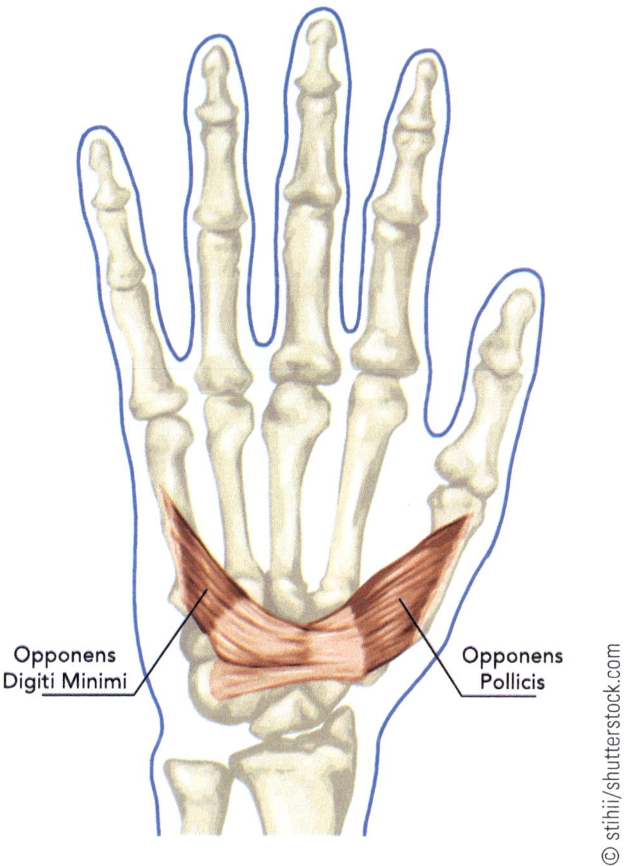

So we have flexed, extended, opposed . . .but we haven't adducted . . . and we like to add things to our lives more than take away---so. . . . how about the adductors?

Adductor Pollicis

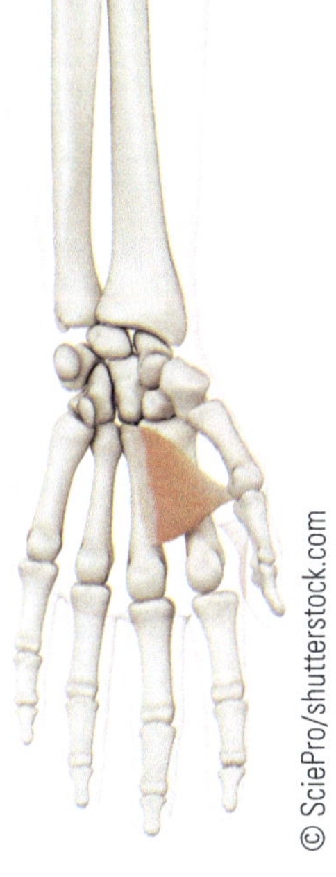

*** Additional intrinsic muscles will help adduct and abduct the remaining digits**

Intrinsic Hand Muscles:

Lumbricales:

- Hang out alongside the Flexor Digitorum Profundus
- Flex MCP 2-5
- Extend IP 2-5

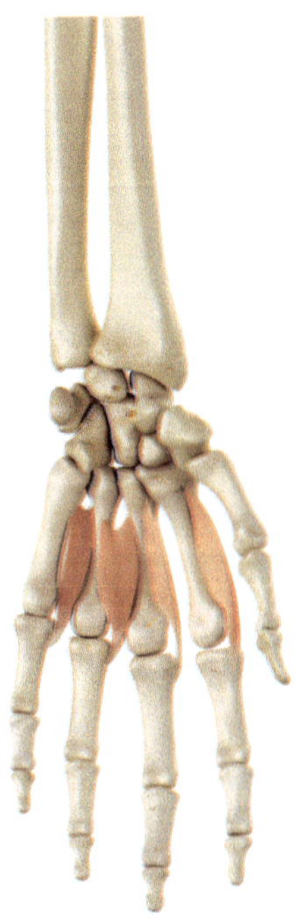

Palmar Interossei (PAD)

- PAD=Palmar ADDuction
- Palmar 2, 4, 5
- Unipennate (one sided muscles)
- Adduction 2, 4, 5 (add to reference)

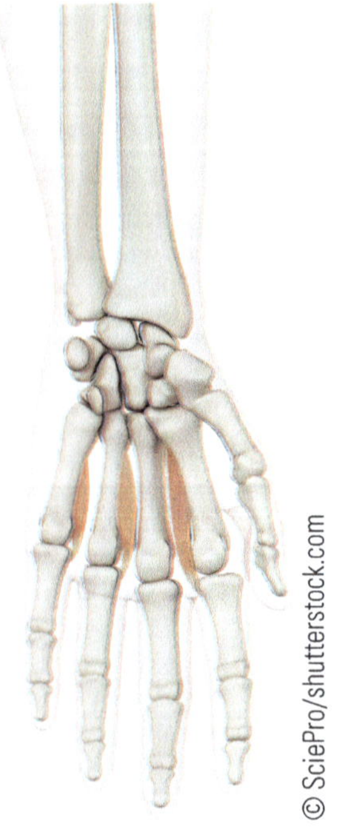

First Remember:

The reference finger for ABduction and ADduction is the 3rd digit.

Dorsal Interossei (DAB)

- DAB= Dorsal ABDuction
- Dorsal (1-4)
- Bipennate (2 sided muscle)
- Abduction 2-4 (away from reference)

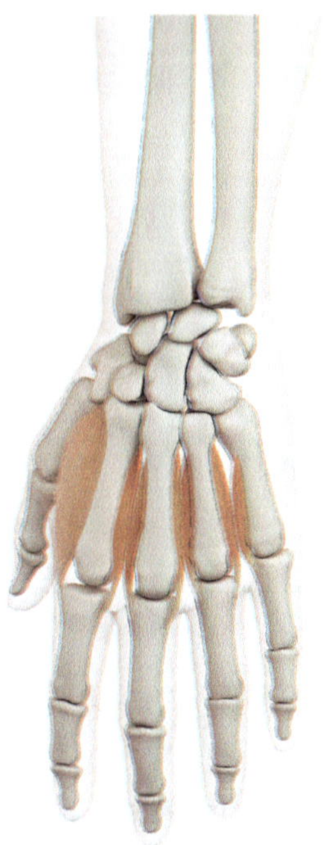

Nerves of Upper Extremity

Trust Me Moment

Make sure you do not make this harder than it has to be . . . pay attention to the names of the nerves as they will usually provide a clue to the location and muscles they innervate. Also, knowing the muscles is still the key-know how to identify the muscles first and foremost. Knowing the muscle compartments will also help tremendously.

Samples:

Dorsal Scapular Nerve: Think posterior scapula—Rhomboids!!!

Long Thoracic Nerve: Think longitudinal along the rib cage- Serratus Anterior!!!

OK!!! LET's DO THIS!!!!

© Leonard Zhukovsky/shutterstock.com

Brachial Plexus

- Forming from cervical ventral rami and thoracic ventral rami (C5-T1)
- Ends at the axilla region as "terminal nerves"
- Learning the brachial plexus is FUN!!! **Learn a REFERENCE as this is where typically tagged on cadavers; in bold in "getting there" column**
- Based on the following foundation:

Root	Trunks	Divisions	Cords	Terminal Nerves
C5	Superior	Anterior	Lateral	Musculocutaneous
C6	Middle	Posterior	Medial	Median
C7	Inferior		Posterior	Ulnar
C8				Radial
T1				(Axillary)

Brachial Plexus in Pieces

__C4__

__C5__

__C6__

__C7__

__C8__

__T1__

→

Nerve	Branching directly from (Anatomically)	Root Origin *Ventral Rami*	Innervates *Think functions Affected*	Getting there . . . (typically) *To help in the lab*
Accessory	Cranial Nerve	NA	Trapezius	NA
Dorsal Scapular	Root C4, C5	C4, C5	Rhomboid Minor Rhomboid Major Levator Scapulae	Supraclavicular through the middle scalene, heading posteriorly to **LS and Rhomboids**
Long Thoracic	Root C5, C6, C7	C5,C6, C7	Serratus Anterior	Supraclavicular longitudinally down the **lateral thorax**; superficial to the serratus anterior
Suprascapular	Superior Trunk	C4, C5, C6	Supraspinatus Infraspinatus GH joint	Supraclavicular; **from superior trunk it moves laterally** and posteriorly to scapular notch; it runs under the transverse scapular ligament to **S/I muscles**
Lateral Pectoral	Lateral Cord	C5, C6, C7	Pectoralis Major	Infraclavicular; **between the pectoralis major/pectoralis minor**; traced back to a cord
Medial Pectoral	Medial Cord	C8, T1	Pectoralis Major Pectoralis Minor	Infraclavicular; comes from under the pectoralis minor, pops through the under surface of the pectoralis major; *is actually **lateral to the lateral pectoral nerve**
Upper Subscapular	Posterior Cord	C5	Upper subscapularis	Infraclavicular; leaves posterior cord to run on top of **the upper subscapularis** muscle (anteriorly)
Thoracodorsal	Posterior Cord	C6, C7, C8	Latissimus Dorsi	Infraclavicular; between the upper and lower subscapular N.; Will move directly towards the thoracic side of **latissimus dorsi**
Lower Subscapular	Posterior Cord	C6	Lower Subscapularis Teres Major	Infraclavicular; most lateral of posterior cord nerves; on top of **lower subscapularis** (anteriorly)

Terminal Nerves

Nerve	Branching directly from (Anatomically)	Root Origin *Ventral Rami*	Innervates *Think of functions Affected*	Getting there . . . (typically) *To help in the lab*
Musculocutaneous	Lateral Cord	C5, C6, C7	Anterior Brachium: Coracobrachialis Biceps Brachii Brachial	Leave lateral cord **popping into the coracobrachialis**- will continue in between the biceps and the brachialis; ends in a cutaneous nerve
Median	Lateral Cord Medial Cord	C5, C6-T1	Anterior Antebrachium PFPF . . . Pronator Teres Flexor Carpi Radialis Flexor Digitorum Sup. Thenar muscles To Palmar 1-3 and half 4	Forms the middle of the "M"; Best reference is as it pops through the **pronator teres** to enter the anterior antebrachium; will be seen again at the carpal tunnel
Ulnar	Medial Cord	C7, C8, T1	Ulnar side Flexor Carpi Ulnaris Flexor Digitorum Profundus Hypothenar muscles Deep Hand To Palmar/Dorsal 4,5	Most medial aspect of "M"; Best reference is as it moves through the **ulnar groove** of the humerus; will run just under the flexor carpi ulnaris
Axillary	Posterior Cord	C5, C6	Deltoid Teres Minor GH joint	Best reference: moves directly to the **armpit**
Radial	Posterior Cord	C5-T1	All posterior (brachium and antebrachium); To Dorsal 1-3 and half 4	Best reference: pops into the **triceps brachii;** will continue through in the forearm through the **supinator** to stay posterior and run with the **brachioradialis**

Median Nerve

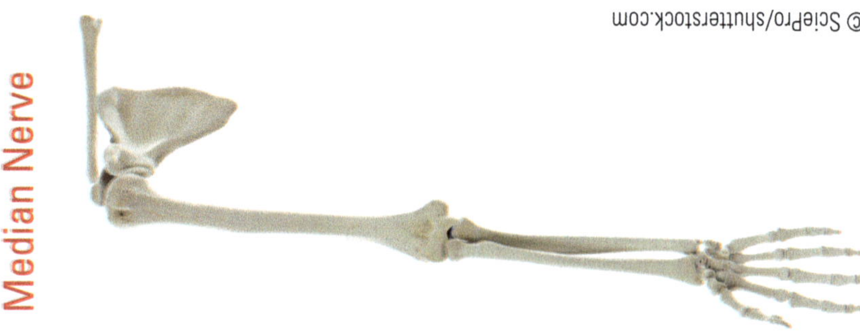

© SciePro/shutterstock.com

Musculocutaneous Nerve

© SciePro/shutterstock.com

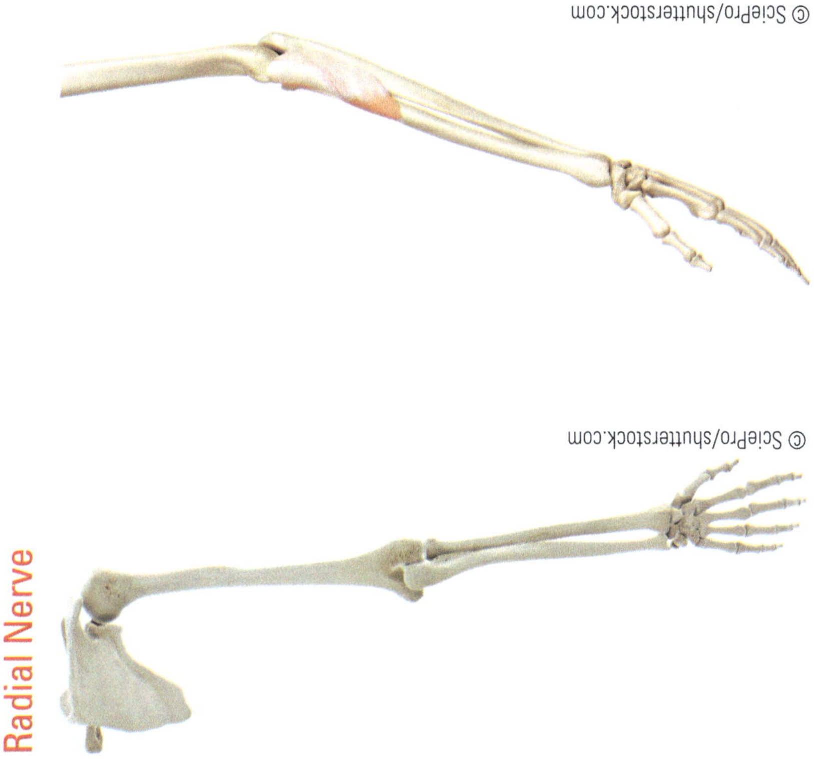

Radial Nerve

© SciePro/shutterstock.com

Ulnar Nerve

© SciePro/shutterstock.com

Sensory Distribution

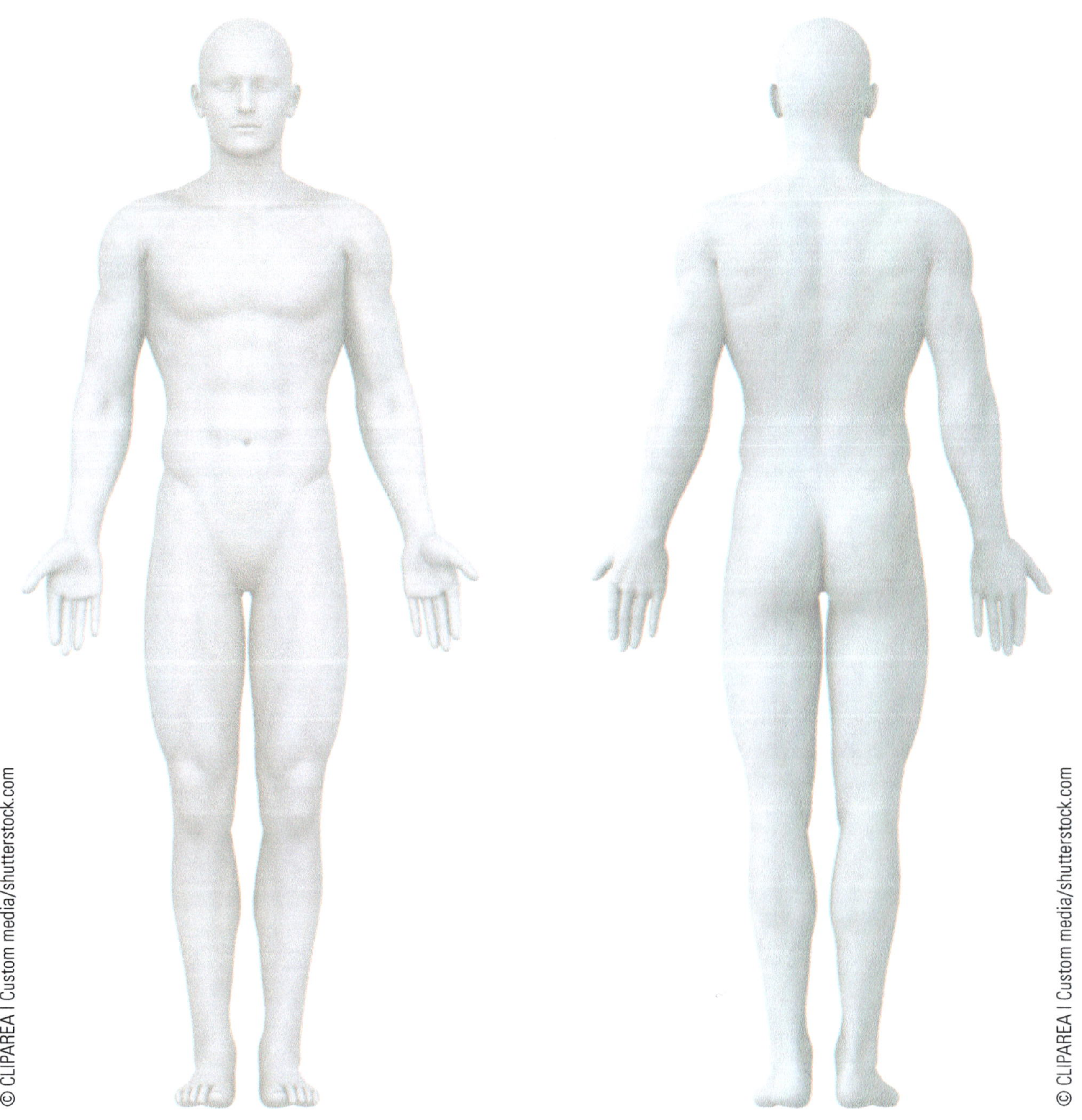

Arteries of the Upper Extremity

Quick Hits:

1. The **Aortic Arch** will have 3 primary branches providing blood supply to the upper extremity:
   a. **Brachiocephalic Trunk** ("brachio" = arm; "cephalic" = head; "trunk" = has branches)
      i. **Right Common Carotid Artery**: branch to right neck/head
      ii. **Right Subclavian Artery:** branch to right brachium/antebrachium
   b. **Left Common Carotid Artery:** to neck/head
   c. **Left Subclavian Artery:** to left brachium/antebrachium
2. The subclavian arteries will transition into the **Axillary Artery**
3. The axillary artery will transition into the **Branchial Artery**
4. The brachial artery will branch into the **Radial Artery and Ulnar Artery**

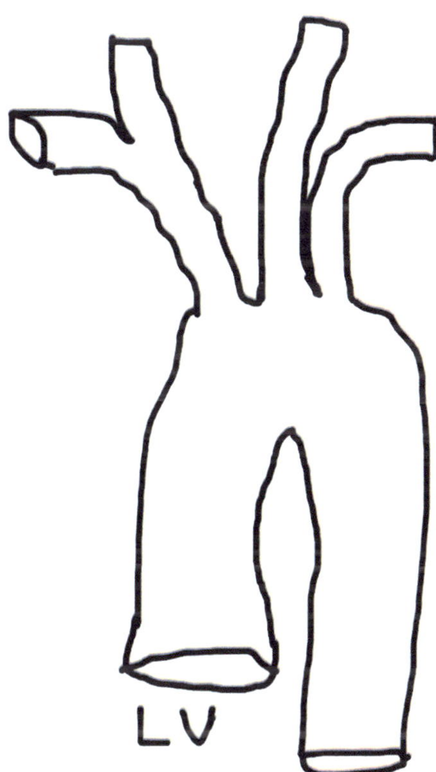

Source: Shireen Rahman

Basic Drawing Right Side-Arteries

And with Skeletal References (Left Side)

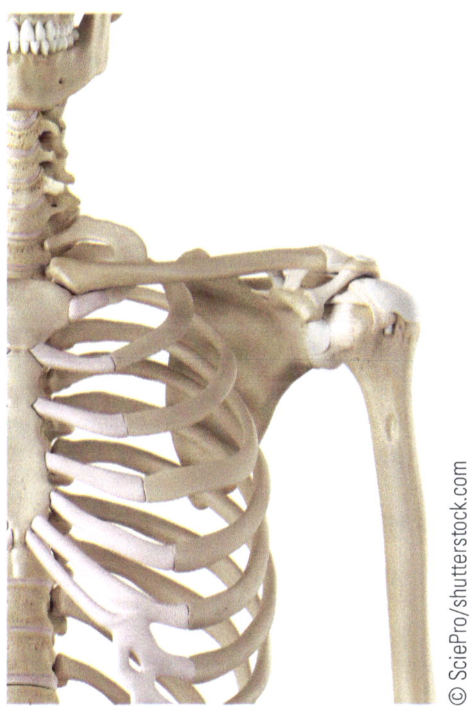

Summary of Shoulder Girdle Arteries

Vic thinks it's smart to let Shireen act pretty bitchy
Arteries ARE NOT exact with supply as they can spread their joy through branches; it helps to think regional!!!

SUBCLAVIAN ARTERY BRANCHES	Supplies	Getting there . . .
Vertebral *Vic*	*Reference artery*	*Ascends from subclavian to travel through* **transverse foramen** *of cervical vertebrae*
Thyrocervical Trunk *thinks*		
1. Inferior thyroid	*Reference artery*	*Ascends to the thyroid gland*
2. Ascending cervical	*Reference artery*	*Ascends cervical region*
3. <u>Transverse cervical</u>	Trapezius	**Moves across supraclavicular** *region to upper trapezius*
4. <u>Dorsal scapular</u>	Rhomboids Levator Scapulae	*May branch from transverse scapular OR come directly from the subclavian artery; will* **loop posteriorly** *to posterior scapula*
5. <u>Suprascapular</u>	Supraspinatus Infraspinatus Some Trapezius	*Moves across the supraclavicular region and moves posterior to* **move over the transverse scapular ligament** *to reach supraspinatus, moves lateral border of the spine of scapula to reach* **infraspinatus**
Internal Thoracic (*it's*)	*Reference artery*	*Descends inside the thorax*
Superior Thoracic (*smart*)	Upper Serratus Anterior	**Descends medial to the pectoralis minor** *to the 1st and 2nd intercostal spaces (short artery)*
Thoracoacromial Trunk (*to***)**		*Cadavers*
1. Clavicular branch	To clavicle	*Are*
2. Acromion branch	To acromion process	*Dead*
3. Deltoid branch	To deltoid	*People*
4. Pectoral branc	To pectorals	

(*Continued*)

SUBCLAVIAN ARTERY BRANCHES	Supplies	Getting there . . .
AXILLARY ARTERY BRANCHES		
Lateral Thoracic (*let*)	Serratus Anterior	Descends **lateral to the pectoralis minor** to the lateral rib cage
Suprascapular Trunk (*Shireen*) 1. <u>Subscapular</u> 2. <u>Circumflex scapular</u> 3. <u>Thoracodorsal</u>	Subscapularis Teres minor Lower infraspinatus Latissimus Dorsi Teres Major	**Descends onto the subscapularis** aspect of the scapula Continues from subscapular artery and **then swings around the lateral border** of the scapula to the infraspinatus; will connect with the suprascapular Continuation of subscapular artery and descends towards the **latissimus dorsi**
Anterior Humeral Circumflex (*act*) **Posterior Humeral Circumflex** (*pretty*)	Think regional Head of humerus GH joint Deltoid Upper Biceps Brachii Coracobrachialis Triceps Brachii Teres Minor (at axilla)	Anterior is smaller than posterior; will pass to the anterior surgical neck and eventually meets up with the posterior humeral circumflex Posterior moves to the posterior surgical neck of humerus

<u>**Anastamoses:**</u> a connection between two; the anterior and posterior humeral circumflex arteries with anastamose with each other as they circle the surgical neck; the circumflex scapular artery with anastamose with the suprascapular artery.

Ok..let's do the rest!!!

Primary arteries only
Easy references (and typically where tagged) are in bold

BRACHIAL ARTERY		
Brachial	Anterior brachium	Continuation of the axillary artery as it leaves the axilla; Lies deep to the Basilic Vein and the "M" of the Brachial Plexus; will continue until the cubital fossa
Deep Brachial (Profunda brachii)	Posterior brachium	Just inferior to the teres major it takes a dive into the **triceps brachii** with the radial nerve
BRACHIAL ARTERY BRANCHES		Branches begin at the cubital fossa
Radial	Lateral antebrachium	Upon leaving the cubital fossa, it will generally follow the outline of the **brachioradialis** as it reaches the wrist. It will be just **lateral to the tendon of the flexor carpi radialis** to enter the hand; Radial Pulse
Ulnar	Anterior and medial anterbrachium (PFPFF)	Upon leaving the cubital fossa, it will travel through the pronator teres to the **ulnar side** of forearm; it will move **under the flexor digitorum profundus and the flexor carpi ulnaris**; it will enter the hand lateral to the flexor carpi ulnaris tendon
Common Interosseous		**Branches from the ulnar artery laterally** to move towards the interosseous region;
1. Anterior interosseous	Deep anterior antebrachium	Runs **anteriorly** between the radius and ulna
2. Posterior interosseous	Posterior anterbrachium; LBDM	Branches from the common interosseous artery to pop posteriorly through the interosseous membrane The anterior and posterior interosseous will eventually anastomose distally; anterior will pop posteriorly

****Hand will be supplied by branches of both Radial and Ulnar arteries- to be shown in drawing****

Cubital Fossa

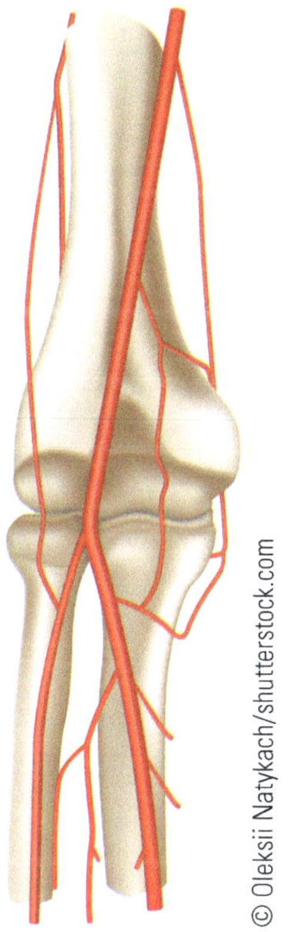

Hand

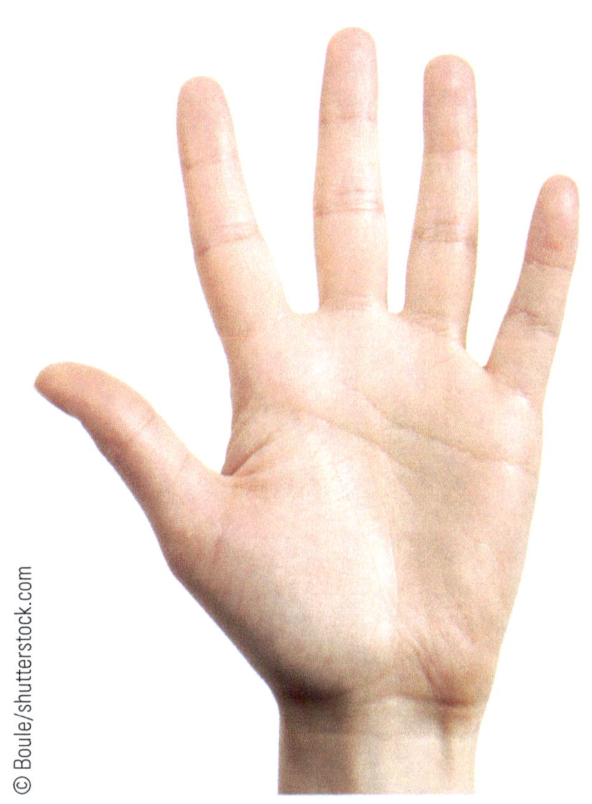

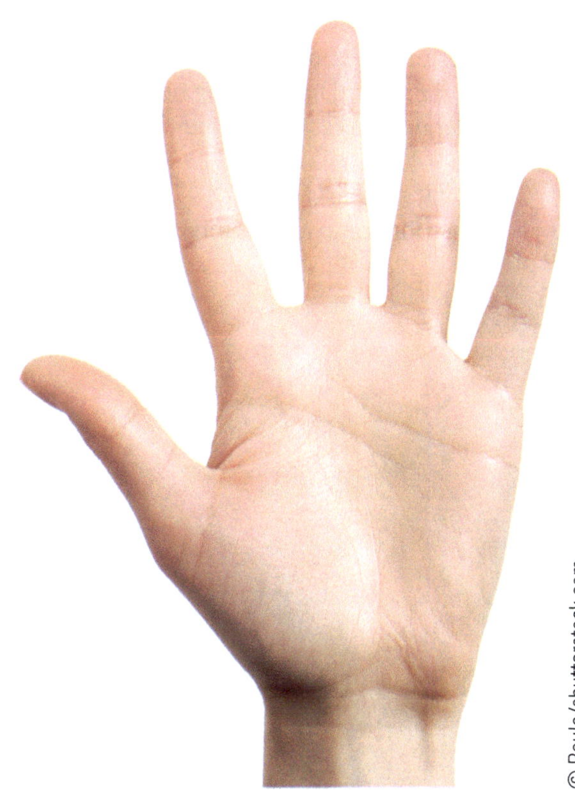

<u>Veins</u> (in general)

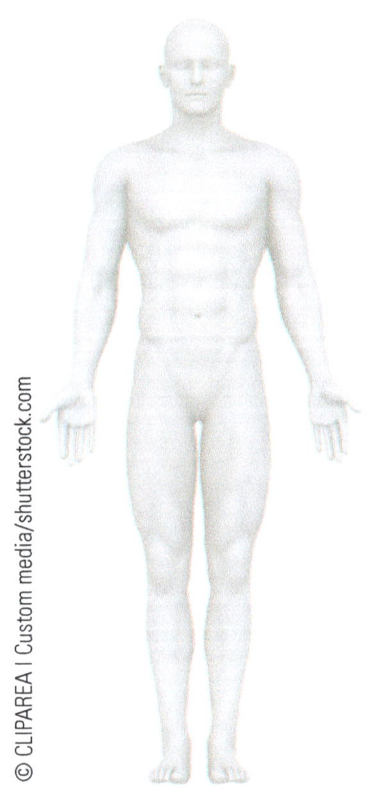

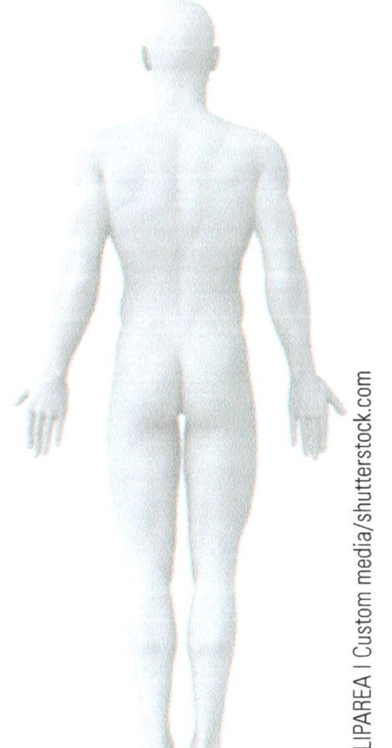

<u>Of Clinical Significance (and place to tag!!!)</u>

Quadrangular Space:

1. Boundaries
    a. Superior: inferior border of teres minor
    b. Inferior: superior border of teres major
    c. Medial: lateral border long head of triceps
    d. Lateral: surgical neck of humerus
2. Contents
    a. Axillary nerve
    b. Posterior humeral circumflex artery and vein

Upper Triangular Space:

1. Boundaries
    a. Superior: inferior border of teres minor
    b. Inferior: superior border of teres major
    c. Lateral: long head of triceps brachii
2. Contents
    a. Circumflex scapular artery/vein

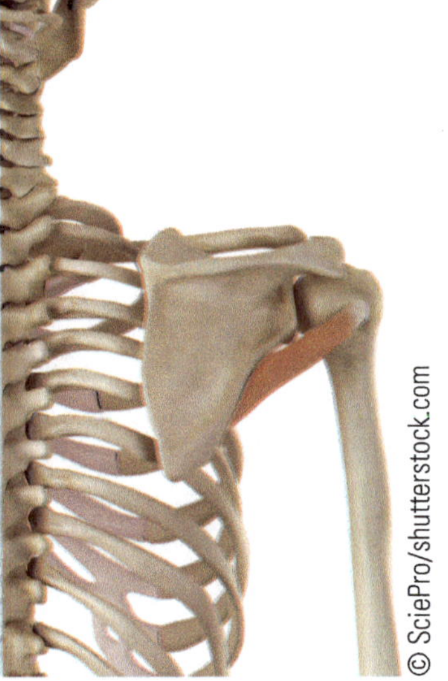

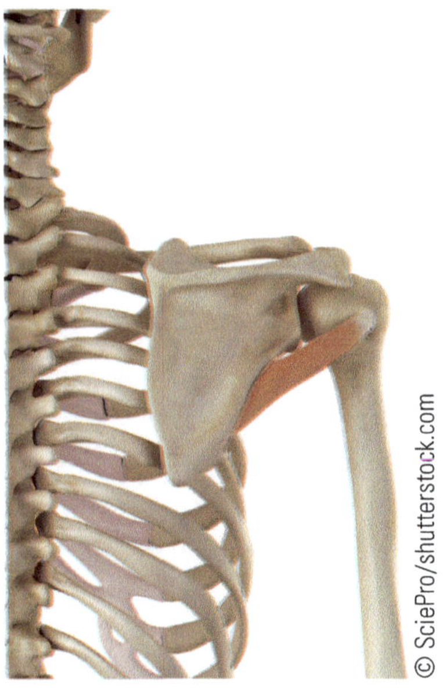

Lower Triangular Space (triangular interval)

1. Boundaries
    a. Superior: inferior border of teres major
    b. Medial: lateral border of long head triceps
    c. Lateral: medial border of lateral head of triceps
2. Contents
    a. Radial nerve
    b. Deep brachial artery

Triangle of Auscultation

1. Boundaries
    a. Medial: inferior/lateral border of trapezius
    b. Lateral: inferior angle/lateral border of scapula
    c. Inferior: superior border of latissimus dorsi

Clinical Importance:

With arms crossed over chest, this is a great location for listening to breath sounds at the inferior lobes of lungs.

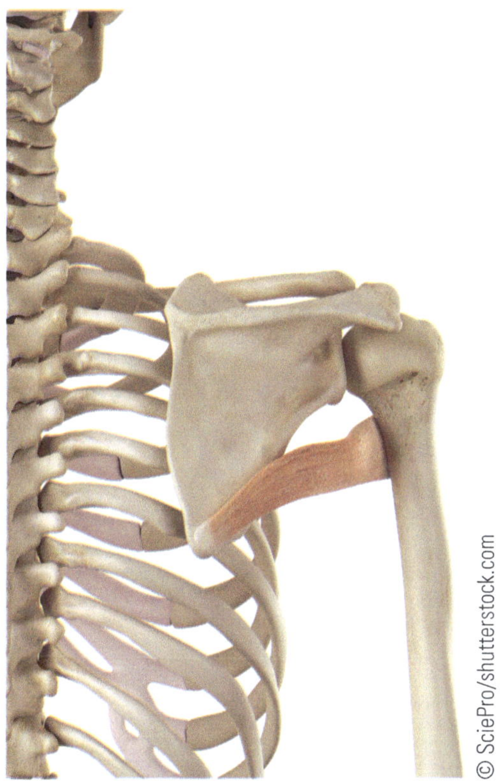

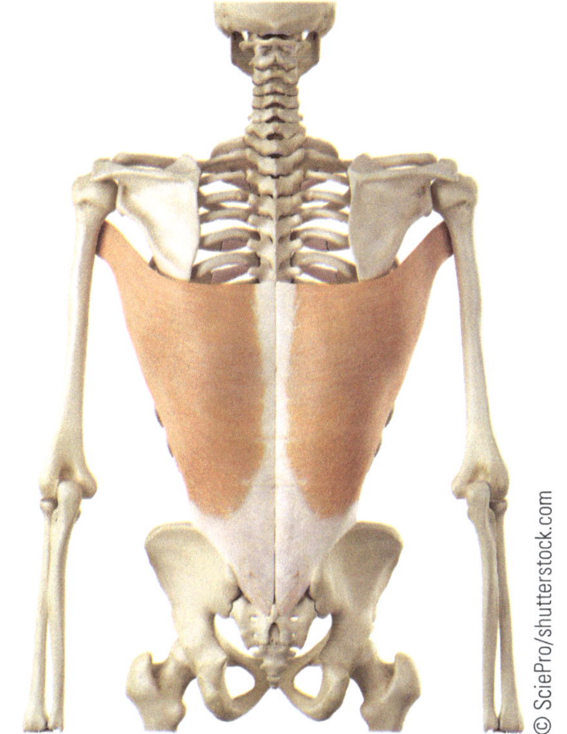

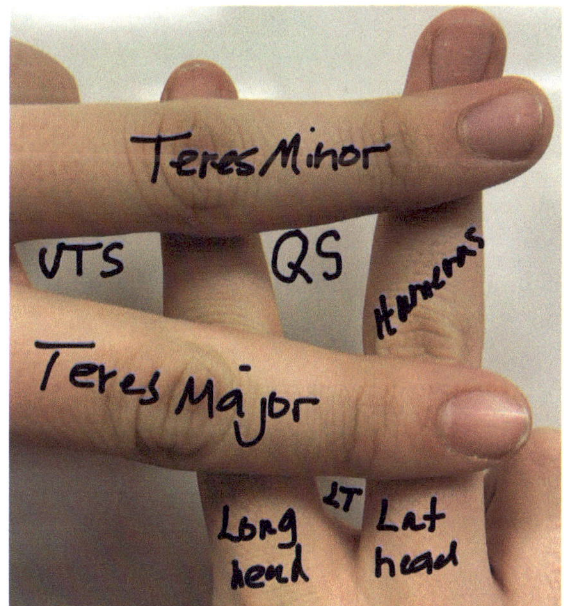

Source: Shireen Rahman

Cubital Fossa

1. Boundaries
   a. Superior: line between both epicondyles of humerus
   b. Medial: pronator teres
   c. Lateral: brachioradialis
2. Contents
   a. Brachial artery to radial and ulnar artery division
   b. Median and radial nerves
   c. Superficial to fossa
      i. Median cubital vein
      ii. Bicipital aponeurosis

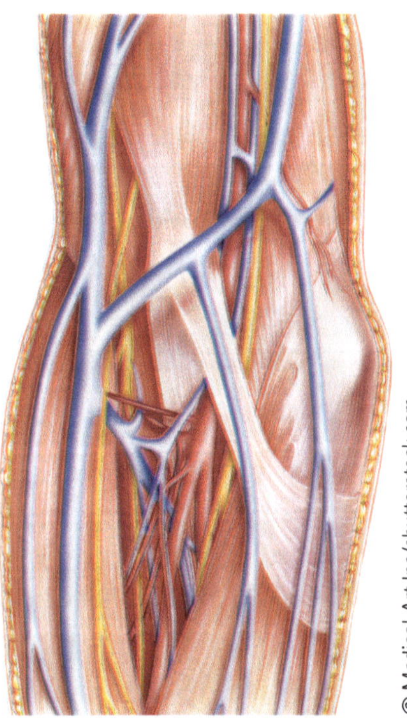

© Medical Art Inc/shutterstock.com

Cubital Tunnel

1. As the ulnar nerve runs through the <u>ulnar groove</u> of the humerus
2. <u>Cubital Retinaculum</u> will hold in place.
3. Compression can lead to sensory deficits to dorsal/palmar digits 4,5

Carpal Tunnel

1. Formed by the carpal bones bridged by the <u>Flexor Retinaculum (transverse carpal ligament)</u>;
2. Contents:
   a. Flexor digitorum superficialis tendons/synovial sheaths
   b. Flexor digitorum profundus tendons
   c. Flexor pollicis tendon
   d. Median nerve

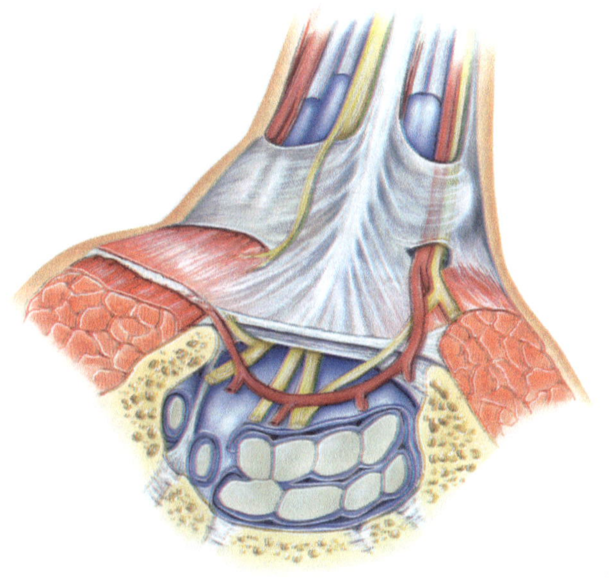

© Medical Art Inc/shutterstock.com

ONE MORE!!!!!

Tunnel of Guyon (Ulnar Tunnel)

1. Bridged by the ulna aspect of the antebrachial fascia which will attach to the pisiform. Just medial to flexor digitorum superficialis tendon
2. Contents:
   a. Ulnar nerve (location of divide into deep (motor) and superficial (sensory)
   b. Ulnar artery

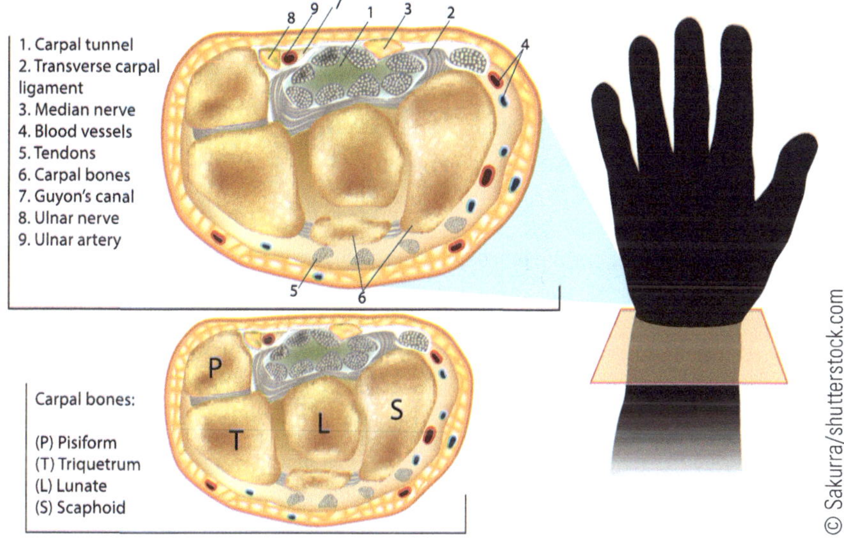

Muscle	Insertion TO	Origin	Function	N	A
ROTATOR CUFF					
Supraspinatus	Greater tuberosity of humerus	Supraspinatus fossa of scapula	<u>Cap Over:</u> Abduction	Suprascapular	Suprascapular
Infraspinatus	Greater tuberosity of humerus	Infraspinatus fossa of scapula	<u>Posterior head:</u> Lateral rotation	Suprascapular	Suprascapular
Teres Minor	Greater tuberosity of humerus	Lateral border of scapula	<u>Posterior head:</u> Lateral rotation	Axillary	Circumflex scapular A
Subscapularis	Lesser tuberosity of humerus	Subscapular fossa of scapula	<u>Anterior head:</u> Medial rotation	Subscapular Upper/ Lower	Subscapular
Other Humeral Movers					
Deltoid	Deltoid tuberosity of humerus	Clavicle Acromion process Spine of scapula	<u>Anterior</u> Flexion <u>Cap over</u> Abduction <u>Posterior</u> Extension	Axillary	Deltoid Humeral circumflex

Muscle	Insertion TO	Origin	Function	N	A
Pectoralis Major	Crest of greater tuberosity of humerus	Clavicle Sternum Rectus Sheath	<u>Anterior</u> Flexion <u>Under arm</u> Adduction <u>Anterior head</u> Medial rotation	Medial pectoral Lateral pectoral	Pectoral
Coracobrachialis	Crest of lesser tuberosity of humerus	Coracoid process of scapula	<u>Anterior</u> Flexion <u>Cross under arm</u> Adduction <u>Anterior head</u> Medial rotation	Musculocutaneous	Humeral circumflex
Latissimus Dorsi	Crest of lesser tuberosity of humerus	Iliac crest (posterior) Thoracolumbar fascia Spinous process of lower thoracic vertebrae Lower ribs Inferior angle of scapula	<u>Posterior</u> Extension <u>Cross under arm</u> Adduction <u>Anterior head</u> Medial rotation	Thoracodorsal	Thoracodorsal
Teres Major	Crest of lesser tuberosity of humerus	Inferior angle of scapula	<u>Posterior</u> Extension <u>Cross under arm</u> Adduction <u>Anterior head</u> Medial rotation	Lower Subscapular	Subscapular Circumflex scap.
Biceps Brachii	Radial tuberosity of radius	<u>Long head</u>: Supraglenoid tubercle of scapula <u>Short head</u>: Coracoid process of scapula	<u>Anterior</u> Flexion	Musculocutaneous	Humeral circumflex Brachial
Triceps Brachii	Olecronon of ulna	<u>Long head</u>: Infraglenoid tubercle of scap, <u>Lateral head</u>: Posterior humerus <u>Medial head</u>: Posterior humerus	<u>Posterior</u>: Extension	Radial	Humeral circumflex Deep brachial
SCAPULAR MOVERS					
Serratus Anterior	Medial border of scapula via sub-scapularis fossa	Lateral ribs 1-9	<u>Anterior</u> Protraction	Long thoracic	Superior thoracic Lateral thoracic
Pectoralis Minor	Coracoid process of scapula	Ribs 3-5	<u>Anterior</u> Protraction	Medial pectoral	Pectoral
Rhomboid Minor and Major	Medial border of scapula	Spinous process C6-T4	<u>Post/Medial</u> Retraction	Dorsal scapular	Dorsal scapular
Trapezius	Lateral clavicle Spine of scapula Acromion process	Occipital Spinous process Cervical through Thoracic	Elevation Retraction Depression Rotation	Accessory CN 11	Transverse cervical Suprascapualar

Muscle	Insertion TO	Origin	Function	N	A
Levator Scapulae	Superior angle of scapula	C1-C4 transverse process	Elevation	Dorsal scapular	Dorsal scapular
BRACHIUM					
Brachialis	Ulnar tuberosity of ulna	Mid anterior shaft	<u>Anterior</u> Flexion	Musculocutaneous	Brachial
Anconeus	Olecronon process	Lat. Epicondyle of humerus	<u>Posterior</u> Extension	Radial	Deep brachial
Biceps/ Triceps Above					
ANTERIOR ANTEBRACHIUM					
Pronator Teres	Radius	Medial epicondyle of humerus	Pronation of prox R/U joint	Median	Ulnar
Flexor Carpi Radialis	Base of 2nd meta-carpal (palmar side)	Medial epicondyle of humerus	Elbow: weak flexion Radiocarpal: flexion abduction	Median	Radial
Palmaris Longus	Palmar aponeurosis	Medial epicondyle of humerus	Elbow: weak flexion Radiocarpal: flexion	Median	Ulnar
Flexor Digitorum Superficialis	Middle phalanges 2-5 (palmar)	Medial epicondyle of humerus	Elbow: weak flexion Radiocarpal: flexion MCP: flexion DIP: flexion	Median	Ulnar
Flexor Carpi Ulnaris	Pisiform, hook of hamate, base of 5th metacarpal	Medial epicondyle of humerus	Elbow: weak flexion Radiocarpal: flexion adduction		
Flexor Pollicis Longus	Palmar distal pha-lange of thumb	Anterior radius	Radiocarpal; flexion 1st CMC: flexion 1st MCP: flexion 1st IP: flexion	Median (Anterior interosseous)	Anterior interosseous
Flexor Digitorum Profundus	Palmar distal pha-lange 2-5	Ulna	Radiocarpal; flexion MCP: flexion PIP/DIP: flexion	Ulnar	Anterior interosseous Ulnar
Pronator Quadratus	Anterior radius	Anterior ulna	Pronation	Median (Anterior Interosseous)	Anterior interosseous
NEUTRAL					
Brachioradialis	Styloid process of radius	Lateral supracondylar ridge of humers	Elbow: Flexion	Radial	Radial
POSTERIOR ANTEBRA-CHIUM					

Extensor Carpi Radialis **L**ongus	Dorsal base of 2nd metacarpal	Lateral supracondylar ridge of humerus	Elbow: weak extension Radiocarpal: extension abduction	Radial	Radial
Extensor Carpi Radialis **B**revis	Dorsal base of 3rd metacarpal	Lateral epicondyle of humerus	Elbow: weak extension Radiocarpal: extension abduction	Radial	Radial
Extensor **D**igitorum	Dorsal distal phalange 2-5	Lateral epicondyle of humerus	Elbow: weak extension Radiocarpal: extension MCP/PIP/DIP: extension *Abduction of digits*	Radial via Posterior interosseous	Posterior interosseous
Extensor Digiti **M**inimi	Dorsal distal phalange 5	Lateral epicondyle of humerus	Elbow: weak extension Radiocarpal: extension MCP/PIP/DIP: extension *Abduction of 5th*	Radial via Posterior interosseous	Posterior interosseous
Extensor Carpi **U**lnaris	Dorsal base of 5th metacarpal	Lateral epicondyle of humerus	Elbow: weak extension Radiocarpal: extension adduction	Radial via Posterior interosseous	Posterior interosseous
Extensor Indicis	Dorsal 2nd distal phalange	Distal ulna	Radiocarpal: extension 2nd MCP/PIP/DIP: extension	Radial via Posterior interosseous	Posterior interosseous

Putting it All Together

<u>Putting it All Together</u>

<u>**Putting it All Together**</u>

<u>Putting it All Together</u>

CHAPTER 3

Lower Extremity

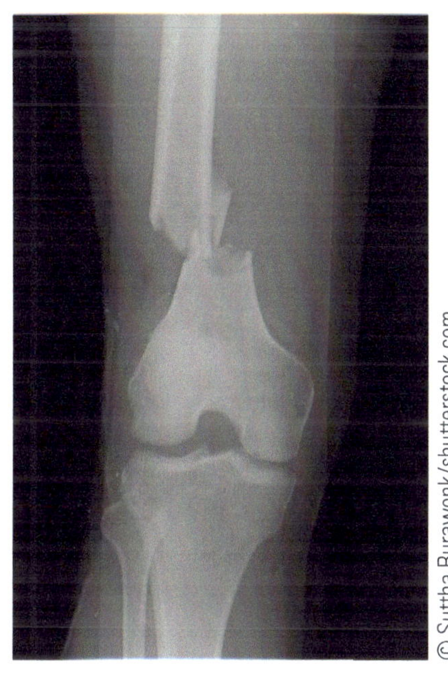

Bony Landmarks
Joints and Ligaments
Muscles
Innervations
Blood Supply

© Suttha Burawonk/shutterstock.com

Bony Landmarks of Lower Extremity

All bony landmarks are listed in a logistical assessment or-der- <u>***trace, label and/or color accordingly***</u>*. Knowing the bony landmarks will help you soon understand muscle location, muscle compartments, and muscle function.*

Anterior Ox Coxae

*Moving from the **anterior superior iliac spine (ASIS) (a)**, move posteriorly along the **iliac crest (b)**. From the middle of the iliac crest drop inferiorly to find the **iliac fossa (c)**. Return to the ASIS and move inferiorly. You will reach a bump, the **anterior inferior iliac spine (AIIS) (d)**. Continue and reach the **acetabular rim (e)**. Now move medially along the **superior pubic ramus (f)**. On the inner border of the ramus, you will find the **pectineal line (g)**. If you follow the pectineal line up as it curves, you find the **arcuate line (h)**. Now return to the ramus to follow it medially. You will feel a small bump, the **pubic tubercle (i)**.*

*The ox coxae comes together at the **pubic symphysis(j)***

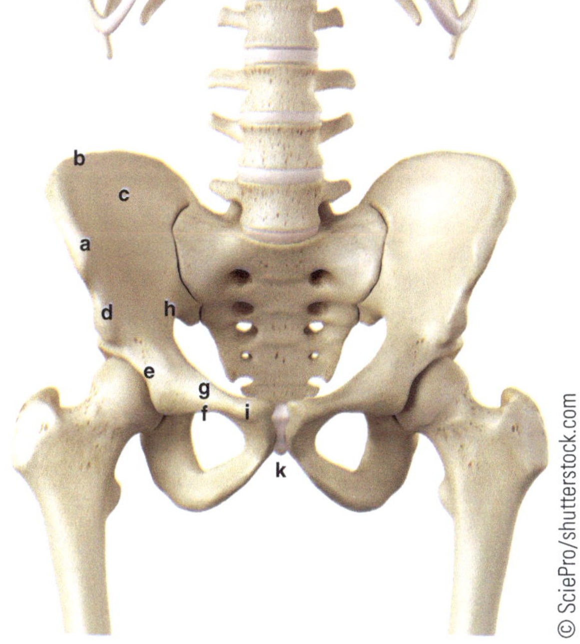

© SciePro/shutterstock.com

Muscle insertion points
Select abdominal muscles and back muscles

Posterior Ox Coxae

Moving from the **iliac crest (a)**, move posteriorly until you come to the first bump, the **posterior superior iliac spine (PSIS) (b)**, if you move just inferiorly, you find the **posterior inferior iliac spine (PIIS) (c)**. Move inferiorly into a "u" shaped structure, the **greater sciatic nerve (d)**, which will allow for passage of the sciatic nerve from anterior nerve roots to posterior muscles. Coming out of the notch, you come across a pointy **ischial spine (e)**. As you move inferiorly, you find a small "u", the **lesser sciatic notch (f)**, where the pudendal nerve will pass. Soon you find a rough wide surface, the **ischial tuberosity (g)**, or the butt bone. As you leave the tuberosity, you move on to the **ischial ramus (h)** to the **inferior pubic ramus (i)** to **pubic tubercle**, not shown here.

The round hole between is the **obturator foramen (j)**.

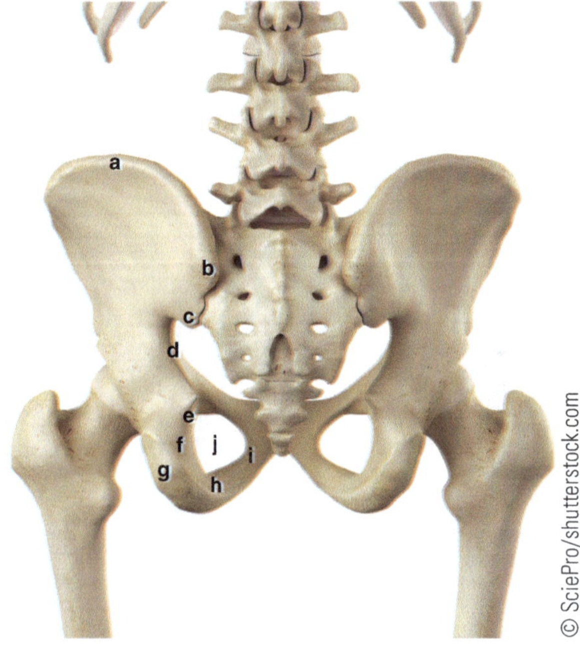

Muscle insertion points
Select abdominal muscles and back muscles

Anterior Proximal Femur

Start at the **head of the femur (a)**. Leaving the head is the **neck of the femur (b)**. The big bump on the lateral aspect is the **greater trochanter (c)**. From the greater trochanter, you will find a line, **intertrochanteric line (d)** that moves down and medially until it reached the small medial bump, **lesser trochanter (e)**. The **shaft (f)** leads us toward the distal femur.

Posterior Proximal Femur

Start at the head, to the neck, superiorly to the **greater trochanter (a)**. From the greater trochanter, you will find a line, **intertrochanteric crest (b)** to the **lesser trochanter (c)**. Directly leaving the lesser trochanter inferiorly, you find the **pectineal line (d)** then into the **gluteal tuberosity (e)**, to the **linea aspera (f)**.

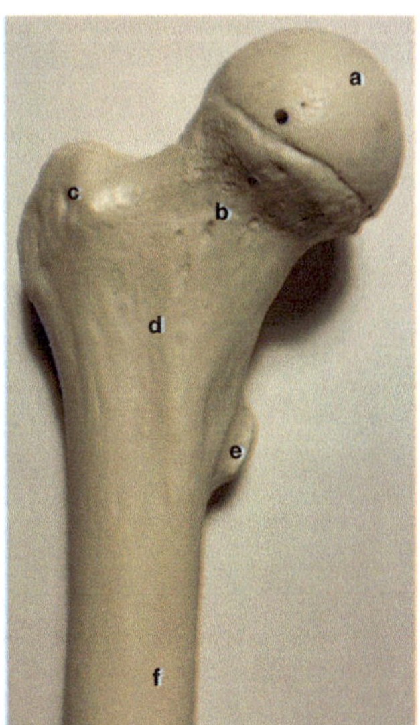

Muscle insertion for gluteals and iliospoas
Source: Shireen Rahman

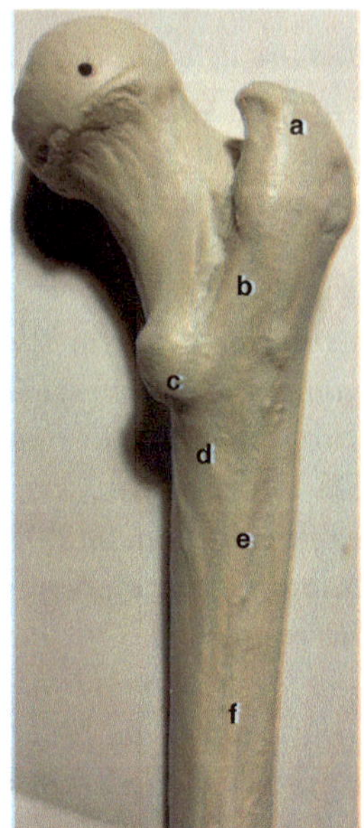

Muscles insertions for gluteals and iliospoas
Source: Shireen Rahman

Distal Femur

The **patella (a)** will sit within the **femoral trochlea (b)**. The big rounded structures, **medial femoral condyle (c)** and **lateral femoral condyle (d)** are separated by the "u"-shaped **intercondylar notch (e)**. To each side of the condyles are the **medial and lateral epicondyles of the femur (f)**.

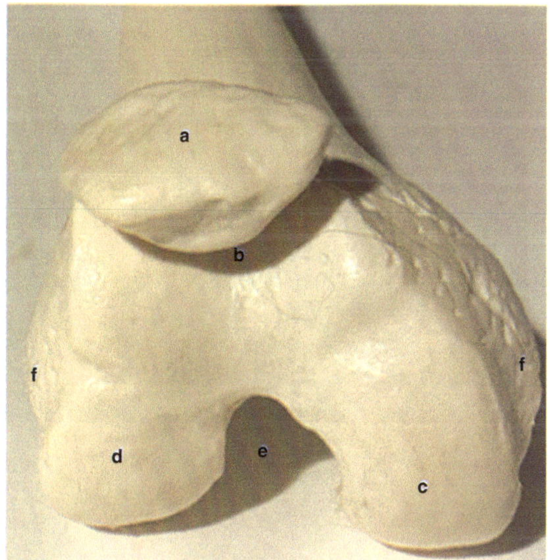

Source: Shireen Rahman

Posterior Distal Femur

Start at the **medial and lateral supracondylar ridges (a)**. As you exit the medial supracondylar ridge, there is small bump, **the adductor tubercle (b)**, an insertion site for the adductor magnus. This will empty into the **medial epicondyle (c)** to **medial condyle (d)**. See the intercondylar notch again. On the lateral side you will again see the lateral epicondyle and lateral condyle of the femur.

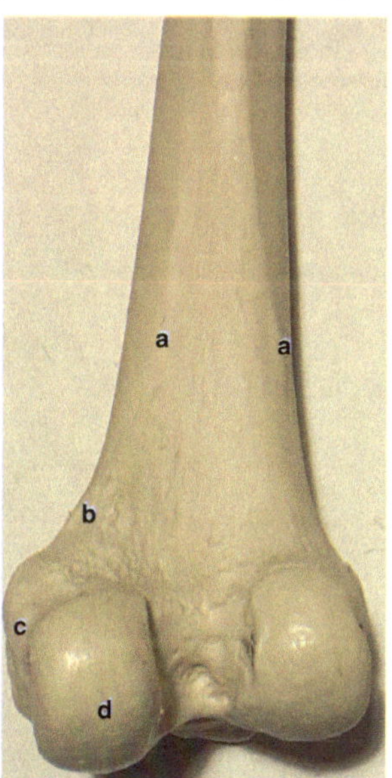

Muscle insertions for adductor magnus, gastrocnemius
Source: Shireen Rahman

Anterior Proximal Tibia/Fibula

At the most superior surface of the tibia find the **tibial plateau (a)**. Moving inferiorly find the **medial and lateral condyle (b)**. The bump in the middle is the **tibial tuberosity (c)**. The **head of the fibula (d)** will articulate with the lateral tibia.

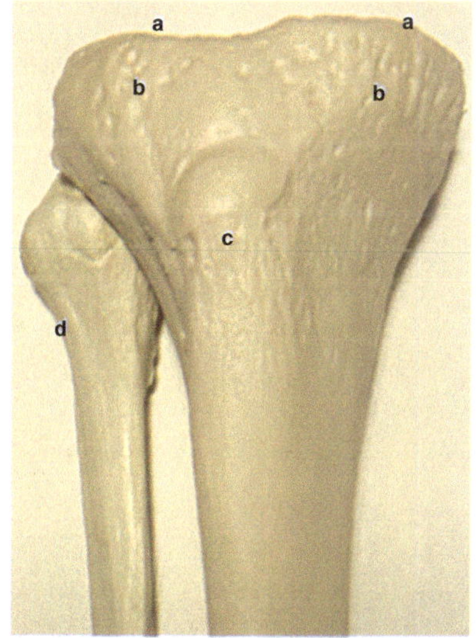

Muscle insertions for Quadriceps and Hamstrings, Sartorius, IT band
Source: Shireen Rahman

Posterior Proximal Tibia/Fibula

At the most superior surface, find the **tibial plateau (a)**, in the middle of the plateau, there is a uplift, **the intercondylar eminence (b)**, serving to anchor menisci and cruciate ligaments. Again, find the lateral and medial condyle (c). There is a line that starts level to the fibular head and leads to the medial tibia, **the soleal line (d)**.

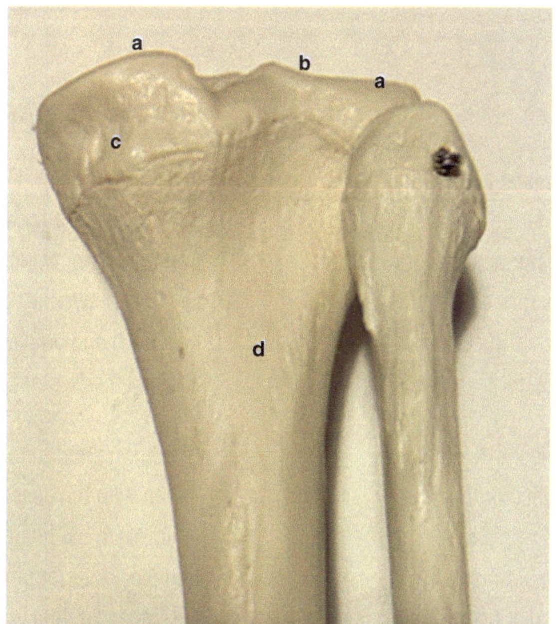

Source: Shireen Rahman

Distal Tibia/Fibula

The **lateral malleolus (a)** will be at the distal end of the fibula. The **medial malleolus (b)** will be at the distal tibia. The tibia has an **inferior articulating surface (c)** at its ditsal end to articular with the talus.

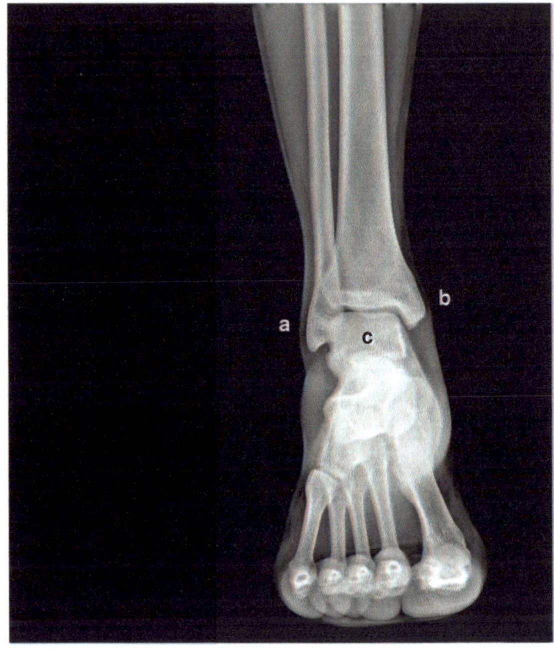

Foot/Tarsal Bones

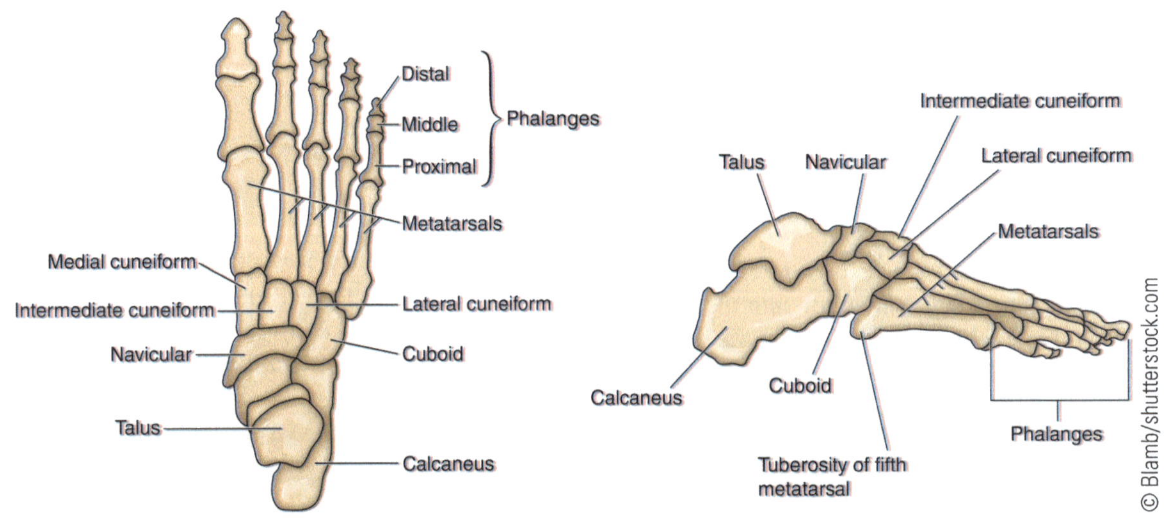

Landmarks Of Clinical Significance

Highest point of iliac crest	L3-L4 to best locate L4-L5 interspace for lumbar puncture
Anterior superior iliac crest	Reference point for assessment of hip
Posterior superior iliac crest	Surface anatomy landnmark; look for posterior dimples
Sacrum	Find just above butt crack
Ischial Tuberosity	Flex thigh and palpater at the inferior buttocks
Greater trochanter of femur	Thumb on iliac crest, move point finger down In line with pubic tubercle
Epicondyles of femur	In line with mid patella
Adductor tubercle	In line with the base of patella, just above the medial epicondyle
Gerdy's tubercle	In line with the head of the fibula—move to the anterior, lateral tibia
Tibial tuberosity	Bump on anterior proximal tibia
Fibular head	Lateral and in line with the tibial tuberostiy

<u>Femoral Head Angles:</u>

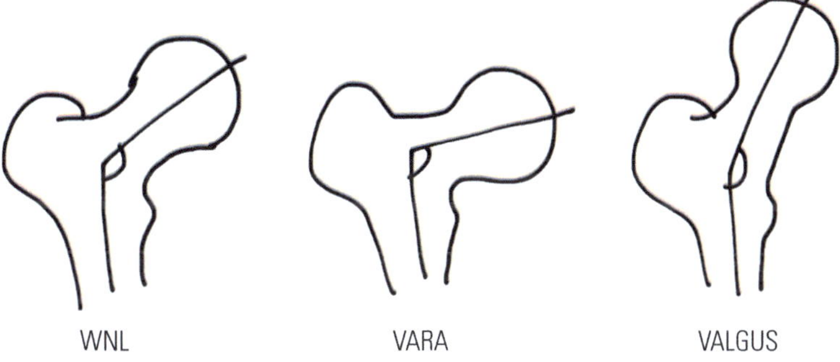

Source: Shireen Rahman

Joints and Ligaments of Lower Extremity

*Joints will provided information about the movement. Ligaments will **restrict** a movement as it lengthens.*

JOINTS

Use the bony landmarks to help you with joints

Joint	Bone to	Bone	Joint Type	Movement Type	Function
Femoroacetabular	Head of femur	Acetabulum	Ball and socket	Multi-axial	F/E Abd/Add LR/MR
Knee Joint	Femoral condyles	Tibial plateau	Condyloid *Modified hinge*	Biaxial	F/E Rotation
Patellofemoral	Articular surface of patella	Patellar surface of femur	Saddle *Gliding*	Non-axial	Glide
Talocrural	Inferior articulating surface of tibia	Talar dome	Hinge	Uniaxial	Dorsiflexion Plantar flexion
Subtalar	Under surface of talus	Calcaneous	Gliding *Hinge*	Nonaxial *Uniaxial*	Inversion Eversion
Talonavicular	Anterior talus	Navicular	Gliding	Nonaxial	Glide
Calcaneocuboid	Calcaneus	Cuboid	Gliding	Non-axial	Glide
Cuneonavicular	Cuneiform	Navicular	Gliding	Non-axial	Glide
Intercuneiform	Cuneiform	Cuneiform	Gliding	Non-axial	Glide
Tarsometatarsal	Tarsals	Metatarsals	Gliding	Non-axial	Glide
Metatarsophalangeal	Metatarsals	Phalanges	Condyloid	Bi-Axial	F/E Abd/Add
Interphalangeal (PIP/DIP)	Phalange	Phalange	Hinge	Uniaxial	F/E

The Femoral-Acetabular Joint

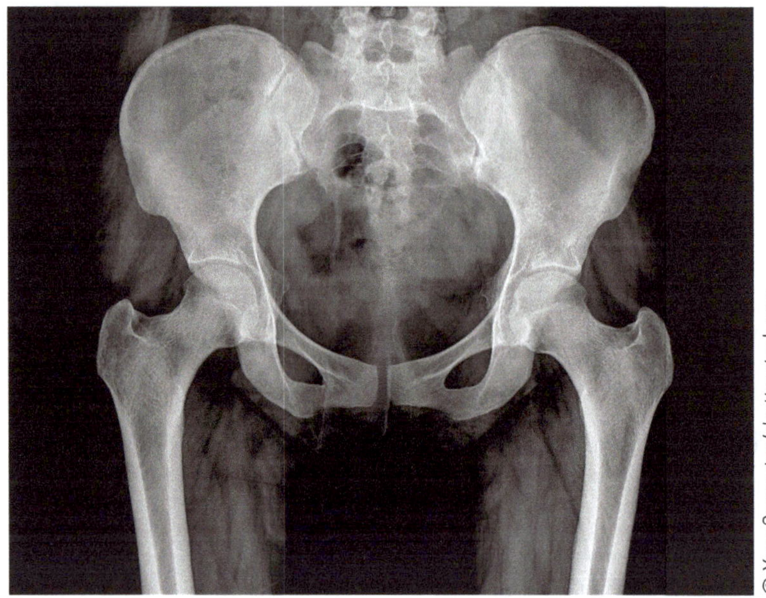

© Xray Computer/shutterstock.com

Integrity of the Joint

Unlike the glenohumeral joint that is known for its mobility via its shallow glenoid fossa, the hip has a deep articulation. Though it is a multi-axial joint, one primary function is to weightbear, over and above mobility.

The integrity of the joint starts at the acetabulum itself. The superior surface of the acetabulum contains incomplete, moon-shaped hyaline cartilage (green, lunate surface); there is a nonarticular surface (brown). The **Transverse acetabular ligament (gray** serves as a hammock at the inferior acetabulum.

The head of the femur is also lined with hyaline cartilage.

Source: Shireen Rahman

Adding to the integrity:

1. Acetabular labrum
2. Ligament of the head of the femur (also carrying blood supply to head of femur via obturator artery)
3. Joint capsule-Synovial membrane
4. Femoral ligaments (discussed below)

HIP JOINT

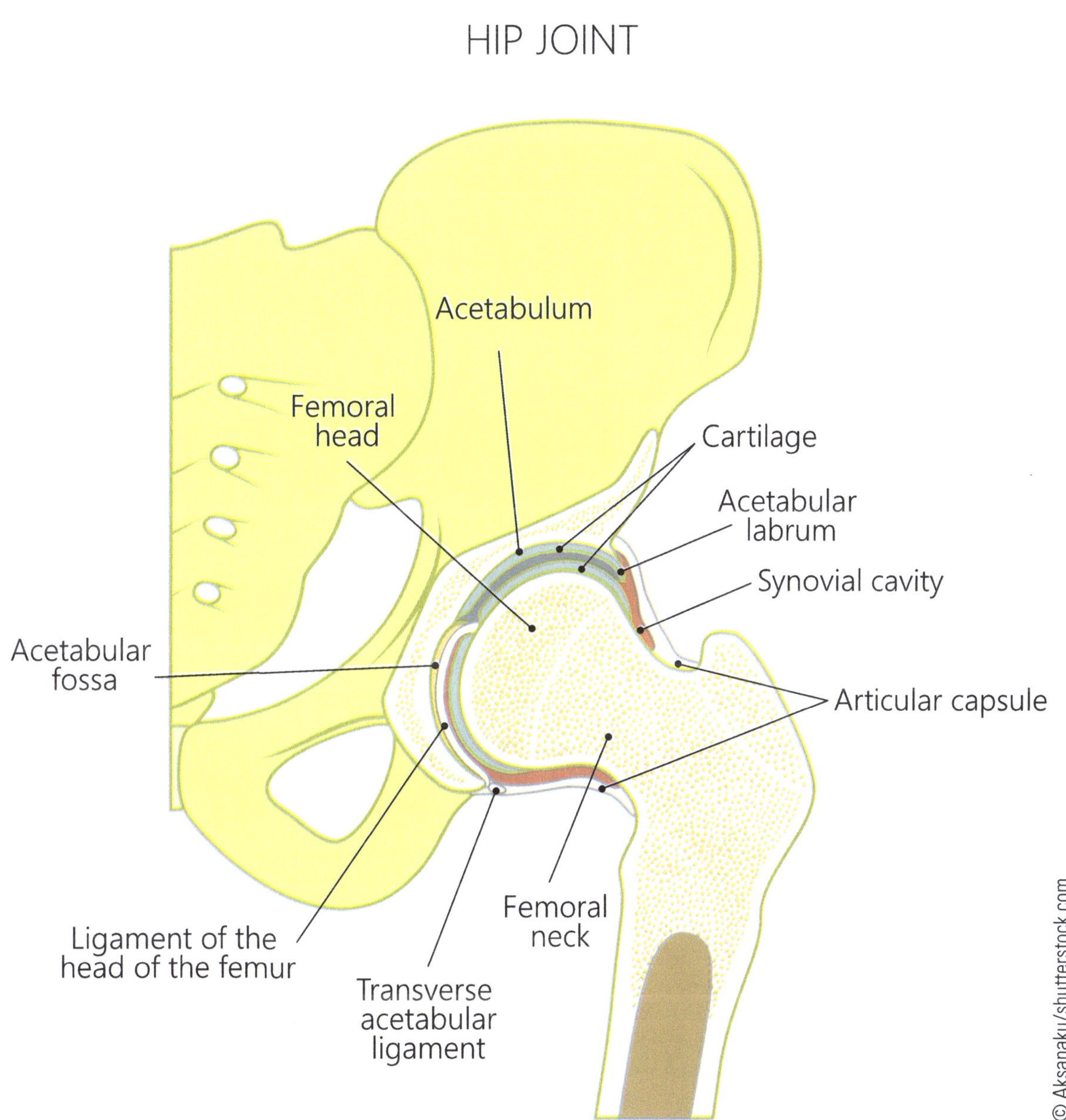

Femoral-acetabular Ligaments
Draw/Outline or Color accordingly

Iliofemoral Ligament: (color accordingly)

From the Ilium → Intertrochanteric line of femur
Y ligament

Restricts:

Ischiofemoral Ligament

From Ischium to Intertrochanteric Line

Restricts:

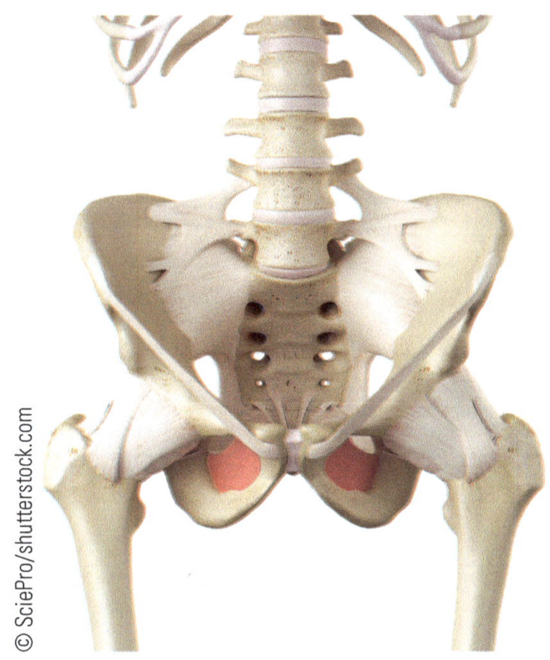

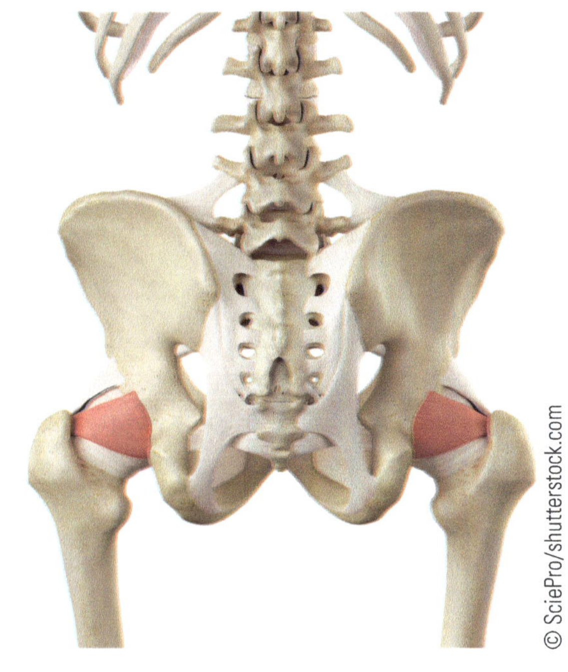

Pubofemoral Ligament (color accordingly)

From the Pubis → Intertrochanteric Line

Restricts:

Inguinal Ligament

From ASIS to Pubic Tubercle

Will function to serve as a passageway for structures
from pelvic region to lower extremity

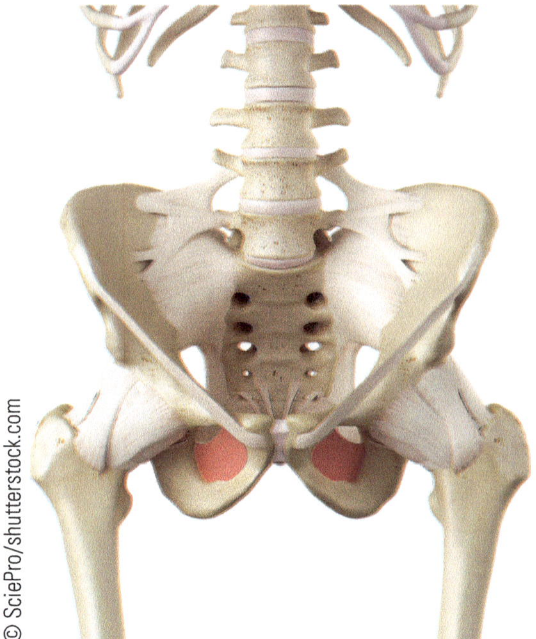

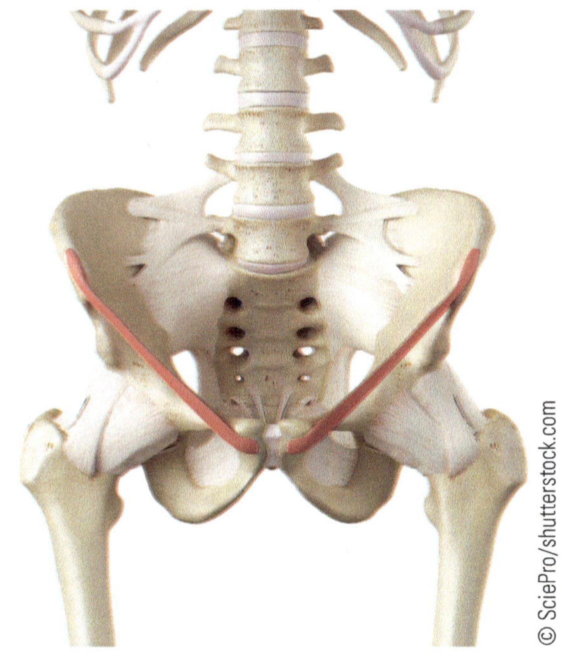

Knee Joint

Much like other synovial joints, the knee joint contains a joint capsule; fibrous outside and synovial inside. It is thin, especially anteriorly and posteriorly.

Articular cartilage lines the femoral condyles and the tibial plateau.

In addition the knee joint has **menisci** (discussed later) and **bursa.**

Extracapsular and Intracapsular ligaments and reticaculum will support the joint capsule.

Go grab a coffee!!!
Maybe skip there. . . ☺

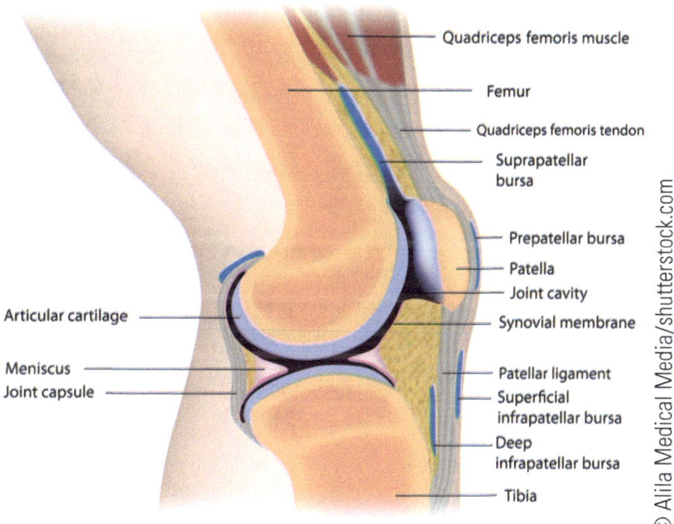

Menisci

Sit on Tibial Plateau
Cushioning to the joint

Medial Meniscus:
1. C-shaped, pliable fibrocartilage
2. Inside joint capsule BUT outside the synovial membrane
3. Attachment with the medial collateral ligament
4. Increased injury frequency

Lateral Meniscus:
1. Circular shaped pliable fibrocartilage
2. Inside joint capsule BUT outside the synovial membrane

The **Transverse Meniscus Ligament (blue)** will attach the anteior aspects (horns) of the menisci

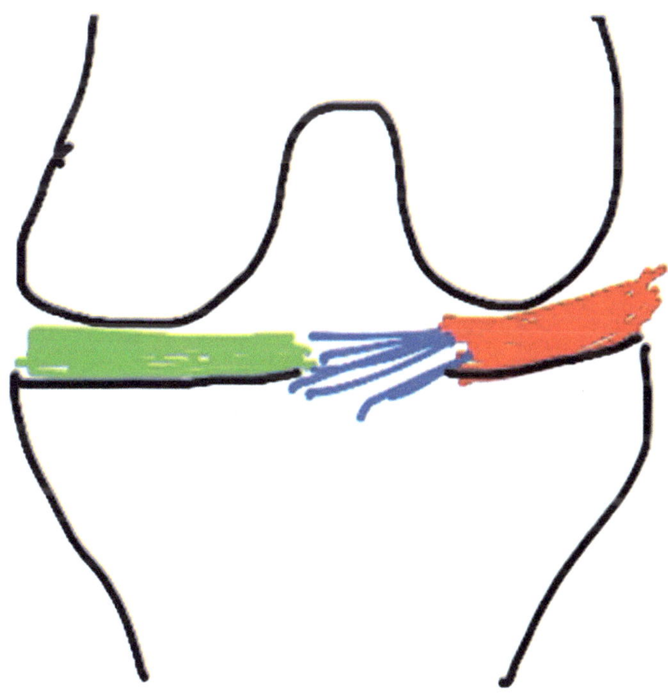

Source: Shireen Rahman

Superior (top) view of the right knee

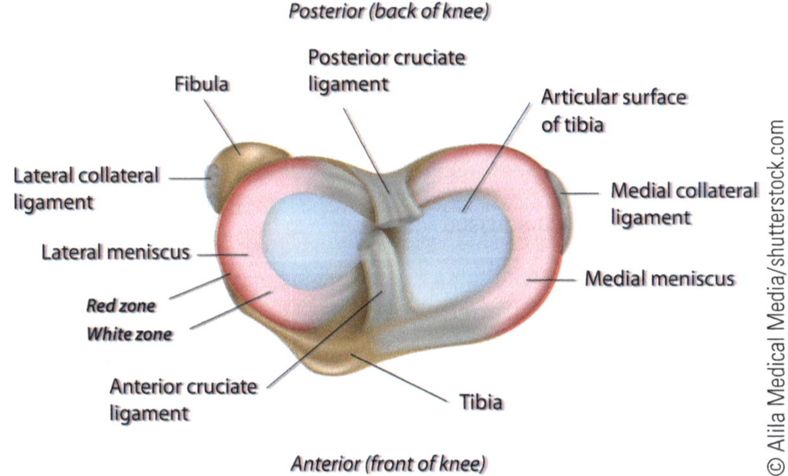

Zones of the Menisci

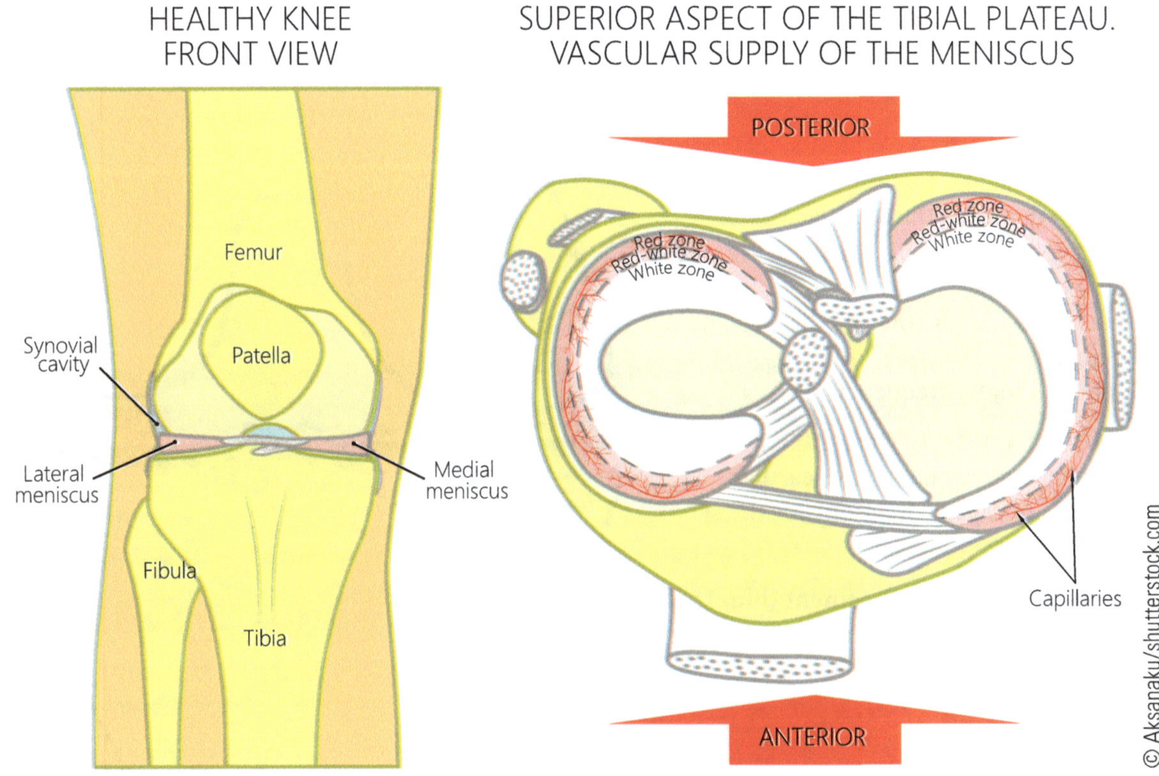

Bursa

Color/Label Accordingly

A	Prepatellar Bursa	Subcutaneous Between skin and patella
B	Infrapatellar Bursa	Subcutaneous Just inferior to patella and superficial to patellar ligament
C	Suprapatellar Bursa	Superior to patella Deep to the quadricep tendon
D	Semimembranosus Bursa	Between semimembranosus tendon and bone (posterior, not shown
E	Pes Anserine Bursa (Subsartorial)	Between medial collateral ligament and pes anserine tendons (sartorius, gracilis, semitendinosus)

KNEE JOINT

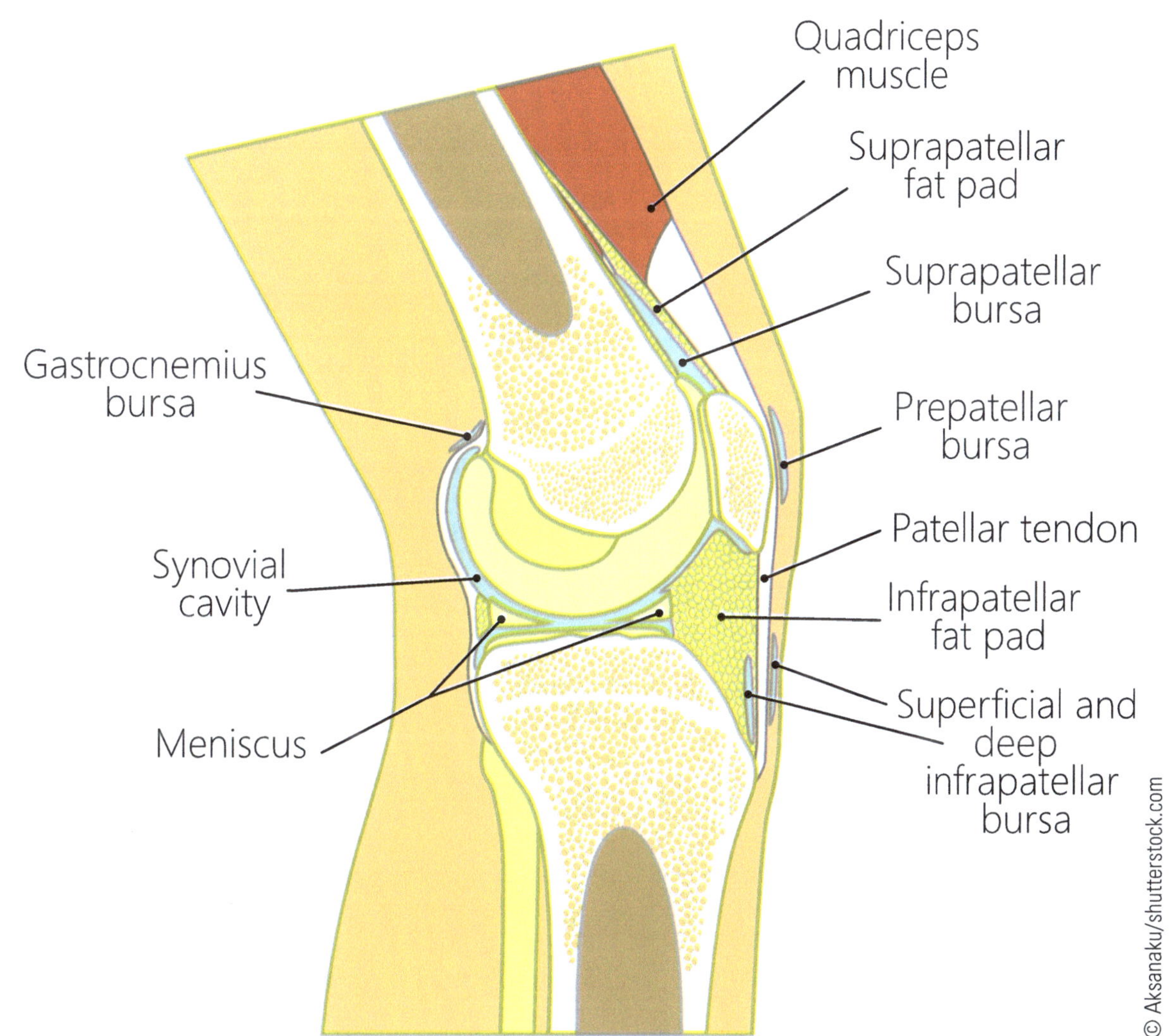

© Aksanaku/shutterstock.com

Quickly…

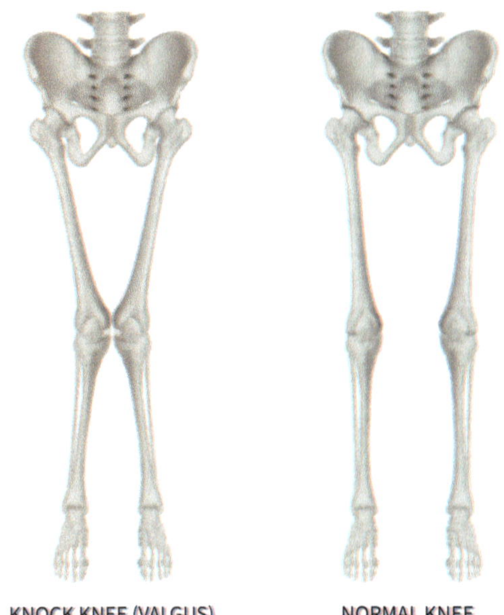
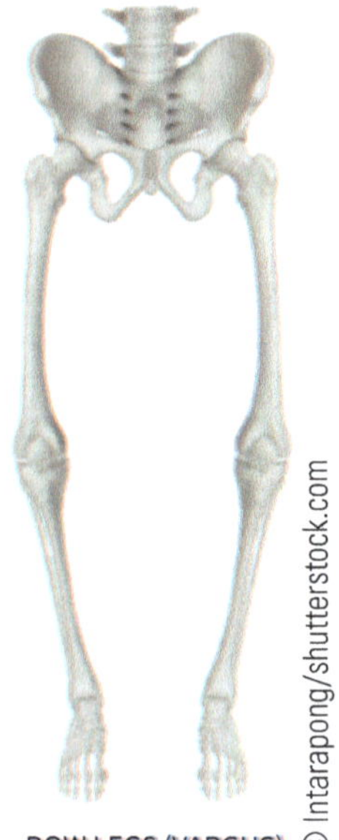

Genu Valgum	Knock-knee	Lateral force →Valgus stress Think injury to the medial aspect
Genu Varum	Bowleg	Medial force →Varus Stress Think injury to the lateral aspect

Knee Ligaments
Draw/Outline and color accordingly

Extracapsular Ligaments

Patellar Ligament
Apex of patella to Tibial Tuberosity of Tibia
- Continuation of quadriceps tendon
- Location for <u>patellar reflex</u> (L2-L4)

Lateral Collateral Ligament (LCL)
Lateral epicondyle of femur to the head of the fibula
- Cordlike
- No direct contact with the meniscus or capsule
- Restricts extension and medial force (varus stress)

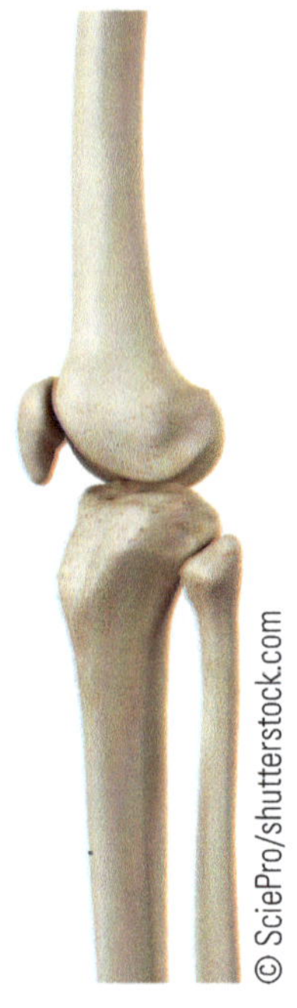

Medial Collateral Ligament (MCL)

Medial epicondyle of femur to medial tibial condyle

1. Broad
2. Attached with the medial meniscus *(often injured together)
3. Restricts extension and lateral force (valgus stress)

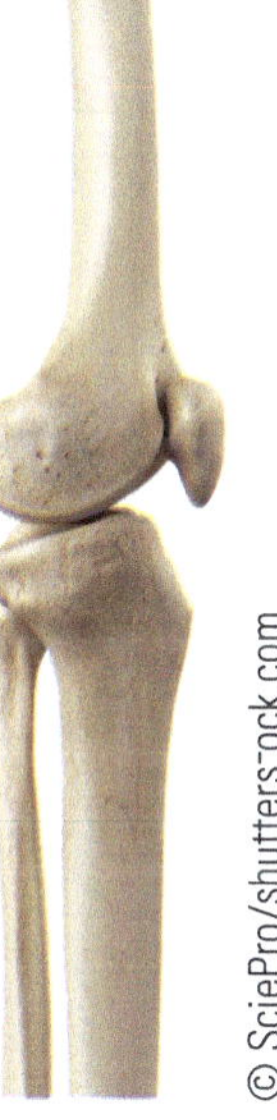

Clinical Quick Hits:

Intracapsular Ligaments

Anterior Cruciate Ligament (ACL)

All Time Luscious Female

Passes posteriorly from the anterior tibia to the lateral femoral condyle

1. Restricts anterior translaton of the tibia on the femur
2. Restricts posterior translation of the femur on the tibia
3. Restricts hyperextension; relaxed in flexion
4. Blood supply limited upon injury

Posterior Cruciate Ligament (PCL)

Part Time Malicious Female

Passes anteriorly from the posterior tibia to the medial femoral condyle

1. Resticts posterior translation of tibia on the femur (dashboard)
2. Restricts anterior translation of femur on tibia
3. Tight in flexion; relaxed in extension
4. Restricts hyperflexion
5. Shorter and stronger than ACL; blood supply limited upon injury

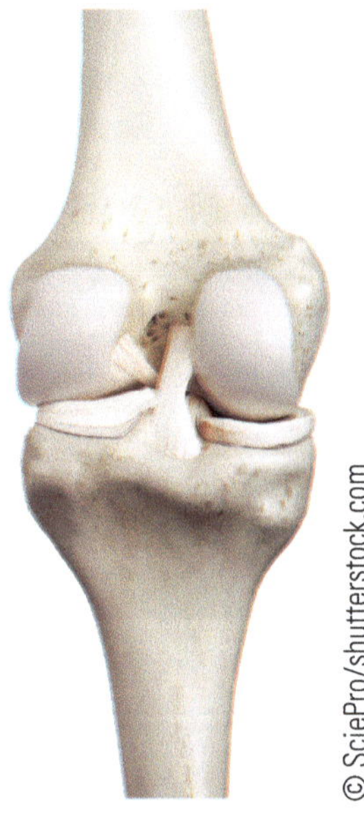

© SciePro/shutterstock.com

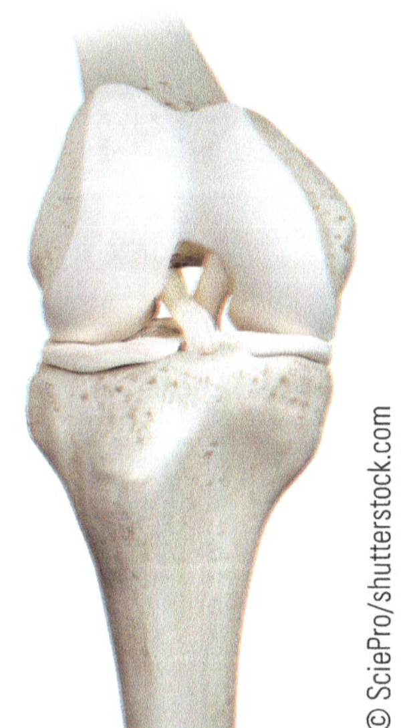

© SciePro/shutterstock.com

TEAR OF THE LATERAL COLLATERAL LIGAMENT (LCL)
FRONT VIEW OF THE KNEE

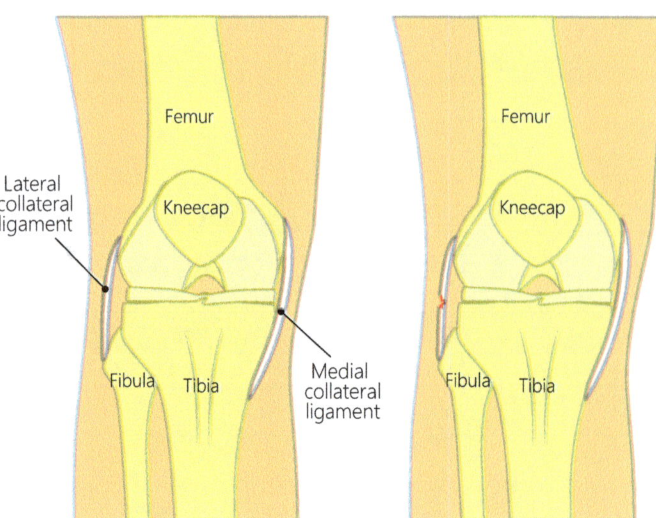

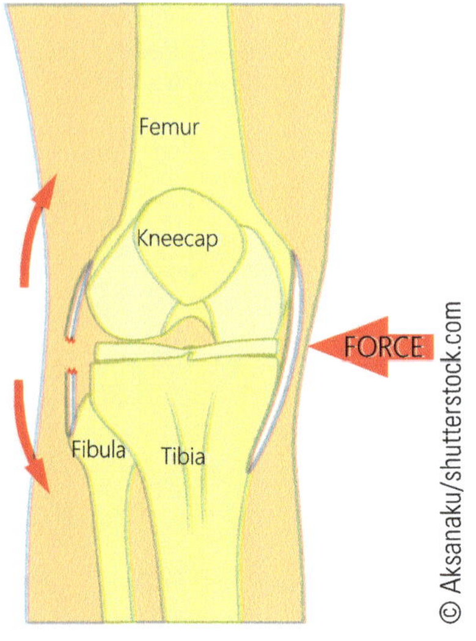

© Aksanaku/shutterstock.com

TEAR OF THE MEDIAL COLLATERAL LIGAMENT (MCL)
FRONT VIEW OF THE KNEE

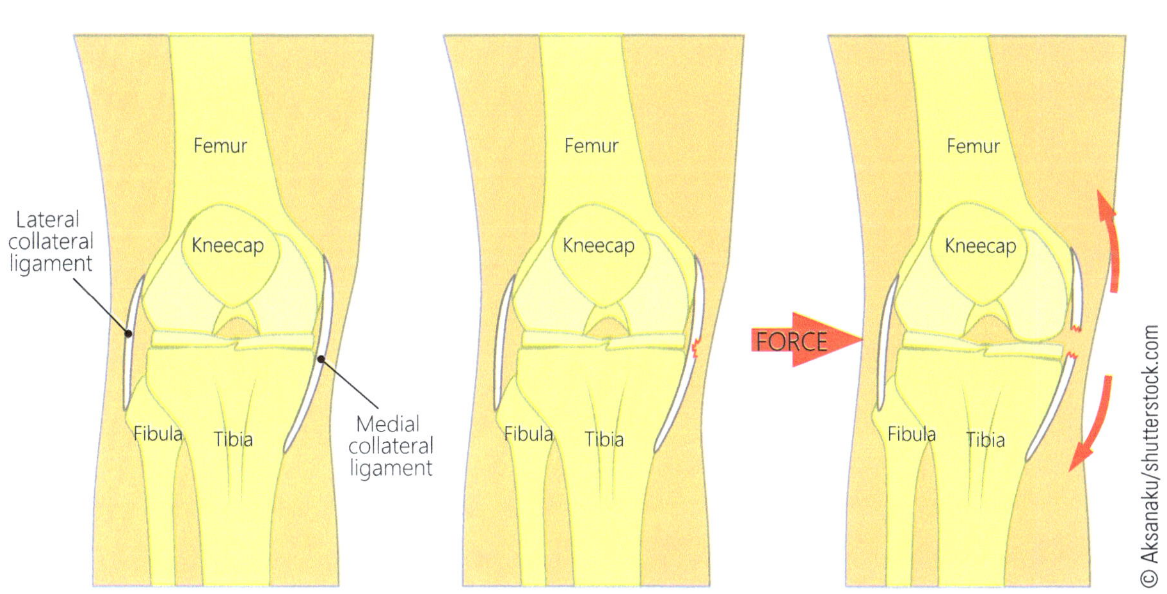

TEAR OF THE ANTERIOR CRUCIATE LIGAMENT (ACL)
MEDIAL VIEW OF THE KNEE

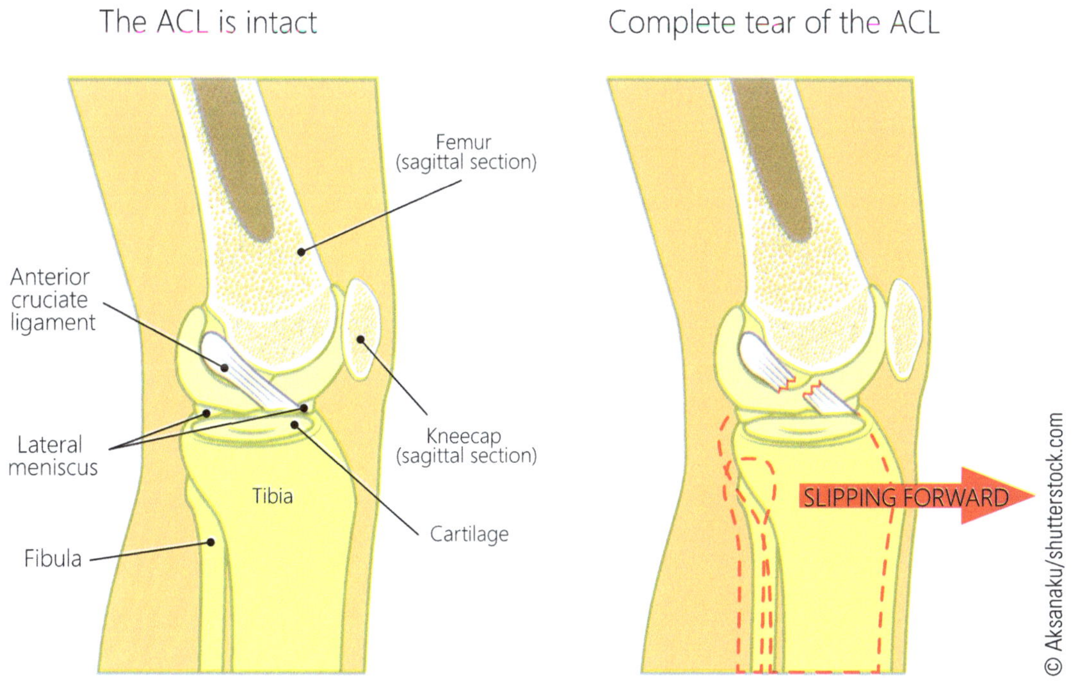

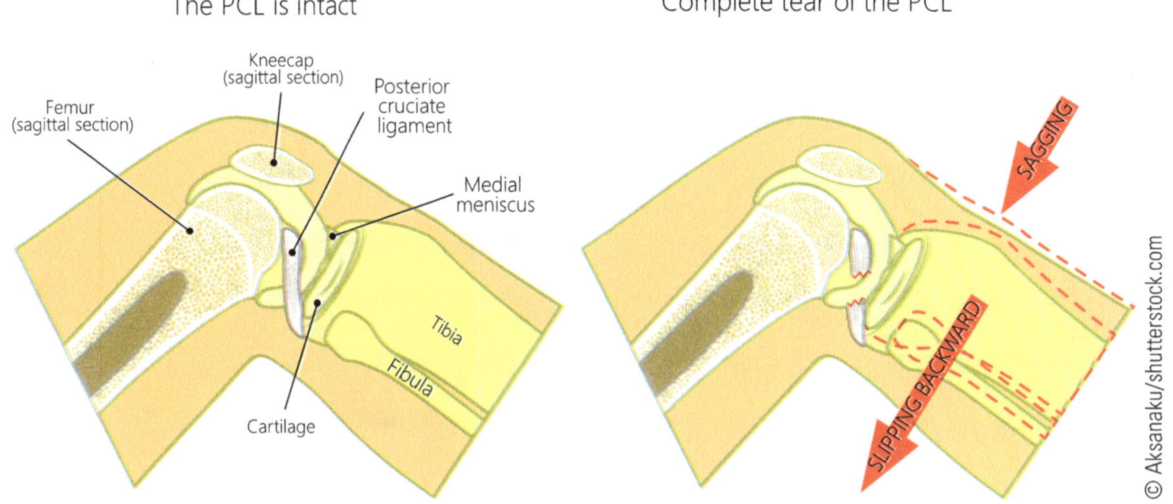

Clinical Quick Hit (or doodle page)

Talocrural Ligaments

Draw and color accordingly

Lateral reinforcement: Restrict excessive **inversion**

Anterior Talofibular *Most likely to be injured with inversion sprain*	Anterior fibular malleolus to	Anterior talus
Calcaneofibular	Mid fibular malleolus to	Lateral calcaneus
Posterior Talofibular	Posterior fibular malleolus to	Posterior talus

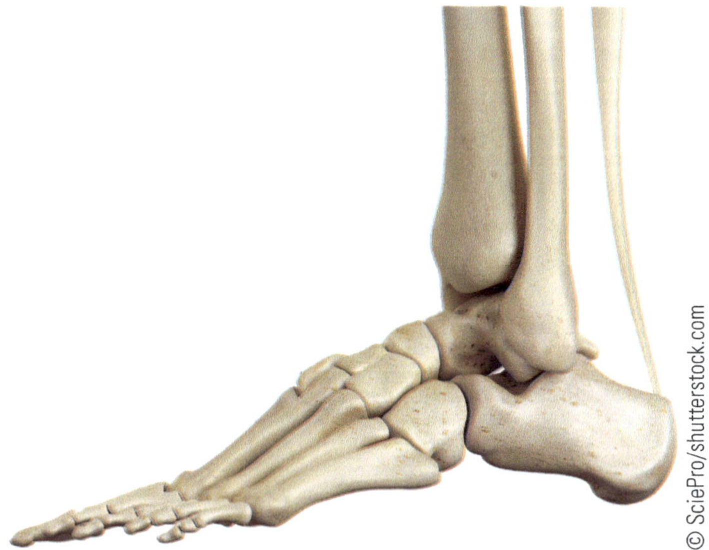

© SciePro/shutterstock.com

Medial reinforcement: (Deltoid Ligament) Restrict excessive **eversion**

Anterior Tibiotalar	Anterior tibial malleolus to	Anterior Talus
Tibionavicular	Mid tibial malleolus to	Navicular
Tibiocalcaneal	Mid tibal malleolus to	Calcaneus
Posterior Tibiotalar	Posterior tibial malleolu to	Posterior talus

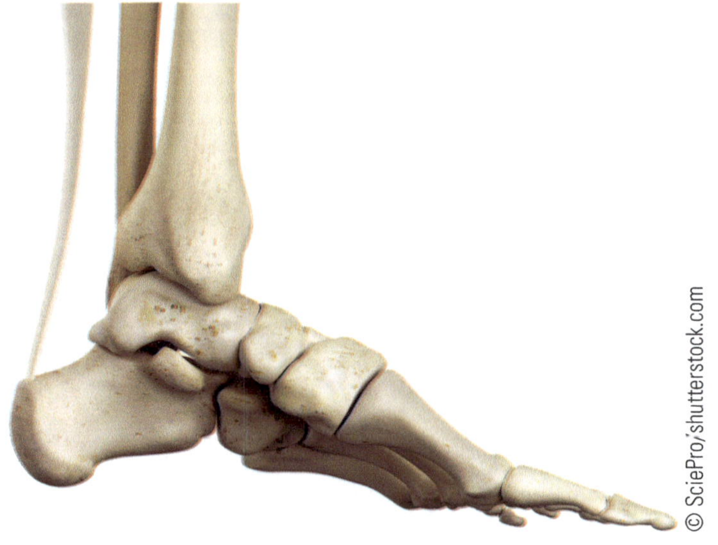

© SciePro/shutterstock.com

<u>Notes/Doodle Page</u>

Plantar Ligaments

Plantar Aponeurosis

1. Fascia that protect the plantar surface of the foot
2. Helps maintain arches
3. From Calcaneus to metatarsal heads 2-5
4. *Think plantar fasciitis

Plantar Calcaneonavicular (spring) Ligament

From the sustentaculum tali to the navicular

1. "spring"; transfer of weight
2. Helps maintain the longitudinal arch of foot

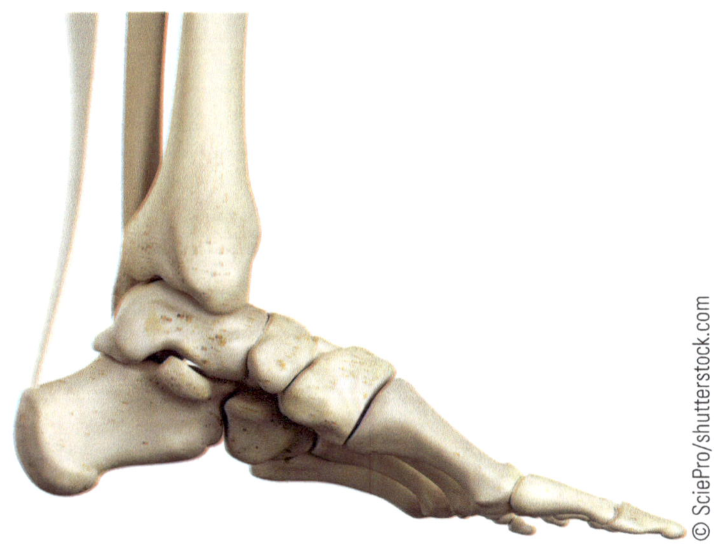

Short Plantar Ligament

From plantar calcaneus to plantar cuboid (calcaneocuboid)

1. Helps maintain the longitudinal arch

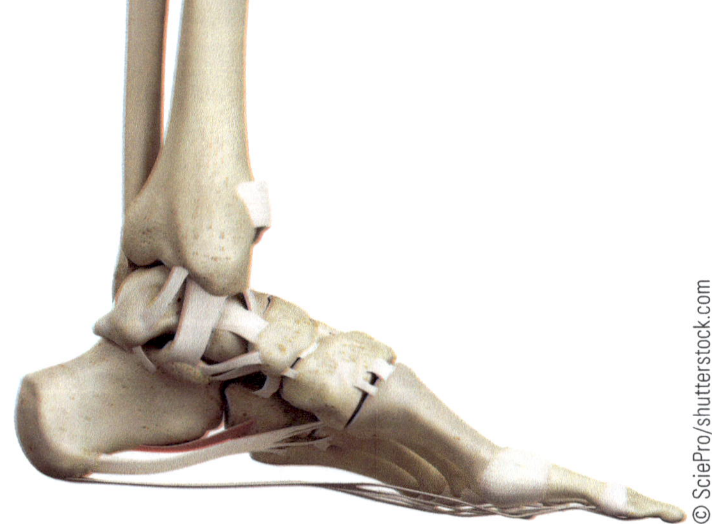

Long Plantar Ligament

From plantar calcaneus to base of metatarsals

1. Helps maintain the longitudinal arch

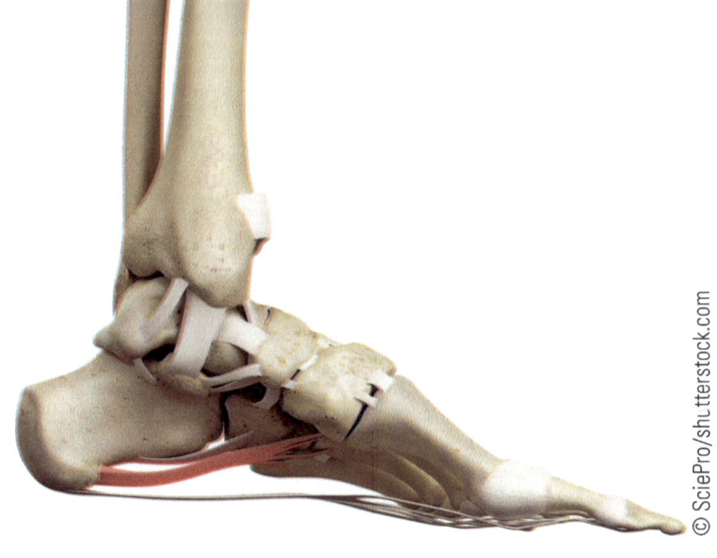

Muscles of the Lower Extremity

First and foremost, make sure you can ID all muscles in general

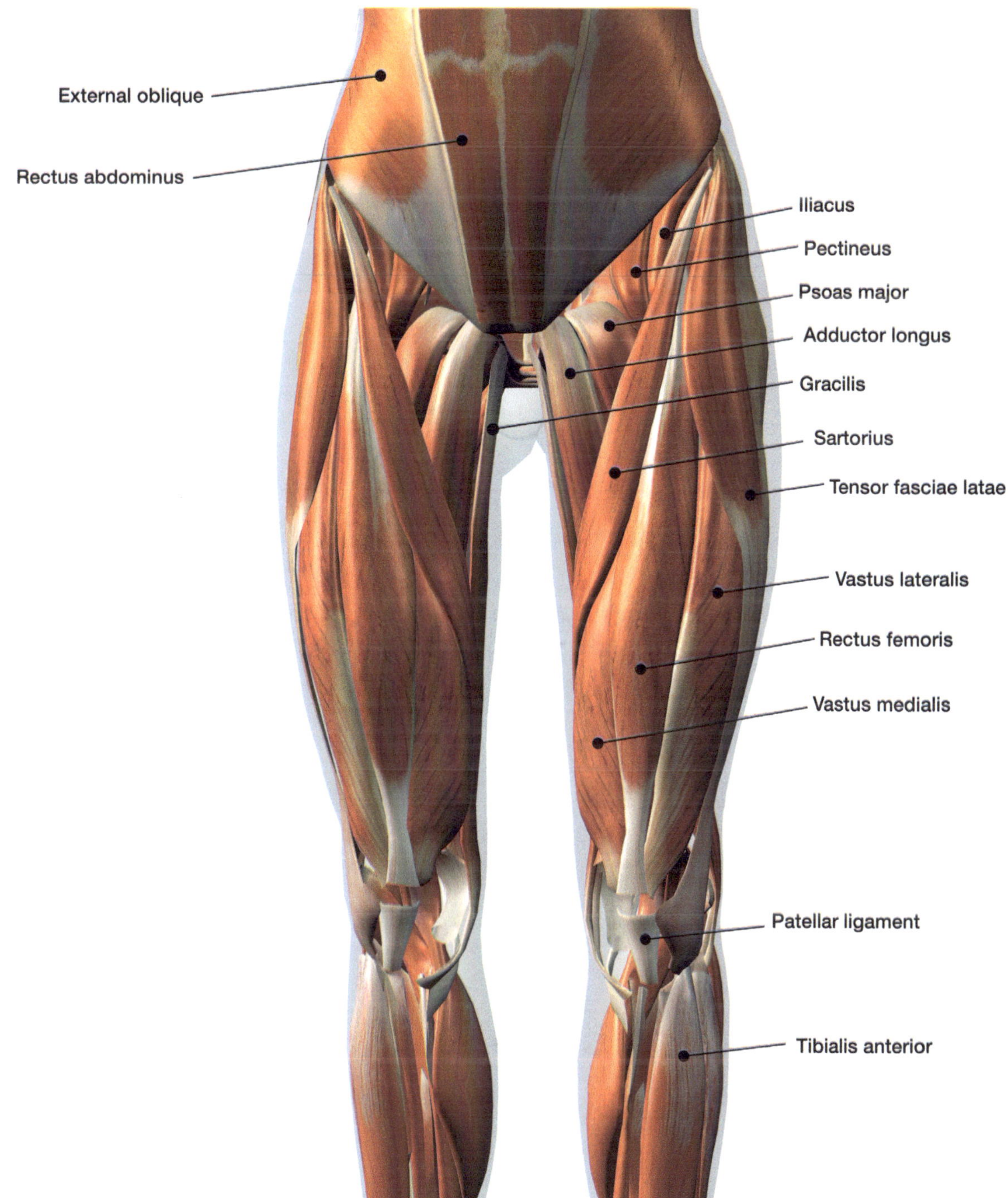

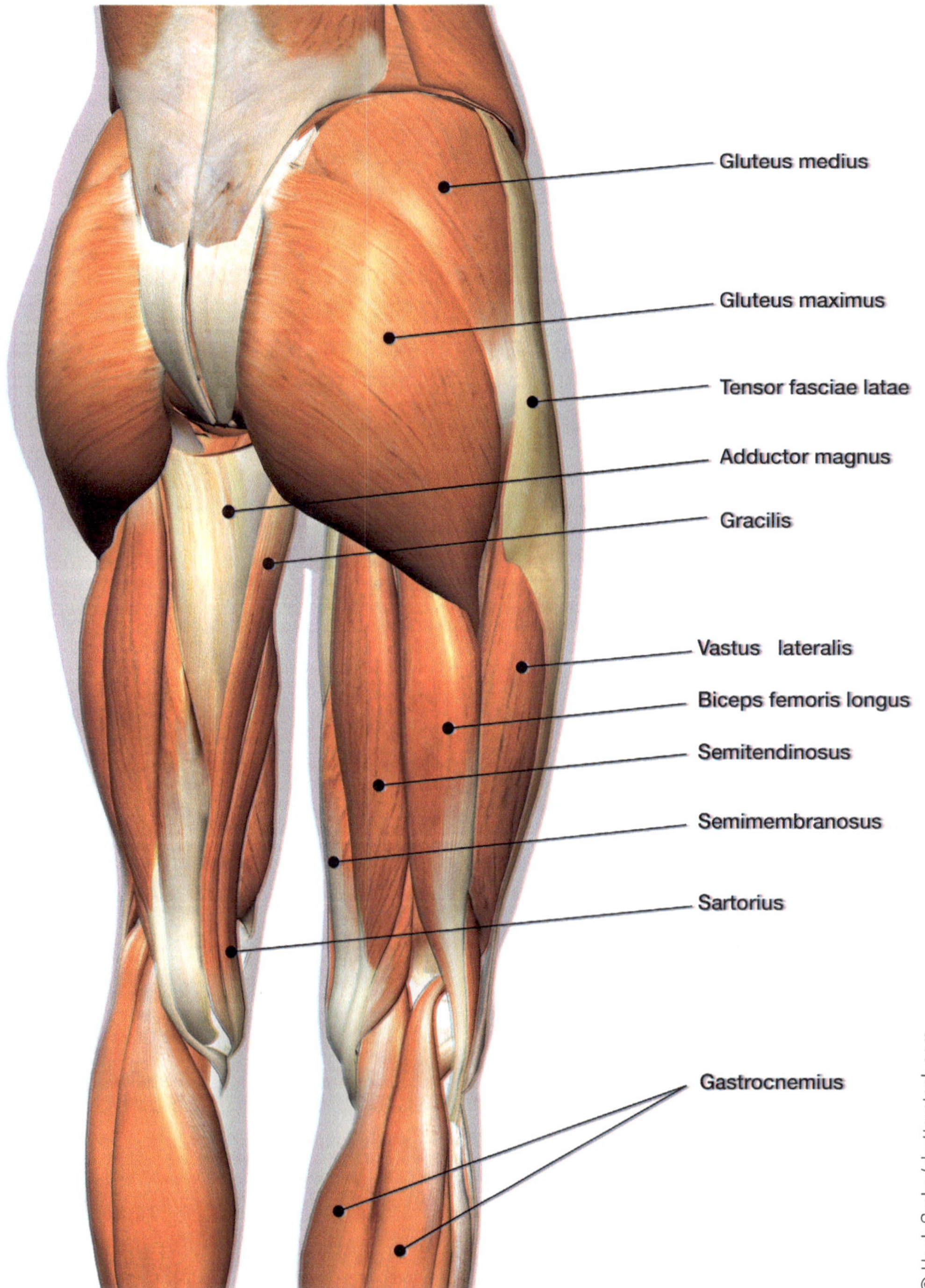

Gluteus medius
Gluteus maximus
Tensor fasciae latae
Adductor magnus
Gracilis
Vastus   lateralis
Biceps femoris longus
Semitendinosus
Semimembranosus
Sartorius
Gastrocnemius
© Hank Grebe/shutterstock.com

FUNCTION COMPARTMENT GOLDEN RULES

<u>Step #1</u>: Be able to identify the muscles- see them in your brain.

<u>Step #2</u>: Compartmentalize do not memorize.

<u>Step #3</u>: Remember some muscles can fit into multiple compartments

Hip Joint Compartments

Compartment	Function	Example
Caps **over** the hip joint to get to femur		Gluteus Medius
Crosses **under** the hip joint to get to humerus		Adductor Longus
Crosses hip anteriorly to get to humerus		Rectus Femoris
Crosses hip posteriorly to get to humerus		Biceps Femoris
Inserts close to the anterior greater trochanter		Gluteus Minimus
Inserts close to the posterior greater trochanter		Gemellus Inferior

Knee Joint Compartments

Crosses the joint anteriorly to reach tibia/fibula   Extension
Crosses the joint posteriorly to reach tibia/fibula   Flexion

Talocrural Joint Compartments

Crosses the joint anteriorly to reach tarsals   Dorsiflexion
Crosses the joint posteriorly to reach tarsals Plantar   Flexion

Subtalar Joint Compartments

Crosses the joint medially   Inversion
Crosses the joint laterally   Eversion

MTP/IP Joint Compartments

Crosses the joint dorsally (top of foot)   Extension
Crosses joint on plantar surface (under foot)   Flexion

© VGstockstudio/shutterstock.com

____ Remember when?

As you learn the muscles to functions, remember your patient will mostly complain of pain or weakness during a movement versus naming a specific muscle. It is important to be able to relate the movement with the muscle.

FA Muscles- Anterior Compartment   FA FLEXORS

If flexion is weak or elicits pain, think of which muscles fall within the anterior compartment.
All the following muscles are FA movers (insert on the femur-and some below)
Note how the following muscles cross the FA joint anteriorly:

Draw from insertion to origin to understand function

Psoas Major/Iliacus

	Insertion moves to ….	Origin	Anterior Function
Psoas Major	Lesser trochanter of femur	Bodies/transverse processes of lumbar	FA flexion
Iliacus	Lesser trochanter of femur	Iliac fossa	FA flexion

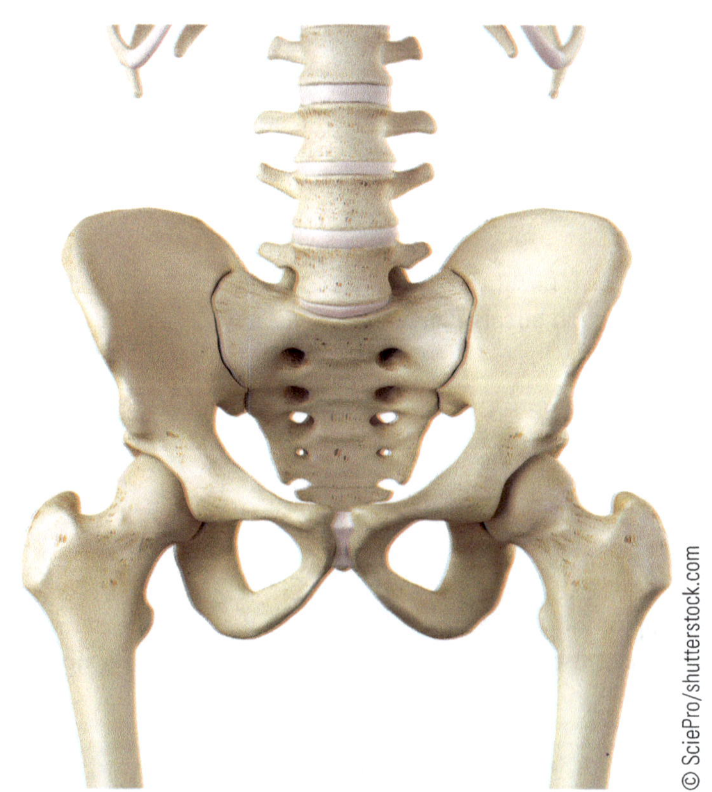

© SciePro/shutterstock.com

Draw one on the following on each leg; from insertion to origin to understand function

Rectus Femoris/Sartorius

Muscle	Insertion moves to ….	Origin	Anterior Function
Rectus Femoris *only quadriceps crossing hip joint	Tibial tuberosity via patellar tendon	AIIS	FA Flexion
Sartorius	Pes anserine (medial to tibial tuberosity)	ASIS	FA Flexion

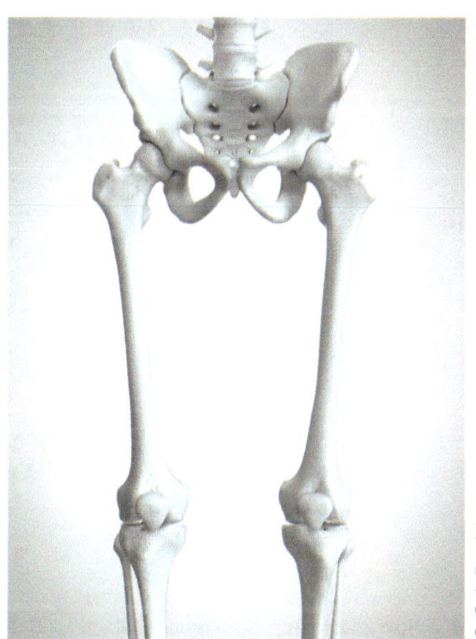

© SciePro/shutterstock.com

Tensor Fascia Latae

Insertion moves to	Origin	Anterior Function
IT band (to Gerdy's tubercle; lateral to tibial tuberosity)	ASIS	FA Flexion

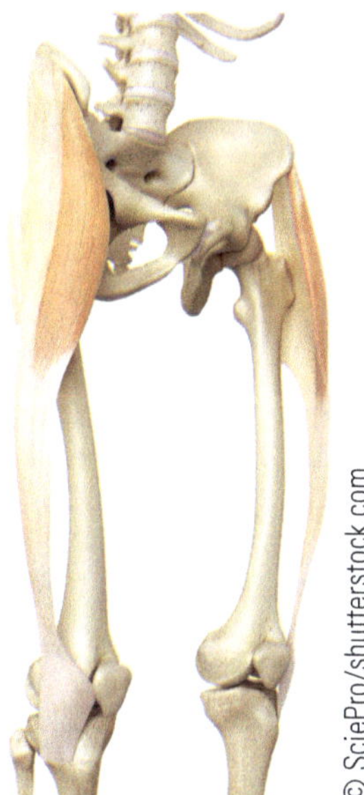

© SciePro/shutterstock.com

*<u>**Pectineus muscle**</u> will also assist in FA flexion (see adductors)

FA Muscles- Posterior Compartment   FA EXTENSORS

If extension is weak or elicits pain, think of which muscles fall within the posterior compartment.

All the following muscles are FA movers (insert on the femur-and some below)

Note how the following muscles cross the FA joint posteriorly:

The Hamstring Group

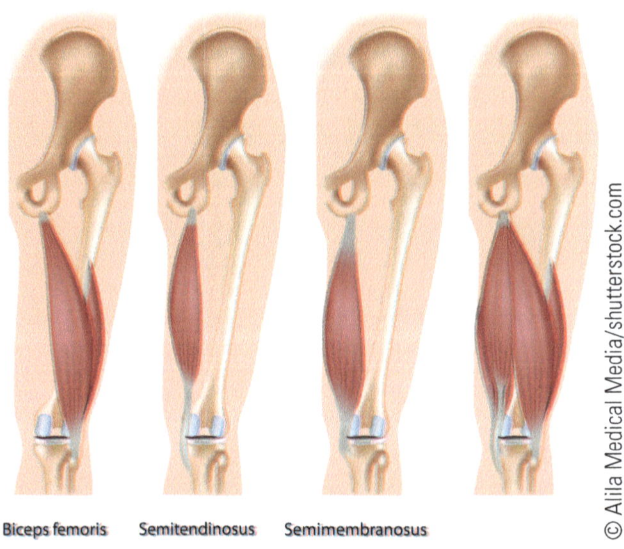

Color/Outline from insertion to origin

Gluteus Maximus

Insertion **moves to ….**	Origin	Posterior Function
Gluteal tuberosity of femur and IT band	Sacrum, gluteal surface of ilium	FA extension

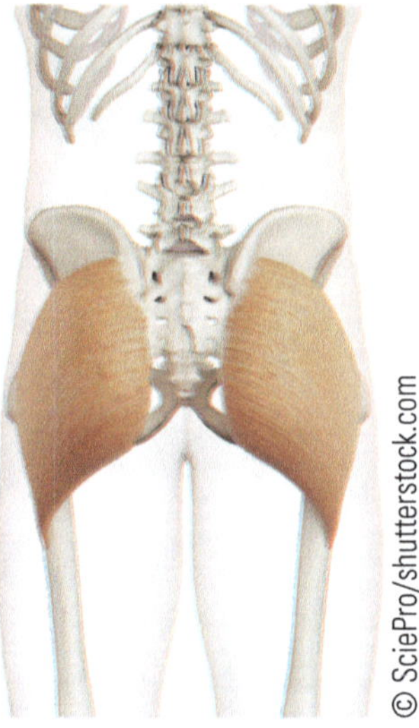

Color/Outline from insertion to origin to understand function

Biceps Femoris, Semimembranosus, Semitendinosus

Muscle	Insertion moves to ….	Origin	Posterior Function
Biceps Femoris (a); only LATERAL hamstring; two heads	Head of fibula	Ischial tuberosity Mid-posterior femur	FA extension
Semimembranosis (b); MORE MUSCLE	Medial tibial condyle	Ischial tuberosity	FA extension
Semitendinosus (c); MORE TENDON; (sits on semimembranosus)	Pes anserine of tibia (medial to tibial tuberosity)	Ischial tuberosity	FA extension

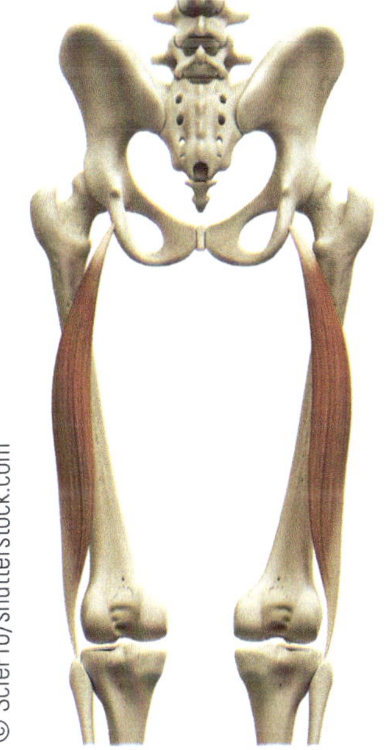

© SciePro/shutterstock.com

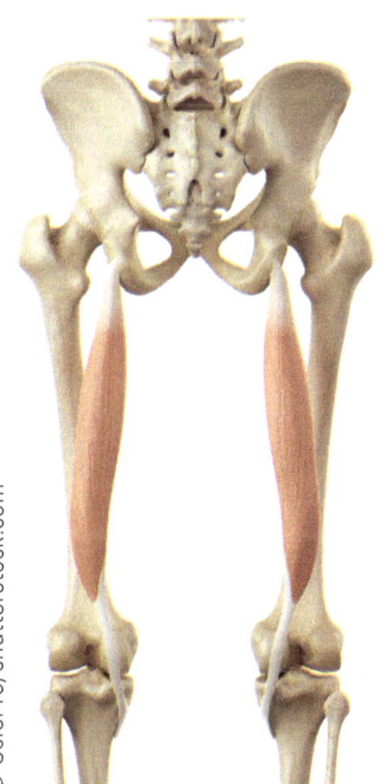

© SciePro/shutterstock.com

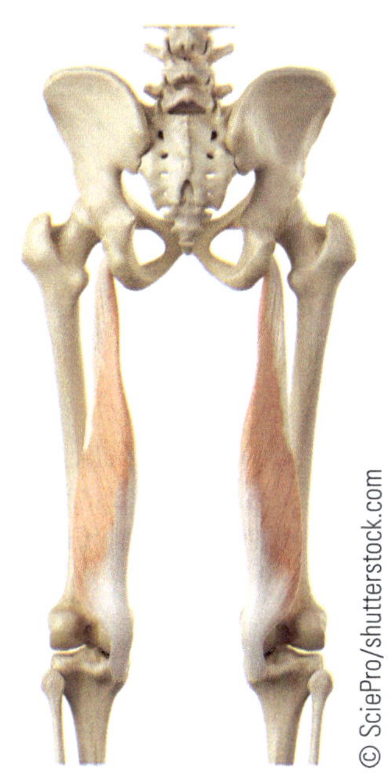

© SciePro/shutterstock.com

Adductor Magnus Color/Outline insertion to origin

Insertion moves to ….	Origin	Posterior Function
Mid and distal linea aspera Adductor tubercle of femur	Ischial tuberosity of pelvis *Inferior pubic ramus*	FA extension

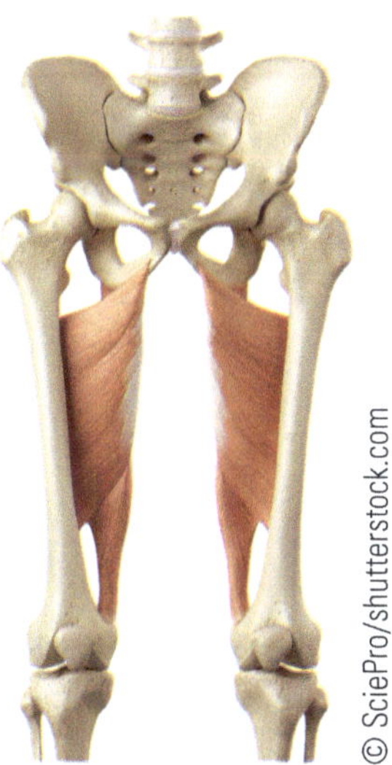

© SciePro/shutterstock.com

Gluteus Medius and Minimis will contribute to extension due to their posterior origin

© Masha Minaeva/shutterstock.com

FA Muscles- Medial (cross under) Compartment   FA ADDUCTORS

If adduction is weak or elicits pain, think of which muscles fall within the medial compartment.
All the following muscles are FA movers (insert on the femur-and some below)
Note how the following muscles cross the FA joint under/medially:

Poor Bills Lose Many Games
Superior Inferior Superior Inferior POW!
Draw from insertion to origin to understand function

Pectineus/Add.uctor Brevis

	Insertion **moves to** ….	Origin	Medial Function
Pectineus (right)	Pectineal line of femur	Pectineal line of pelvis (superior pubic ramus)	FA adduction
Adductor Bevis (left)	Upper linea aspera	Inferior pubic ramus	FA adduction

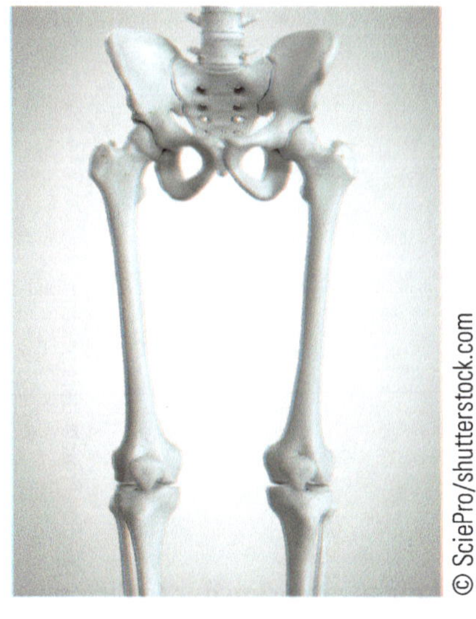

© SciePro/shutterstock.com

Poor Bills Lose Many Games
Superior Inferior Superior Inferior POW!

Adductor Longus/Adductor Magnus/Gracilis

	Insertion **moves to ….**	Origin	Medial Function
Adductor Longus (right)	Mid linea aspera	Superior pubic ramus	FA adduction
Adductor Magnus (left)	Mid to distal linea aspera Adductor tubercle of femur	Inferior pubic ramus Ishial tuberosity	FA adduction
Gracilis (right)	Pes anserine (medial to tibial tuberosity)	Pubic tubercle	FA adduction

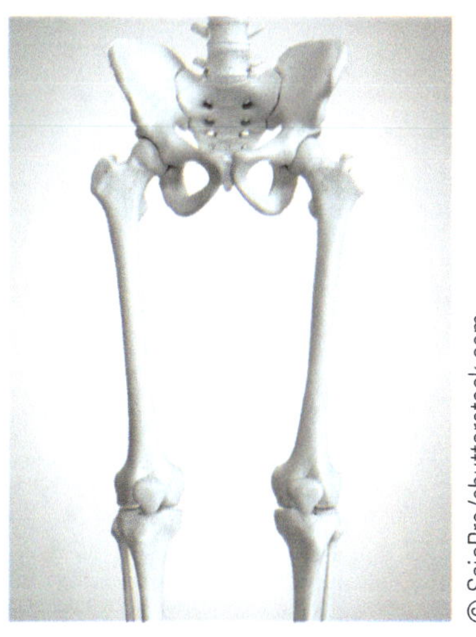

FA Muscles- Lateral (cap over) Compartment   FA ABDUCTORS

If abduction is weak or elicits pain, think of which muscles fall within the lateral compartment.
All the following muscles are FA movers (insert on the femur-and some below)

Note how the following muscles cross the FA joint over/laterally:

Color/Outline from insertion to origin to understand function

Gluteus Medius/Gluteus Minimus

	Insertion **moves to ….**	Origin	Lateral Function
Gluteus Medius	Lateral surface of greater trochanter of femur	Just below iliac crest (posterior)	Abduction ***primary Abductor***
Gluteus Minimus	Anterior/lateral greater trochanter of femur	Just below medius insertion on iliac crest	Abduction

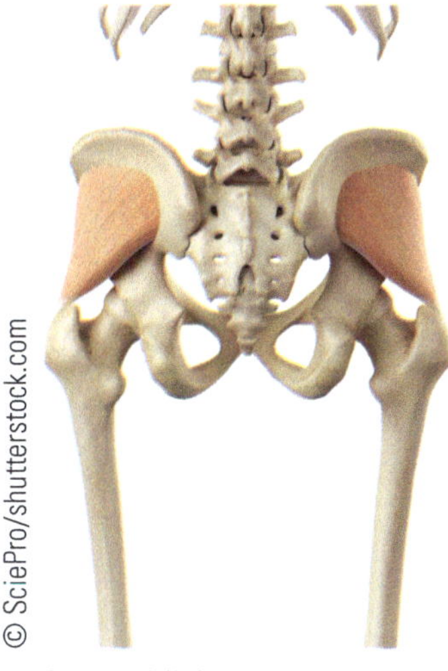

Gluteus Minimus

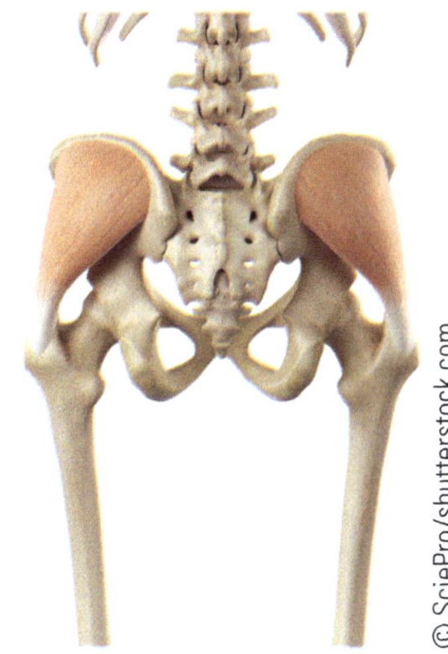

Gluteus Maximus

Tensor Fascia Latae

Insertion **moves to**	Origin	Lateral Function
IT band (to Gerdy's tubercle; lateral to tibial tuberosity)	ASIS	FA Abduction

Color/Outline from insertion to origin

Piriformis

Insertion **moves to**	Origin	Lateral Function
Lip of greater trochanter	Pelvic surface of sacrum	FA Abduction

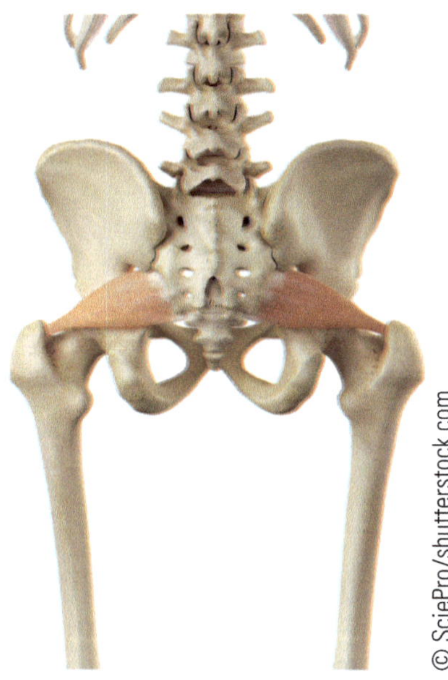

© SciePro/shutterstock.com

Colo/Outline from insertion to origin

Sartorius

Insertion moves to	Origin	Lateral Function
Pes anserine (medial to tibial tuberosity)	ASIS	FA Abduction

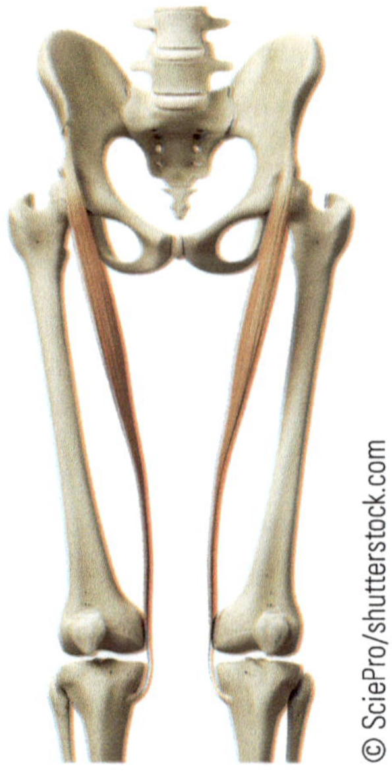

© SciePro/shutterstock.com

FA Movers: Anterior Greater Trochanter Compartment   FA Medial Rotators

1. Gluteus Medius (anterior aspect)
2. Gluteus Minimus (anterior aspect)
3. Tensor Fascia Latae (anterior aspect)

FA Movers: Posterior Greater Trochanter Compartment   FA Lateral Rotators

1. Psoas Major
2. Iliacus
3. Gluteus Maximus
4. Gluteus Medius (posterior aspect)
5. **<u>Primary Lateral Rotators</u>:** *Peeing? Go outside, Go quickly!!!*
   a. Piriformis
   b. Gemellus Superior
   c. Obturator Internus
   d. Gemellus Inferior
   e. Quadratus Femoris

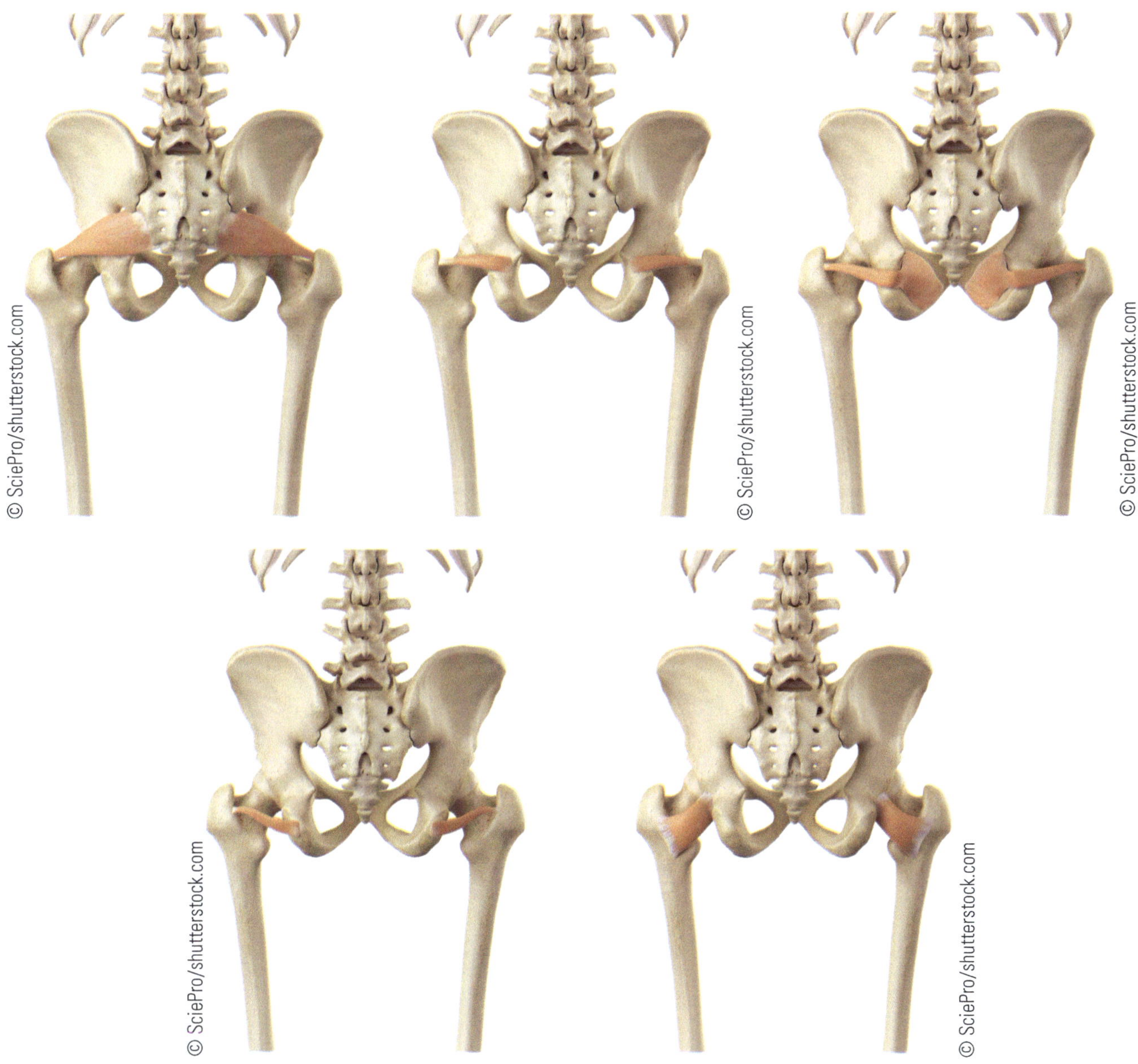

Knee Movers: Crossing the Joint Anteriorly   Knee Extensors

1. Rectus Femoris
2. Vastus Lateralis
3. Vastus Medialis
4. Vastus Intermedius (deep to rectus)

Rectus Femoris is the only quadriceps that crosses the hip and knee; all others only function at the knee

The quadriceps will cross the knee joint through the patellar ligament to the tibial tuberosity

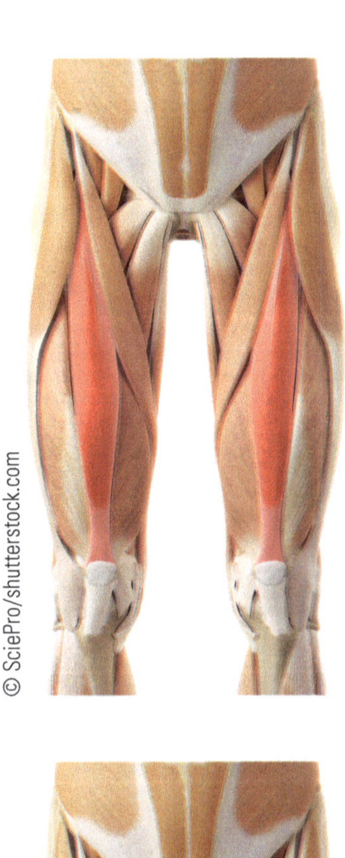

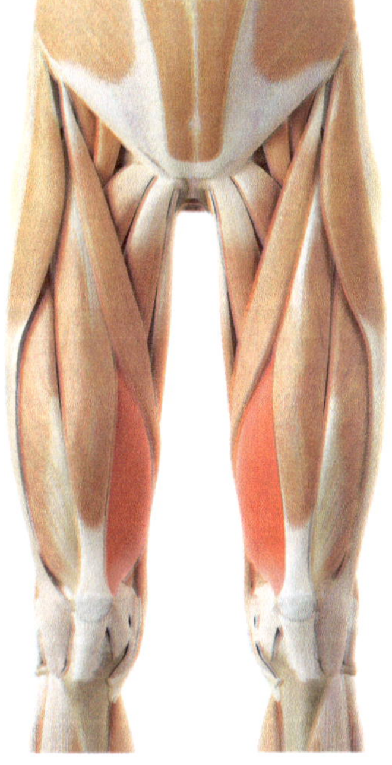

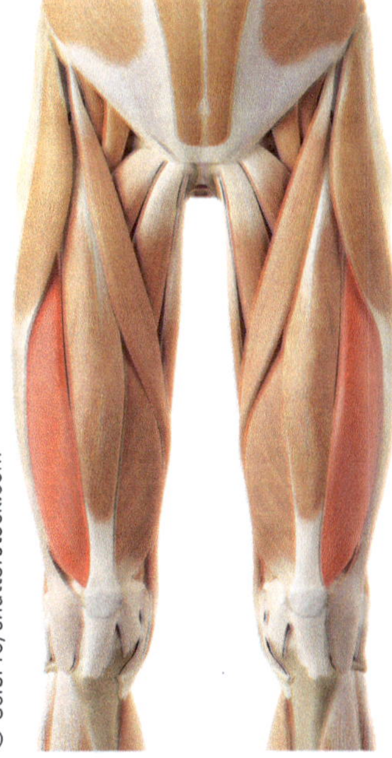

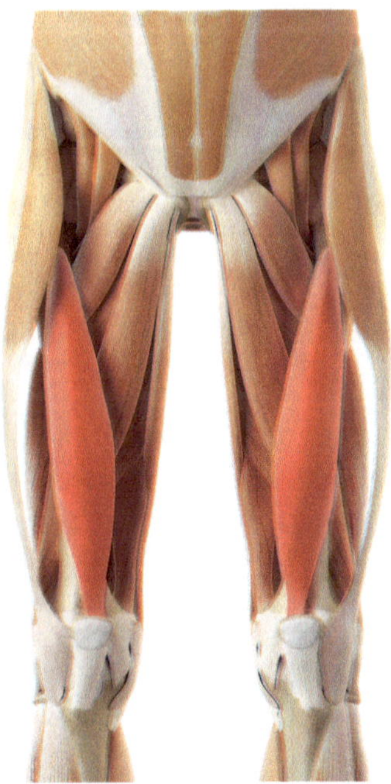

Knee Movers: Crossing the Joint Posteriorly   Knee Flexors

1. Biceps Femoris
2. Semimembranosus
3. Semitendinosus
4. Sartorius (as pass posteriorly to reach insertion)
5. Gracilis (as pass posteriorly to reach insertion)
6. Gastrocnemius
7. Popliteus

Keep in mind that muscles that cross the joint medially will also medially rotate the knee joint, and laterally will laterally rotate the knee.

Gastrocnemius

Insertion **moves to**	Origin	Posterior Function
Achilles tendon to calcaneus	Medial and lateral femoral condyles	Knee flexion

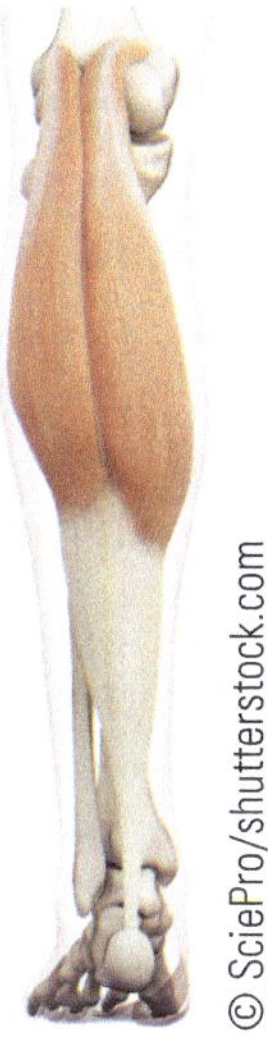

© SciePro/shutterstock.com

Popliteus

Insertion **moves to**	Origin	Posterior Function
Posterior tibia	Lateral femoral condyle of femur	Knee flexion

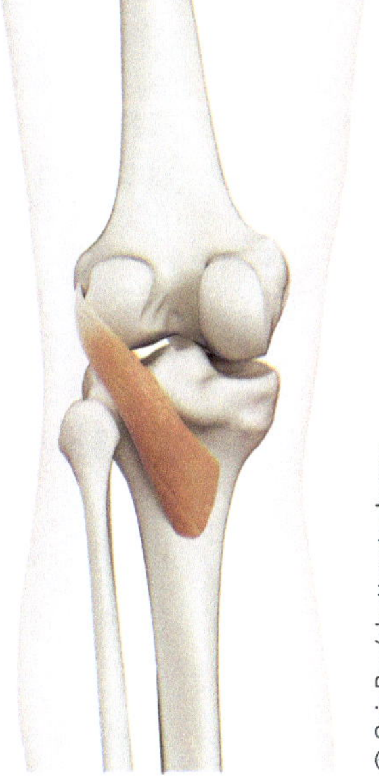

© SciePro/shutterstock.com

Talocrural Joint Movers: Crossing the Joint Anteriorly   TC Dorsiflexors

If dorsiflexion is weak or elicits pain, think of which muscles fall within the anterior compartment.

All the following muscles are TC movers (insert on the tarsal, MTP, IP)

Note how the following muscles cross the TC joint anteriorly:

<u>Color/Outline</u> from insertion to origin to understand function

TOM HARRY DICK   TIB FIB ALL

In the lab, the tendons are the key to locating these muscles-the names will tell you where to look-<u>NEVER FORGET TO SIMPLIFY, LOOK FOR PATTERN, DO NOT MEMORIZE!!!</u>

1. **Tibialis Anterior (TOM)** *My favorite muscle of lower*

Insertion **moves to**	Origin	Anterior Function
Medial cuneiform	Lateral Tibia	TC dorsiflexion
Plantar 1ˢᵗ metatarsal	Interosseous membrane	

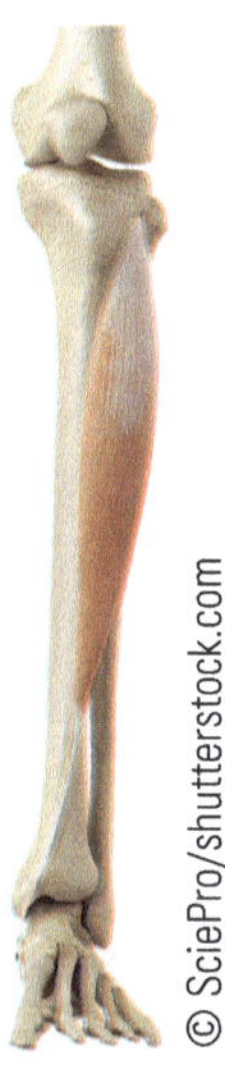

2. **Extensor Hallucis Longus (Harry); think big toe**

Insertion **moves to**	Origin	Anterior Function
Base of 1ˢᵗ distal phalanx	Mid fibula	TC dorsiflexion
	Interosseous membrane	

3. **Extensor Digitorum Longus (Dick); think digits**

Insertion **moves to**	Origin	Anterior Function
Distal phalanx 1-5	Tibia, fibula, interosseous	TC dorsiflexion

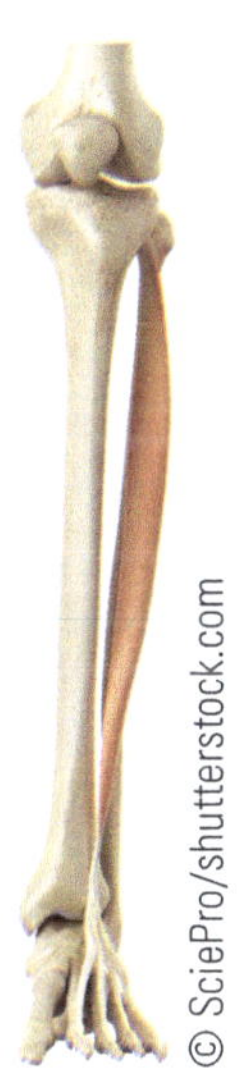

4. **Fibularis Tertius:** *Draw on*

Insertion **moves to**	Origin	Anterior Function
5th metatarsal (dorsal surfaces)	Distal fibular	TC dorsiflexion

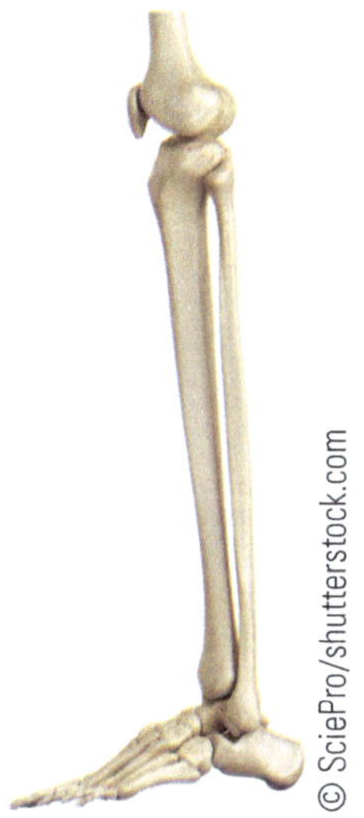

Talocrural Joint Movers: Crossing the Joint Posteriorly   TC Plantar Flexors

If plantar flexion is weak or elicits pain, think of which muscles fall within the posterior compartment.

All the following muscles are TC movers (insert on the tarsal, MTP, IP)

Note how the following muscles cross the TC joint posteriorly:

<u>**Color/Outline**</u> from insertion to origin to understand function

TOM DICK & HARRY   ALL TIB FIB

In the lab, the tendons are the key to locating these muscles-GO TO YOUR MEDIAL MALLEOLUS and plantar aspect for an excellent reference location to tag!

1. Tibialis Posterior (TOM)

Insertion **moves to**	Origin	Anterior Function
Navicular, cuneiforms, base of 2-4 metarsals	Tibia, fibula, interosseous membrane	TC plantar flexion

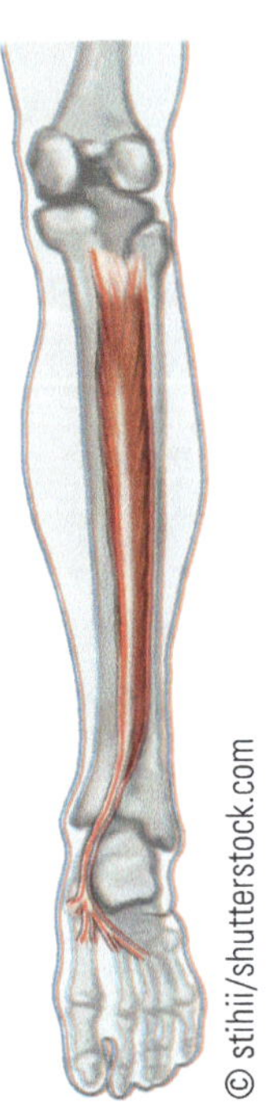

2. Flexor Digitorum Longus (DICK)

Insertion **moves to**	Origin	Anterior Function
Distal phalanges 2-5	Tibia (posterior)	TC plantar flexion

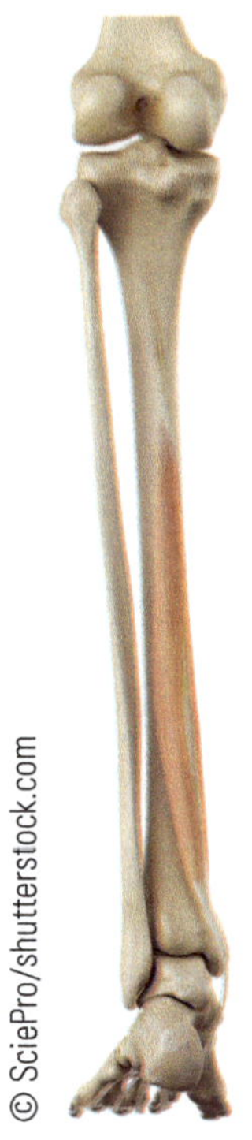
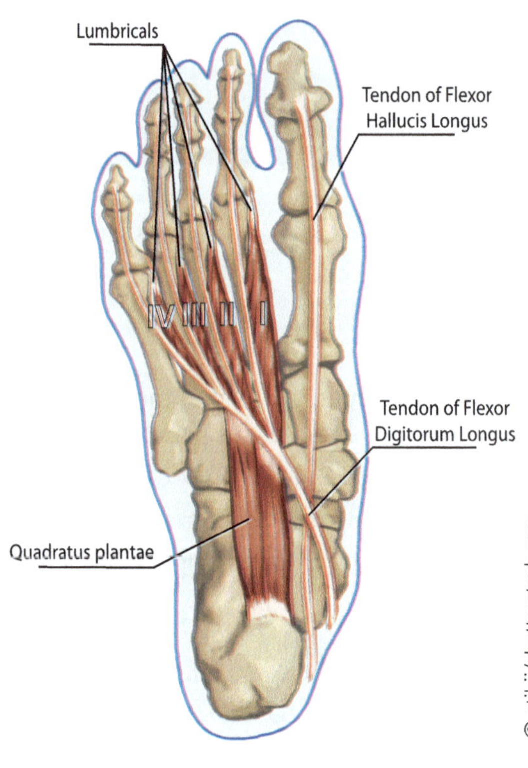

3. Flexor Hallucis Longus (HARRY)

Insertion **moves to**	Origin	Anterior Function
Distal phalange 1	Fibula	TC plantar flexion

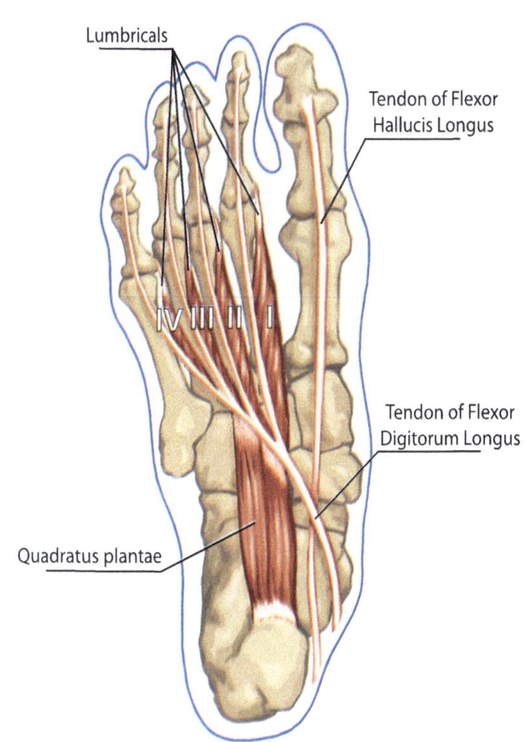

Soleus (GASTROCNEMIUS will also Plantar Flex)

Insertion **moves to**	Origin	Posterior Function
Achilles tendon to calcaneus	Fibula Soleal line on tibia	Knee flexion

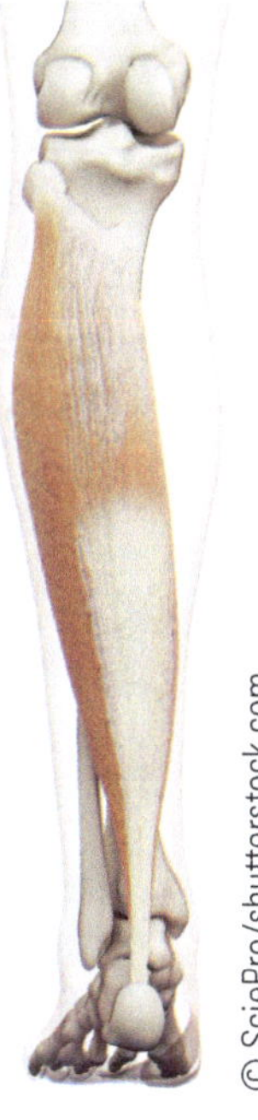

Subtalar Joint Movers: Crossing the Joint Medially   Subtalar Invertors

If inversion is weak or elicits pain, think of which muscles fall within the medial compartment.

All the following muscles are ST movers (insert on the tarsal, metatarsal)

Note how the following muscles cross the ST joint medially:

**Reminder: Subtalar is the Talocalcaneal joint*

1. Tibialis Anterior
2. Tibialis Posterior
3. Flexor Digitorum Longus
4. Flexor Hallucis Longus

Extensor Hallucis Longus will help in both inversion/eversion (secondary-FYI)

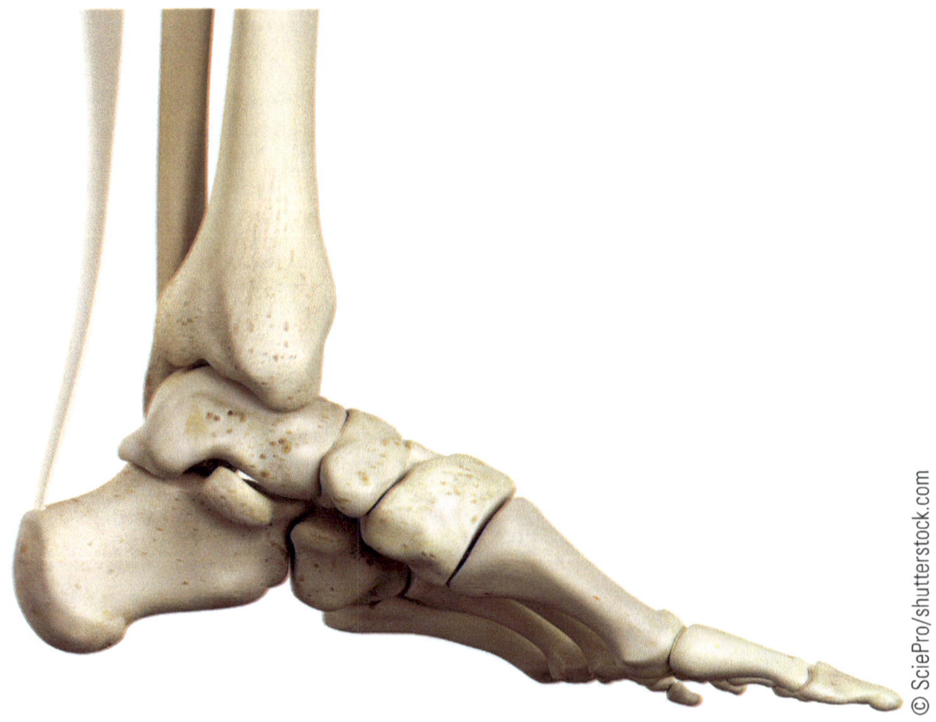

Subtalar Joint Movers: Crossing the Joint Laterally   Subtalar Evertors

If eversion is weak or elicits pain, think of which muscles fall within the lateral compartment.

All the following muscles are ST movers (insert on the tarsal, metatarsal)

Note how the following muscles cross the ST joint laterally:

Wait for it……☺

Fibularis Longus/Fibularis Brevis (Fibularis=Peroneals)

Muscle	Insertion **moves to** ….	Origin	Anterior Function
Fibularis Longus	Plantar medial cuneiform Plantar base of 1ˢᵗ metatarsal	Head of fibula	ST eversion
Fibularis Brevis	Tuberosity of 5ᵗʰ metatarsal	Mid-fibula	ST eversion

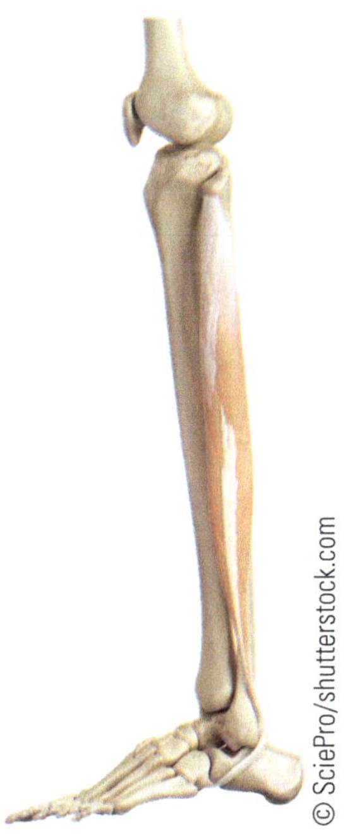

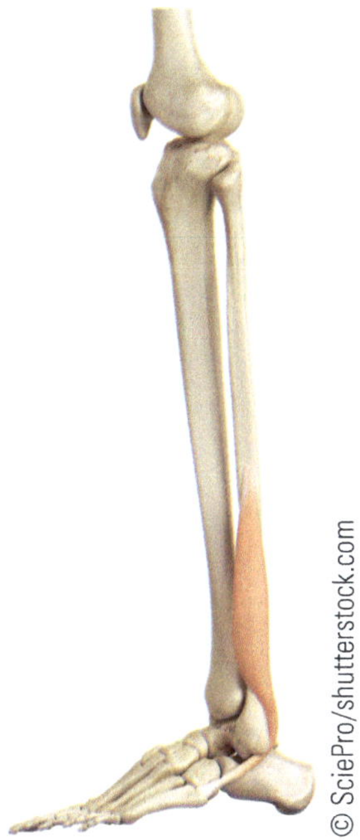

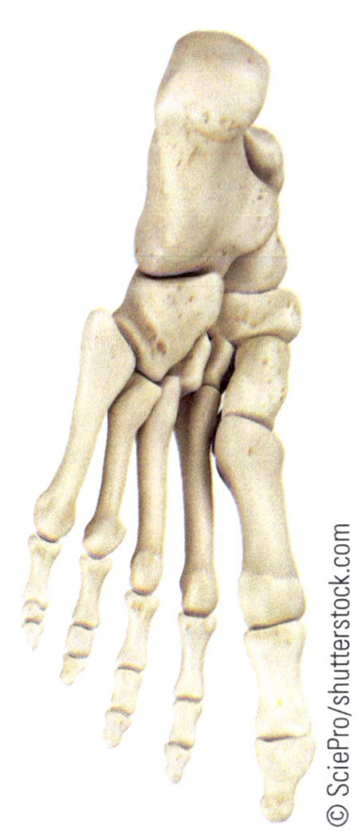

Other Evertors:

1. Extensor Digitorum Longus
2. Fibularis Tertius

Extensor Hallucis Longus will help in both inversion/eversion (secondary-FYI)

And..Finally the FOOT (to fill in some gaps)

Dorsal Foot:
Extensor Hallucis Brevis

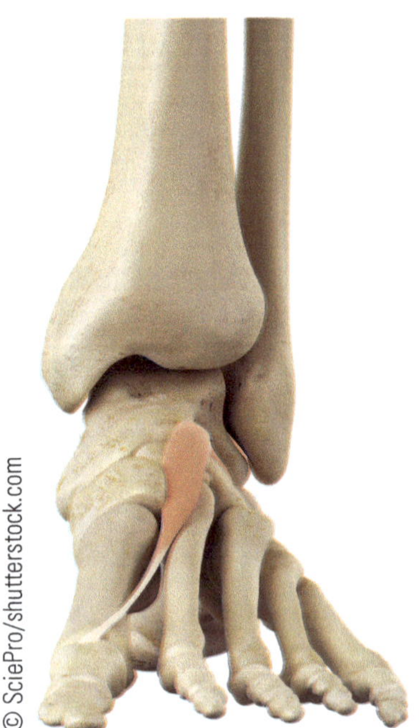

Extensor Digitorum Brevis

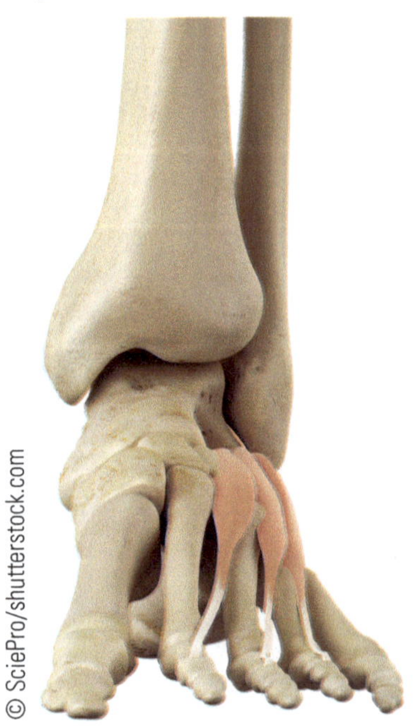

Plantar Foot:
Flexor Hallucis Brevis

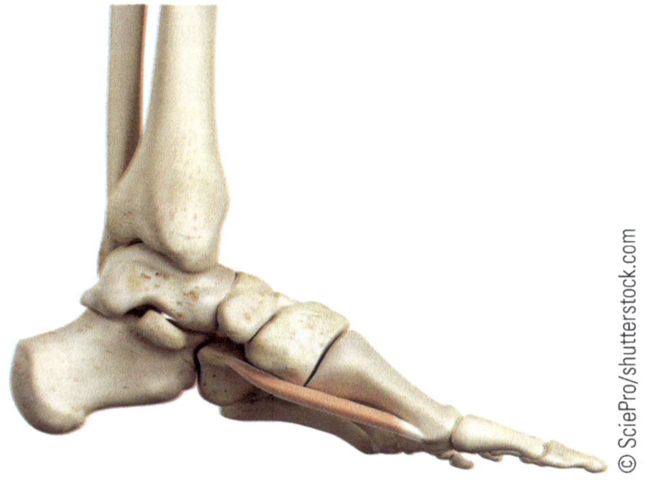

Flexor Digitorum Brevis, Abductor Hallucis, Abd Digitorum

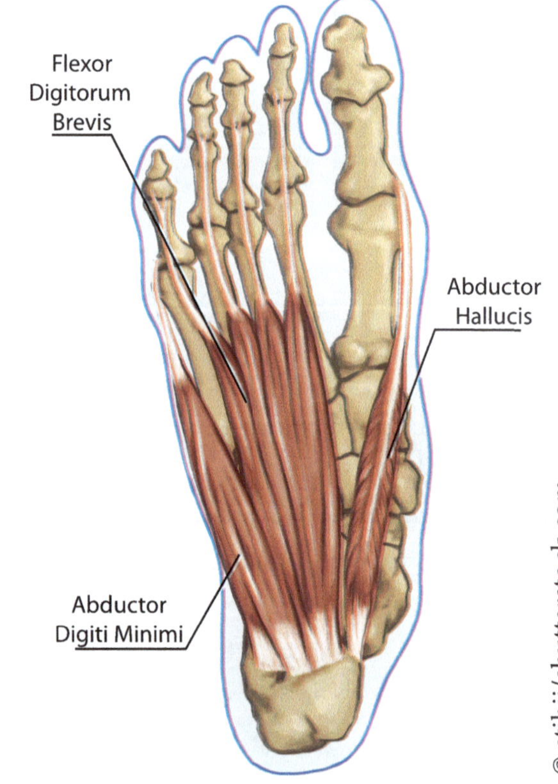

Plantar Foot Continued:
Adductor Hallucis Brevis

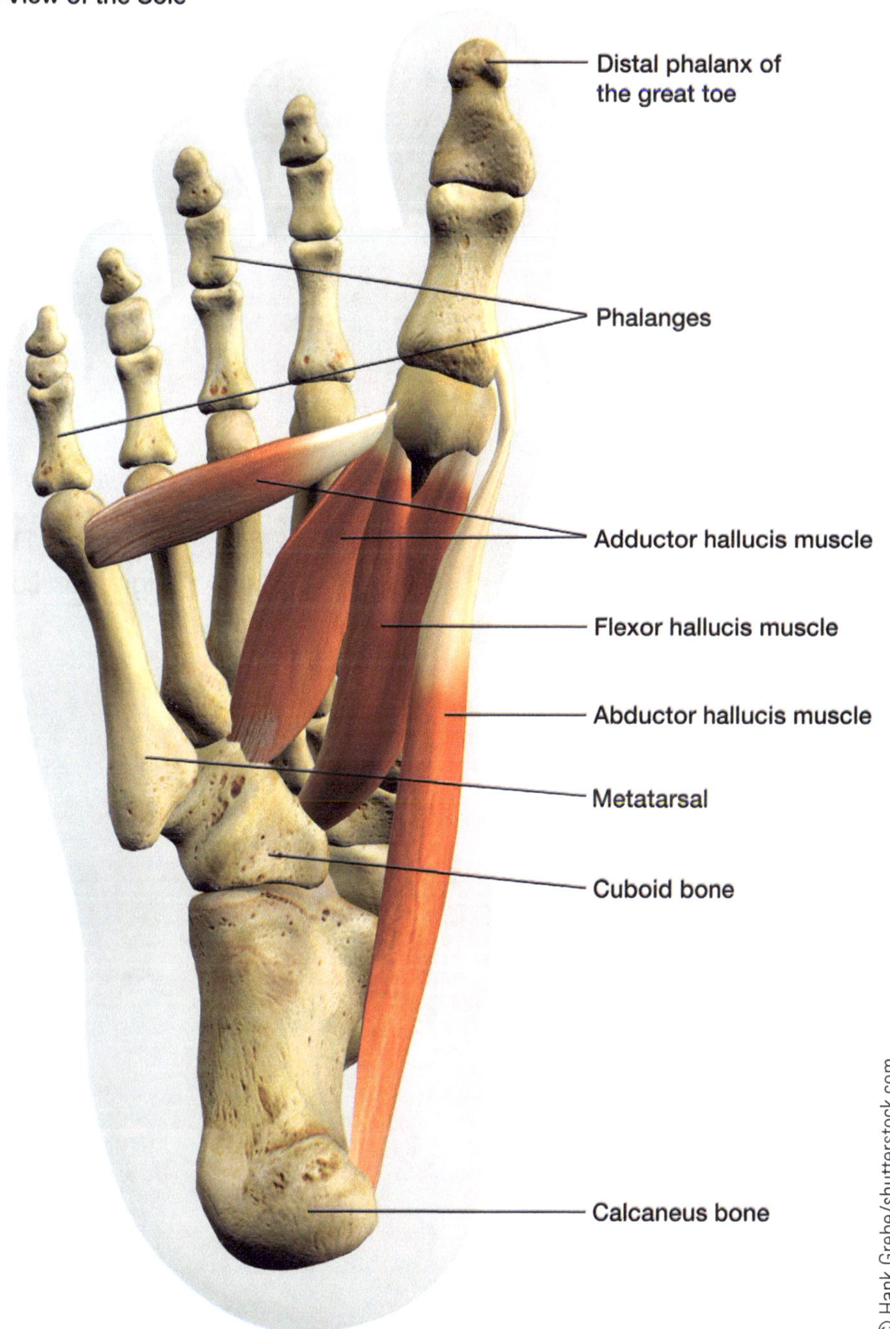

Quadratus Plantae/Lumbricales

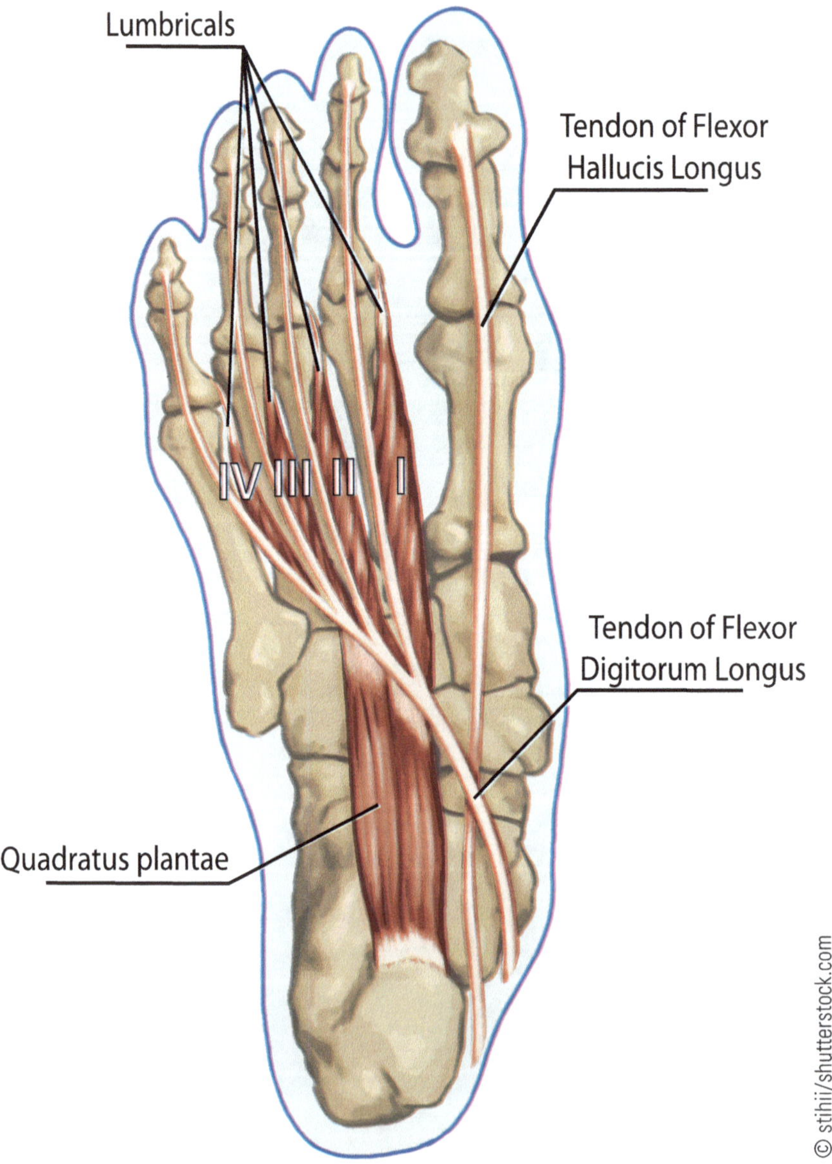

<u>Interosseous Muscles:</u>
Plantar Interossei (PAD; 3, 4, 5)- adduction
Dorsal Interossei (DAB; 1,2,3,4,5)- abduction

Nerves of Upper Extremity

Trust Me Moment
Lower extremity nerves are compartmentalized much like the muscles so keep it simple!! Knowing the muscle compartments will also help tremendously.

Lumbrosacral Plexus

- Supplies the lower limb; much like the brachial plexus in the upper limb
- **Ventral rami** of T12-Sacral; **all nerves of lower extremity will begin anteriorly**
- Reference point is the 12th rib at the level of spinous process T12
- This section will only discuss those nerves that innervate the lower extremity muscles discussed previously

Femoral Nerve/Obturator Nerve

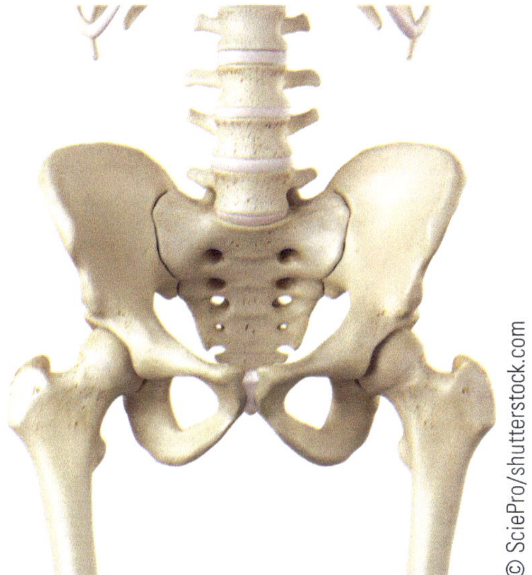

© SciePro/shutterstock.com

Sciatic Nerve, Pudendal Nerve

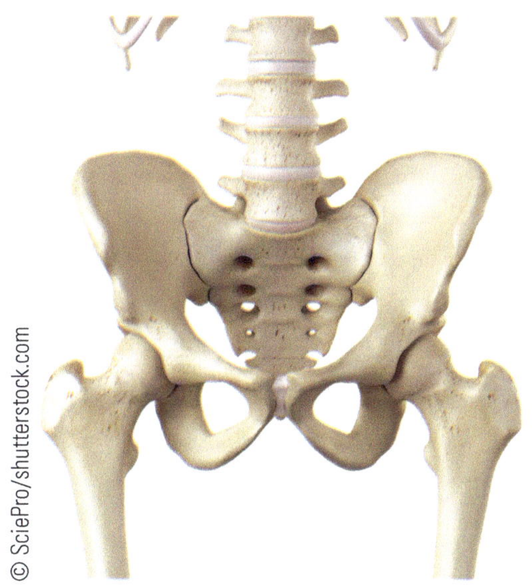

© SciePro/shutterstock.com

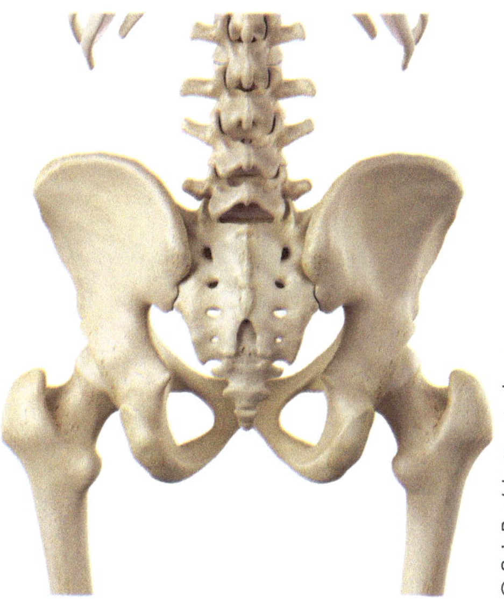

© SciePro/shutterstock.com

Superior Gluteal N/Inferior Gluteal N

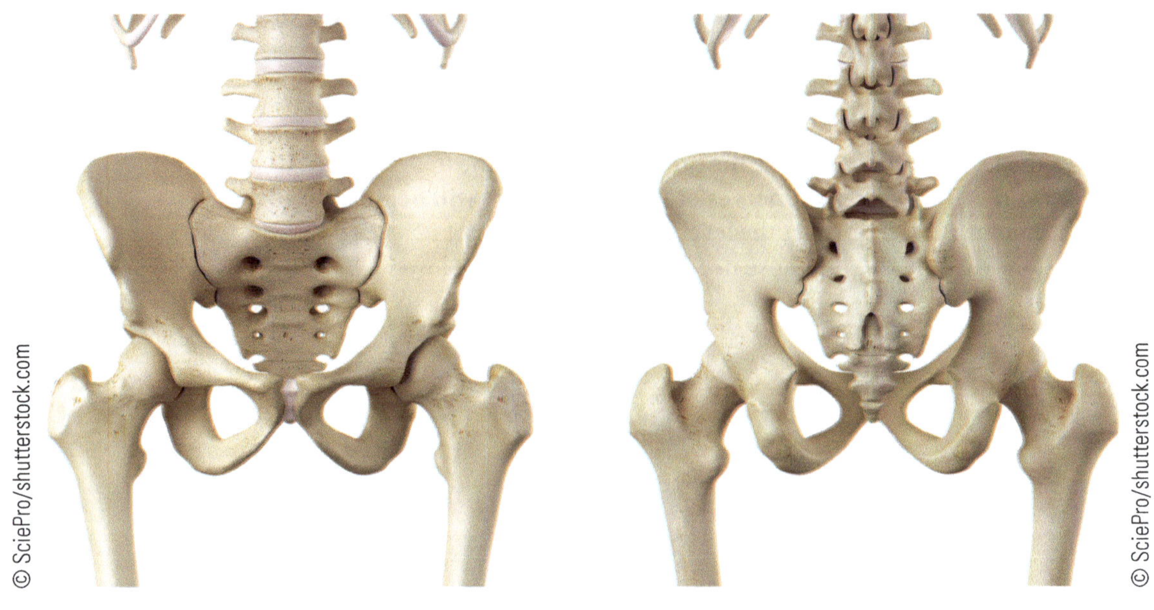

Common Fibular N

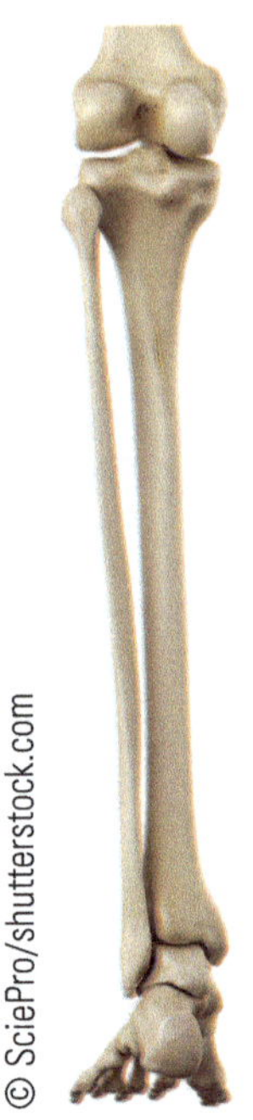

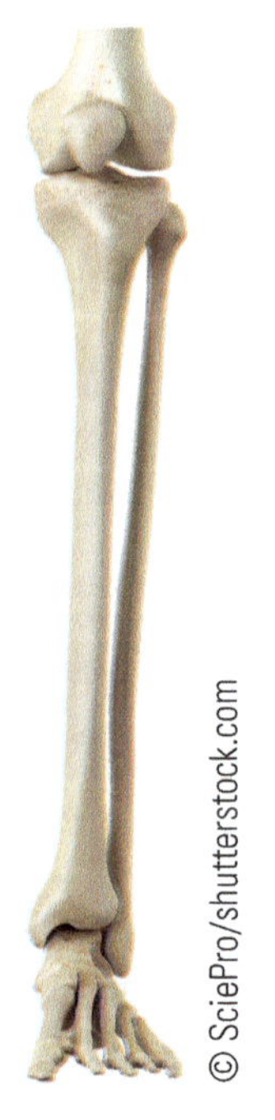

Tibial N to M/L Plantar N

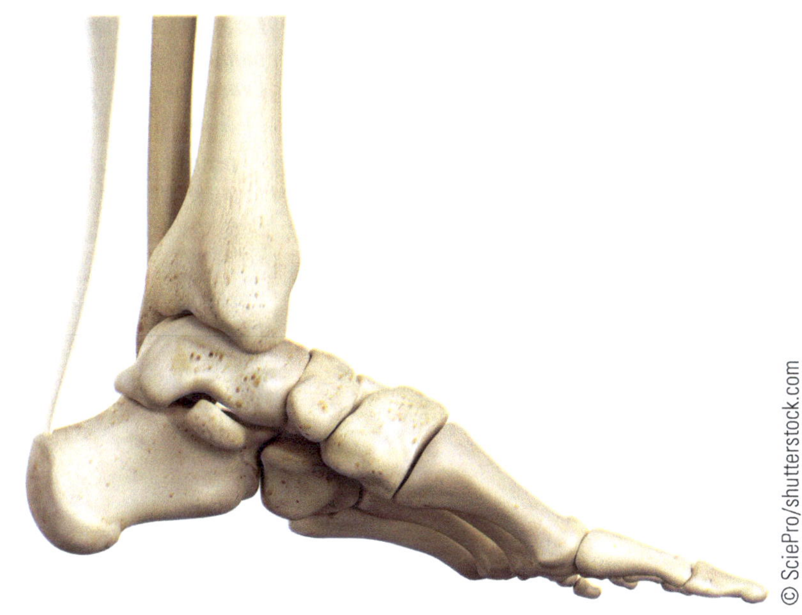

Nerve	Branching directly from (Anatomically)	Root Origin *Ventral Rami*	Innervates *Think functions Affected*	Getting there…(typically) *To help in the lab*
Femoral N	Roots L2-L4	L2-L4	**Anterior FA/Knee compartment** *Hip Flexors* *Knee Extensors*	Comes out from under the lateral border of the psoas major → under the inguinal ligament → to anterior thigh through the femoral triangle *(see it here superficially before it goes deep to rectus femoris)* **Patellar Tendon Reflex**
Obturator N	Roots L2-L4	L2-L4	**Medial FA compartment** *Hip Adductors* *Poor Bills Lose Many Games*	Comes out from the medial border of the psoas major and heads through the obturator membrane (obturator foramen) → medial compartment; *look over the pectineus and under adductor longus*
Superior Gluteal N	Root L4-S1	L4-S1	**Lateral FA compartment** *Hip Abductors* *G med, G min, TFL*	*Moves posteriorly through the greater sciatic notch → pops out above the piriformis and travels laterally across the posterior ilium to reach lateral muscles*
Inferior Gluteal N	L5-S2	L5-S2	*Gluteus Maximus*	*Moves posteriorly through the greater sciatic notch → pops out below the piriformis and travels down into the gluteus maximus*
Sciatic N	S4-S3	S4-S3	**Posterior FA compartment** *Hip Extensors*	*Moves posteriorly through the greater sciatic notch → pops out below the piriformis (large) and travels down the posterior thigh; will branch into* **common fibular and tibial n.**
Common Fibular N	From Sciatic N	L4-S2	**Lateral and Anterior TC** *TC dorsiflexors/evertors* *TDH, Fib Longus/Brevis*	*Branches from Sciatic nerve and heads towards the head of the fibula, once at the head of the fibula, it will branch into the* **Deep and Superficial Fibular N**

Nerve	Branching directly from (Anatomically)	Root Origin *Ventral Rami*	Innervates *Think functions Affected*	Getting there…(typically) *To help in the lab*
Superficial Fibular N	*From Common Fibular N*		**Lateral talocrural compartment** **Eversion** *Fib. Longus/Brevis*	*At head of fibula will descend in between the fibularis brevis and longus until it becomes cutaneous at the dorsal foot*
Deep Fibular N	*From Common Fibular N*		**Anterior talocrural compartment** *Dorsiflexion* *TDH longus/brevis, Tertius*	*At head of fibular will move around the head of fibula to move anteriorly deep and between anterior tibialis and extensor digitorum longus to dorsum*
Tibial N	From Sciatic N	L4-S3	**Posterior/medial talocrural compartment** *Plantar/inversion* *THD, G, S*	*From sciatic, moves posteriorly deep to soleus → passing the medial malleolus to the plantar foot where it splits into the* **medial/lateral plantar nerves**

Sensory/Dermatome Distribution

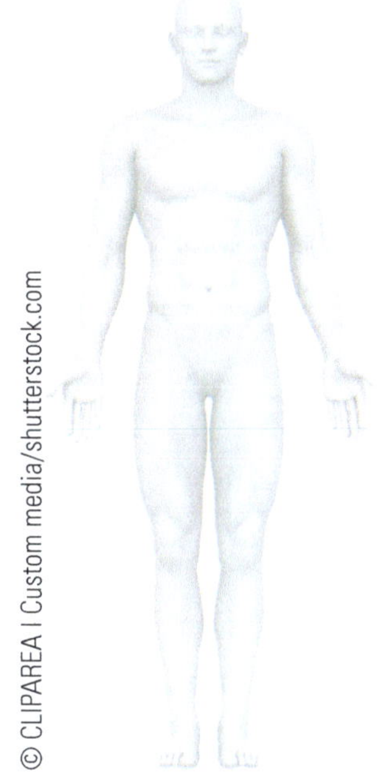

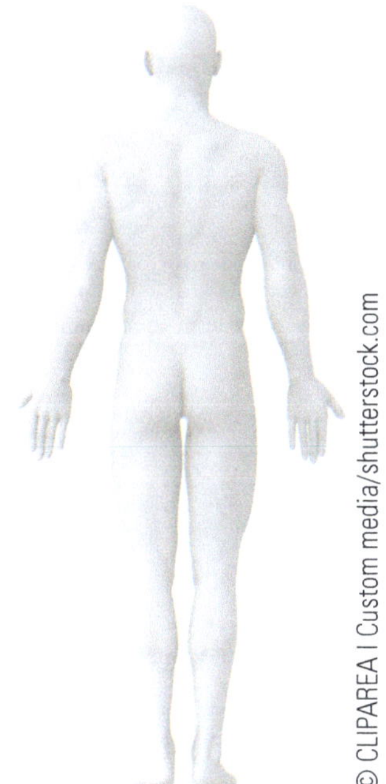

Arteries of the Lower Extremity

Quick Hits:

The **Abdominal Aorta** will have feed the lower extremity through 2 primary branches:
   a. **Right Common Iliac A and Left Common Iliac** branch into: (easy ☺)
      i. Right and Left External Iliac A
      ii. Right and Left Internal Iliac A
      iii. Branches into common iliac arteries at ~L4
Still easy ☺
   b. **Right and Left External Iliac A** will be the primary source for lower extremity
   c. **Right and Left Internal Iliac A** will be the primary source for pelvic/gluteal regions

Pelvic/Gluteal Regions

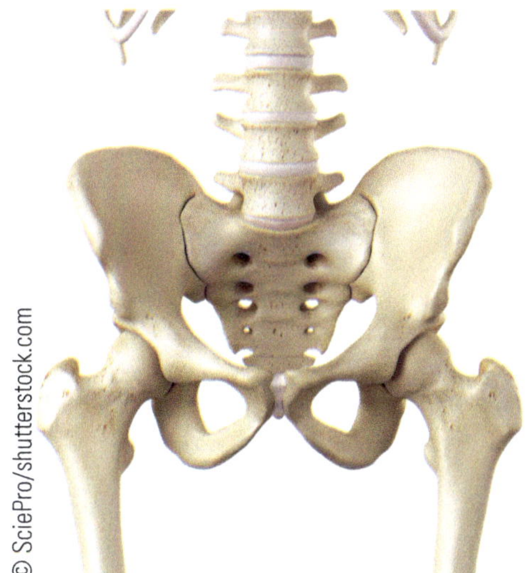

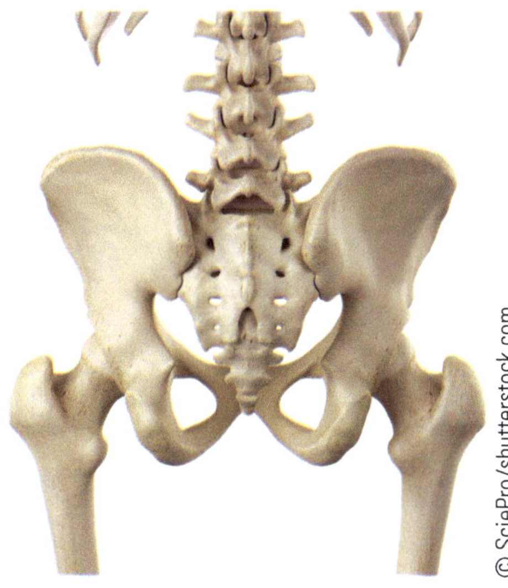

Lower Limb

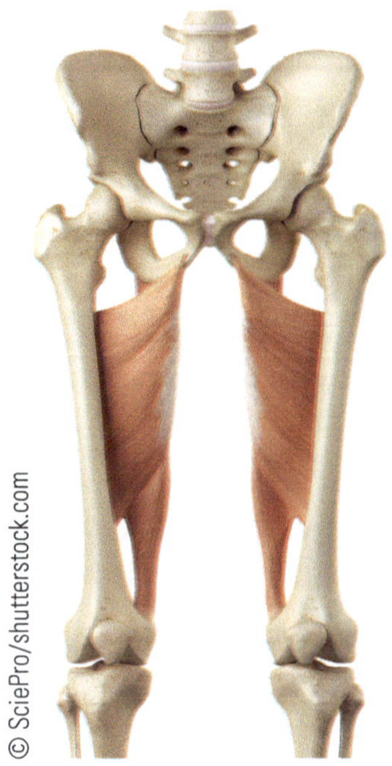
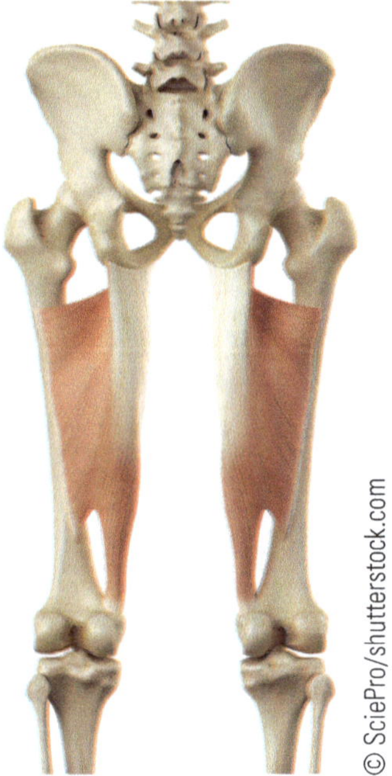

Lower Leg

Closer Look at Foot for 1 second

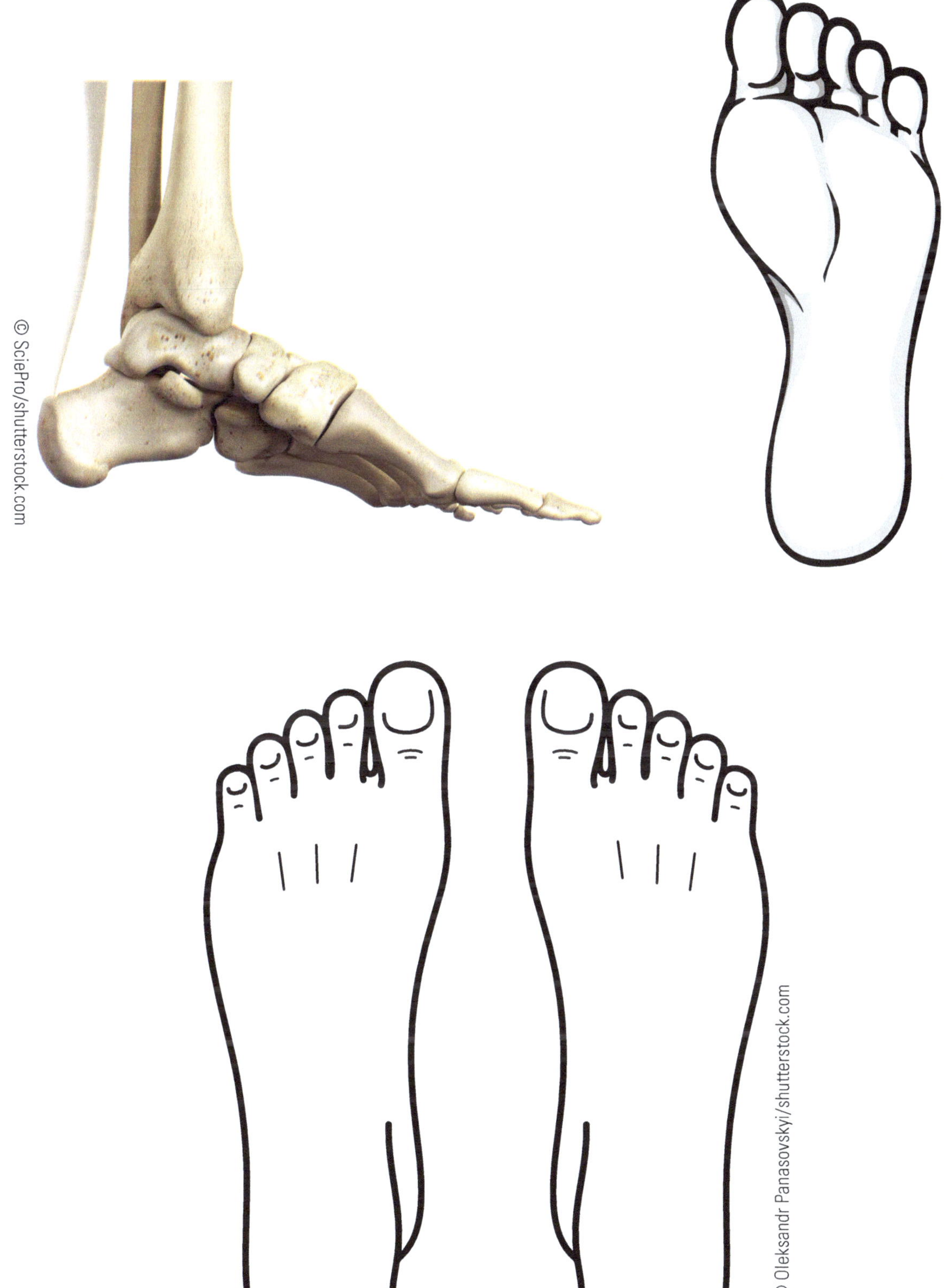

Think WHITE COAT!!!!

Blood Supply: Think Compartments

From	Artery	Getting there?	Supplies (in general)
Internal Iliac	Superior Gluteal	*Through greater sciatic notch to come out above the piriformis; makes its way to the posterior ilium to lateral region; Abductors*	Gluteus maximus (so big, needs 2) Gluteus medius Gluteus minimus TFL Piriformis
Internal Iliac	Inferior Gluteal	*Through greater sciatic notch to come out below piriformis;*	Gluteus maximus (so big, needs 2) Piriformis (so close, why not?) Lateral rotators (P, G, O, G)
Internal Iliac	Obturator	*Descends through the obturator membrane (foramen) to medial compartment; Adductors*	Pectineus Adductor brevis, longus, magnus Gracilis
Internal Iliac	Iliolumbar	*Early branch of posterior division; moves towards the iliolumbar region*	Psoas major Iliacus
External Iliac	Femoral	*Under the inguinal ligament and to the femoral triangle and continues through the adductor hiatus to move posteriorly and end as popliteal artery; has many **muscular branches** to reach quadriceps, etc.*	Anterior compartment of hip and knee *Sartorius* *Quadriceps*
Femoral	Deep Femoral via perforating branches	*Branches from the femoral artery within the femoral triangle; it will stay anterior but run deep (perforate) to posterior compartment via **perforating branches*** *Will branch into **lateral/medial femoral artery***	Posterior compartment of hip and knee *Hamstrings* *Adductor magnus*
Femoral Circumflex	Deep Femoral	*Will move around the femur; think location around proximal femur (trochanter region); there is a **lateral and medial femoral circumflex artery** (think location of muscle)*	Proximal femur TLF Proximal quadriceps Proximal adductors
Femoral	Popliteal	*Upon becoming posterior, the femoral artery becomes the popliteal artery; this will branch into **4 genicular arteries** that supply the knee joint* ****Popliteal Pulse***	Distal hamstrings Gastrocnemius Soleus Plantaris Popliteus
Popliteal	Anterior Tibial	*Branching from the popliteal; heads immediately through the interosseous membrane to enter the anterior compartment; will descend the anterior lower leg lateral to anterior tibialis to dorsal foot* ****Dorsal Pedal Pulse***	Anterior compartment/Dorsal foot *Dorsiflexors*
Popliteal	Posterior Tibial	*Continuation of popliteal as it descends the posterior lower leg with the tibial nerve deep to soleus; heads to the medial malleolus to the plantar surface of the foot to branch into the **medial and lateral plantar arteries***	Posterior/medial compartment/Plantar foot *Plantar flexors/Invertors*
Posterior Tibial	Fibular	*Move toward the fibular region*	Lateral compartment *Evertors*

Need a moment?

Of Clinical Significance

Femoral Triangle

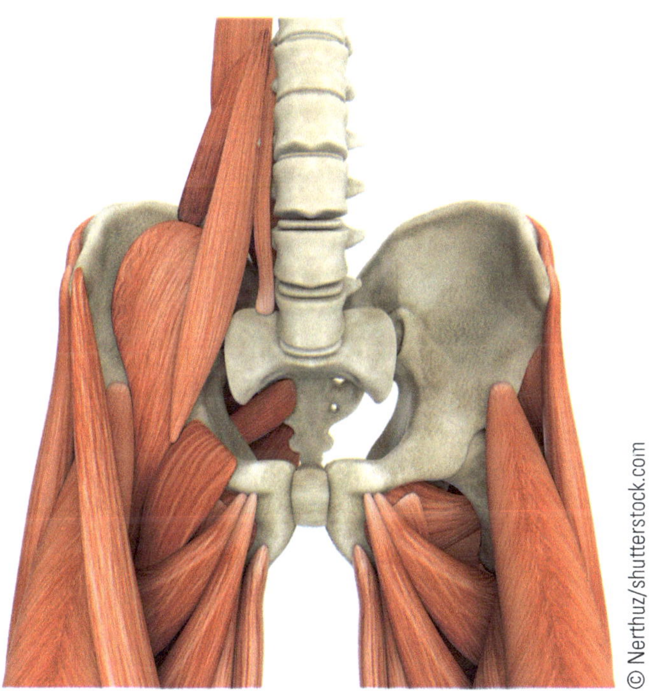

Boundaries: SAIL NAVY

1. Superior: Inguinal ligament
2. Lateral: Sartorius
3. Medial: Adductor longus
4. Floor: Iliospoas, Pectineus

Contents:

1. Femoral N, A, V and lymphatics (femoral canal) **(femoral pulse)**

Significance:

Of Clinical Significance

Popliteal Fossa

Boundaries:

1. Lateral upper: Biceps Femoris
2. Medial lower: Semimembranosus
3. Lower: Both heads of Gastrocnemius

Contents:

1. Popliteal artery **(popliteal pulse)** and geniculars
2. Popliteal vein to small saphenous vein
3. Sciatic nerve to tibial and common fibular n
4. Popliteal lymph nodes

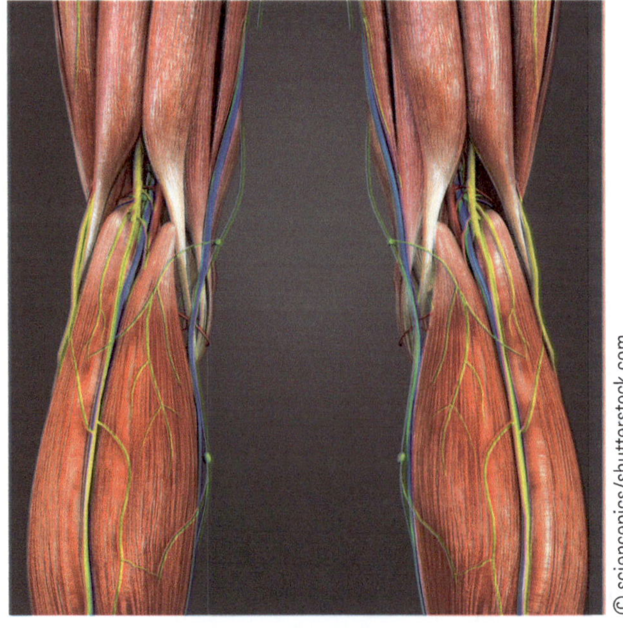

© sciencepics/shutterstock.com

Putting it All Together

Muscle	Insertion TO	Origin	Function	N	A
ANT HIP/KNEE					**THINK REGION**
Psoas Major	Lesser trochanter of femur	Bodies/transverse processes of lumbar	Ant FA: Flexion	Upper ventral rami Lumbar	Iliolumbar
Iliacus	Lesser trochanter of femur	Iliac fossa	Ant FA flexion	Upper ventral rami Lumbar	Iliolumbar
Sartorius	Pes anserine (medial to tibial tuberosity)	ASIS	Ant FA: Flexion Lat FA: Abduction Med Knee: Med rotation Post Knee: Flexion	Femoral	Femoral Femoral Circumflex
Tensor Fascia Latae	IT band	ASIS	Lat FA: Abduction Ant FA: Flexion Anterior Head: Medial rotation	Superior gluteal	Superior gluteal Femoral Circumflex
Rectus Femoris *only quadriceps crossing hip joint	Tibial tuberosity via patellar tendon	AIIS	Ant FA: Flexion Ant Knee: Extension	Femoral	Femoral Femoral Circumflex
Vastus Medialis	Tibial tuberosity via patellar tendon	Intertrochanteric line Medial linea aspera	Ant Knee: Extension	Femoral	Femoral
Vastus Lateralis	Tibial tuberosity via patellar tendon	Intertrochanteric line Lateral linea aspera	Ant Knee: Extension	Femoral	Femoral
Vastus Intermedius	Tibial tuberosity via patellar tendon	Anterior femur	Ant Knee: Extension	Femoral	Femoral
POST HIP/KNEE					**THINK REGION**
Gluteus Maximus	Gluteal tuberosity of femur and IT band	Sacrum, gluteal surface of ilium	Post: Extension Post head of femur: Lateral rotation	Inferior gluteal	Superior gluteal Inferior gluteal *Femoral circumflex*
Gluteus Medius	Lateral surface of greater trochanter of femur	Just below iliac crest (posterior)	Lat FA:Abduction ***primary Abductor***	Superior gluteal	Superior gluteal *Femoral circumflex*
Gluteus Minimus	Anterior/lateral greater trochanter of femur	Just below medius insertion on iliac crest	Lat FA: Abduction Ant head of femur: Medial rotation	Superior gluteal	Superior gluteal *Femoral circumflex*
Piriformis	Lip of greater trochanter	Pelvic surface of sacrum	Lat FA: Abduction Post head of femur: Lateral rotation	N to piriformis	Superior gluteal Inferior gluteal

Muscle	Insertion TO	Origin	Function	N	A
<u>Lateral Rotators</u> Gemellus Superior Obturator Internus Gemellus Inferior Quadratus Femoris			Post head of femur: Lateral rotation	N to obturator internus N to quadratus femoris	Inferior gluteal Femoral circumflex
Biceps Femoris	Head of fibula	Ischial tuberosity Mid-posterior femur	Post FA: Extension Post Knee: Flexion Lateral Knee: Lateral rotation	Sciatic to Tibial	Inferior gluteal Deep femoral Popliteal
Semitendinosus	Pes anserine of tibia (medial to tibial tuberosity)	Ischial tuberosity	Post FA: Extension Post Knee: Flexion Medial Knee: Medial rotation	Sciatic to Tibial	Inferior gluteal Deep femoral Popliteal
Semimembranosus	Medial tibial condyle	Ischial tuberosity	Post FA: Extension Post Knee: Extension Medial Knee: Medial rotation	Sciatic to Tibial	Inferior gluteal Deep femoral Popliteal
ADDUCTORS					**THINK REGION**
Pectineus	Pectineal line of femur	Pectineal line of pelvis (superior pubic ramus)	Med FA: Adduction *Hip Flexion*	Femoral *Obturator*	Obturator Femoral Circumflex
Adductor Brevis	Upper linea aspera	Inferior pubic ramus	Med FA: Adduction *Hip Flexion*	Obturator	Obturator Femoral Circumflex
Adductor Longus (right)	Mid linea aspera	Superior pubic ramus	Med FA: Adduction	Obturator	Obturator Femoral Circumflex
Adductor Magnus	Mid to distal linea aspera Adductor tubercle of femur	Inferior pubic ramus Ishial tuberosity	Med FA: Adduction Ant Hip: Flexion Post Hip: Extension	Obturator Sciatic to Tibial	Obturator Inferior gluteal Femoral Circumflex Deep femoral Popliteal
Gracilis	Pes anserine (medial to tibial tuberosity)	Pubic tubercle	Med FA: Adduction *Flexion* *Med Rotation*	Obturator	Obturator Femoral Circumflex
LOWER LEG					**THINK REGION**
Anterior Compartment	Dorsum	Tib, Fib, All *Interosseous membrane*	TC: Dorsiflexion ST: *If medial; inversion* *If lateral: eversion* MTP/IP: *Extension*	Deep fibular	Anterior tibial

Muscle	Insertion TO	Origin	Function	N	A
Posterior/Medial Compartment	Plantar	All, Tib, Fib *Interosseous membrane*	TC: Plantar Flexion ST: *Med: Inversion* MTP/IP: *Flexion*	Tibial to Plantar	Posterior tibial to Plantar
Gastrocnemius	Achilles to calcaneus	Femoral condyles	Post Knee: Flexion TC: Plantar flexion	Tibial	Popliteal
Soleus	Achilles to calcaneus	Soleal line of tibia	TC: Plantar flexion	Tibial	Popliteal Posterior tibial Fibular
Lateral Comparment	Lateral	Fibula	Posterior TC: Plantar flexion Lateral ST: Eversion	Superficial fibular	Fibular
Plantar foot			Flexion ADD	Med/Lat Plantar	Med/Lat Plantar
Dorsum foot			Extension ABD	Deep fibular	Anterior tibialis through dorsal pedis

Remember that Arteries branch as needed; compartmentalize and consider regional distribution

Putting it All Together

Putting it All Together

<u>Putting it All Together</u>

<u>Putting it All Together</u>

<u>Putting it All Together</u>

Putting it All Together

Back

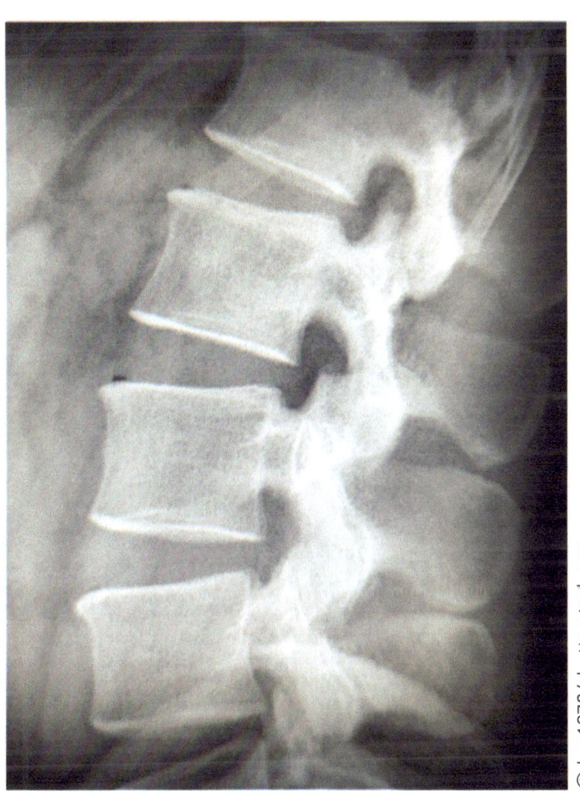

© kos1976/shutterstock.com

Bony Landmarks
Joints and Ligaments
Muscles
Innervations
Blood Supply

The Vertebral Column

1. Protects the spinal cord and spinal nerves
2. Supports the weight of the body
3. Posture and locomotion
4. 33 vertebrae; 7 cervical, 12 thoracic, 5 lumbar, 5 fused sacrum, 4 fused coccyx

Curvatures of the Vertebral Column

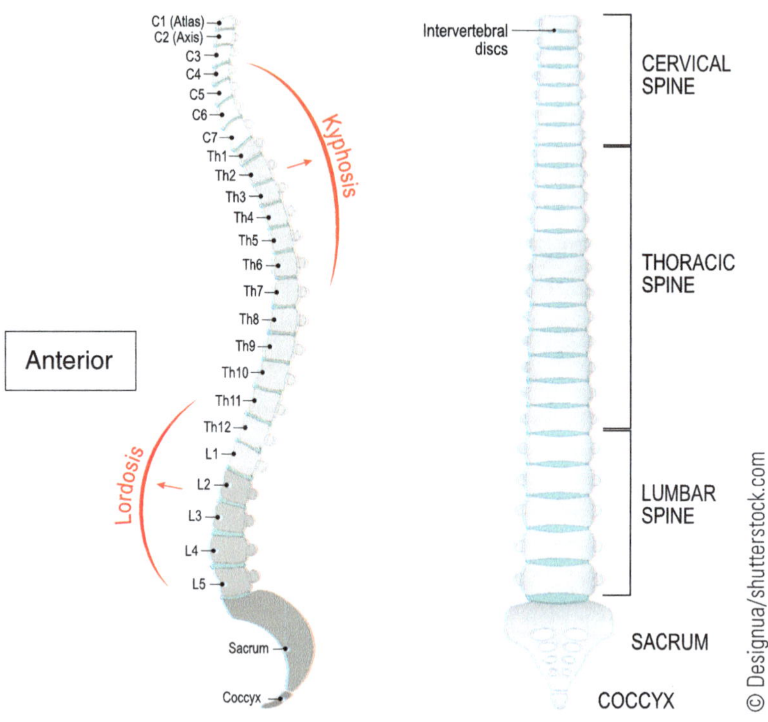

Lordosis	Concave posteriorly Bodies project anteriorly	Cervical Lumbar	If excessive: "Humpback" *Osteoporosis, Text neck*
Kyphosis	Concave anterior; Spinous process projects posteriorly	Thoracic Sacral	If excessive: "Hollowback" *Pregnancy/Obesity*

Clinical Quick Hit

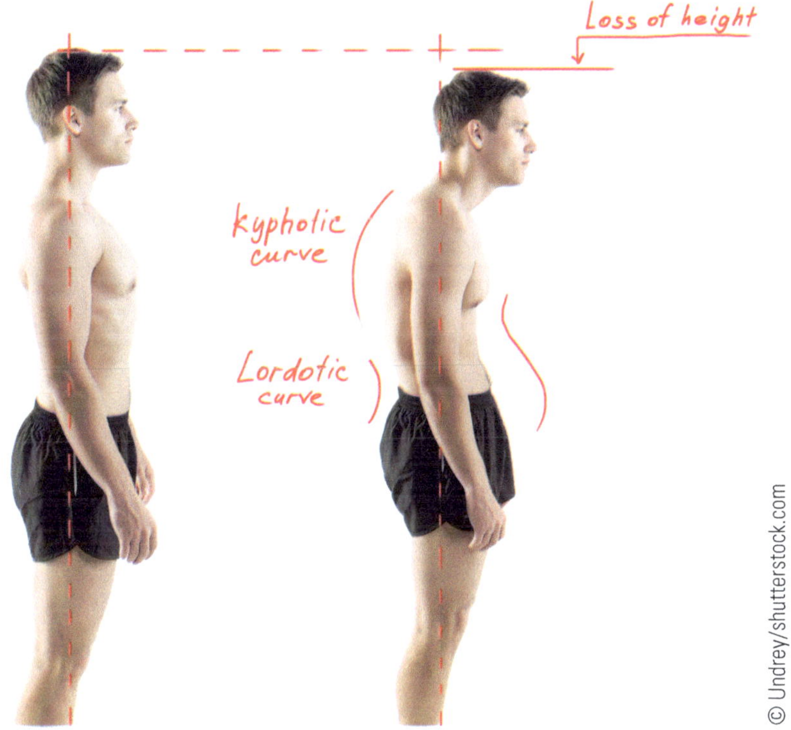

1. <u>Text Neck Syndrome</u>

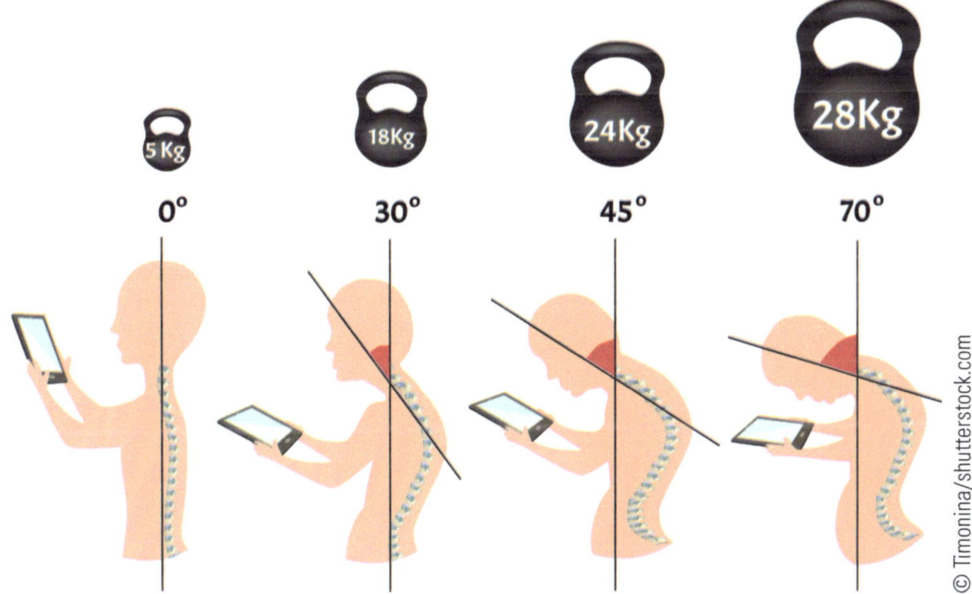

More vertebral column deformities discussed after bony landmarks reviewed

The Vertebrae

Palpation Landmarks

Mandible	Spinous process C2-3
Hyoid	C3
Vertebrae Prominens	Spinous Process C7
Spine of Scapula	Spinous Process T3
Inferior angle of Scapula	Spinous Process T7
Xiphisternal junction	Spinous Process T10
L1	End of Spinal Cord (Conus Medullaris)
Iliac Crest	Spinous Process L4 *Lumbar Puncture*
PSIS	Spinous Process S2

Vertebrae in General

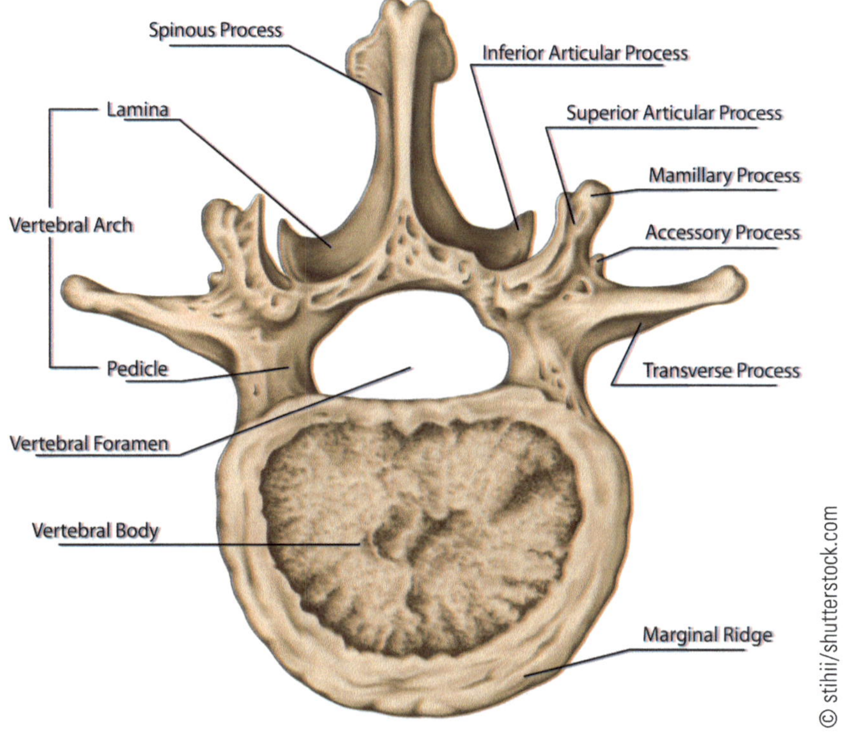

Body of Vertebrae	Anterior aspect; support body weight
Pedicle	Connects body to vertebrae **P**eanut **B**utter (Pedicle with Body)
Lamina	Connect spinous process to vertebrae **L**ittle **S**andwich *Laminectomy*
Transverse Process	Between the pedicle and lamina; horizontal For muscle and ligament attachment

Vertebral Arch	Formed by pedicle and lamina
Vertebral Foramen	Spinal cord passage; all adjacent foramen form the *vertebral canal*
Spinous Process	Posterior aspect; muscle and ligament attachment
Superior Inferior Articulating Process	Have smooth facets for adjacent vertebrae articulation; *the inferior articulating process of L4 will articulate with the superior articulating process of L5*
Intervertebral Foramen (see below)	When 2 vertebrae articulate, a foramen is formed; the posterior wall of the body forms the anterior aspect and the anterior surface of the lamina comprise the posterior aspect; passage for spinal nerve roots, dorsal root ganglion, spinal artery

Intervertebral Foramen

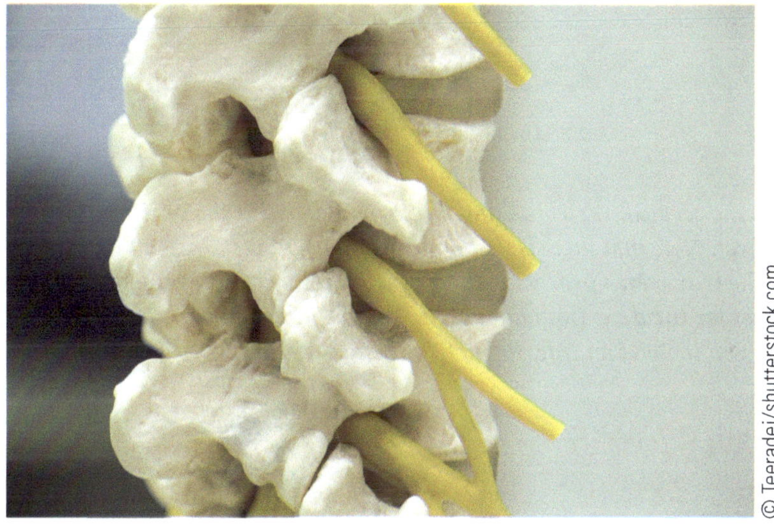

Shown with the spinal nerve root

Each section of vertebrae has its own distinguishing features that correspond with function. Below are the most noteable:

Cervical:

- Have a **transverse foramen** which allows passage of the <u>vertebral artery/vein</u>
- Smaller than other areas.
- Have Bifid spinous process
- Have a Uncinate
- C1 and C2 are different than the others

Thoracic:

- Have extra articulating facets for the ribs (costal facets)
- Long, narrow spinous process (giraffe)

Lumbar:

- Large body compared to others
- Short spinous process (moose)

Cervical Vertebrae C1- ATLAS

Superior View

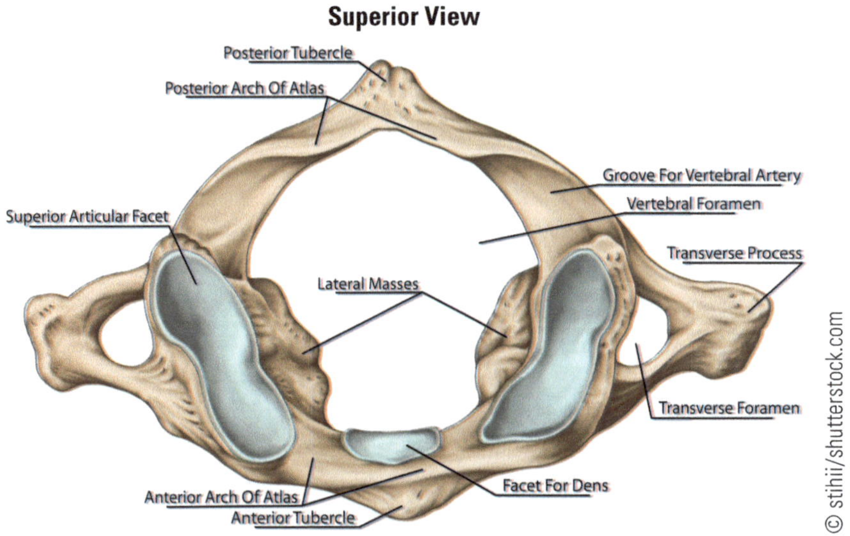

The **Atlas** is the first of the cervical vertebrae and will articulate with the **occipital condyles of the skull** via the **inferior articulating facet.** Note that there is no body or spinous process, instead there is an **anterior tubercle** and a **posterior tubercle.** The **transverse process** is short and contains a **transverse foramen** for the vertebral artery. The atlas also has a **facet for the dens** (found on C2). Note the **lateral masses** on both sides of the **vertebral foramen.** The inferior view will include an **inferior articulating facet** for C2 to articulate.

Cervical Vertebrate C2-AXIS

Superior View

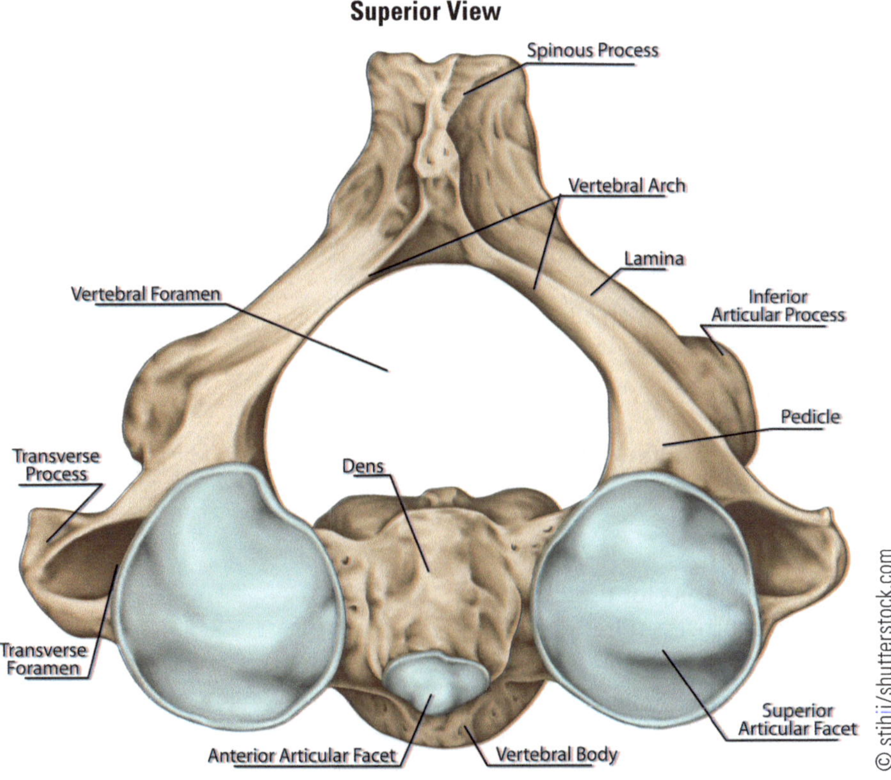

Note that the **Axis** has a **spinous process** and a very small **body.** Projecting from the body is the **dens**, which will articulate with the atlas to help with rotation. The remaining anatomy is the same as other cervical vertebrae, including the **transverse foramen**.

Cervical Vertebrae

Superior View

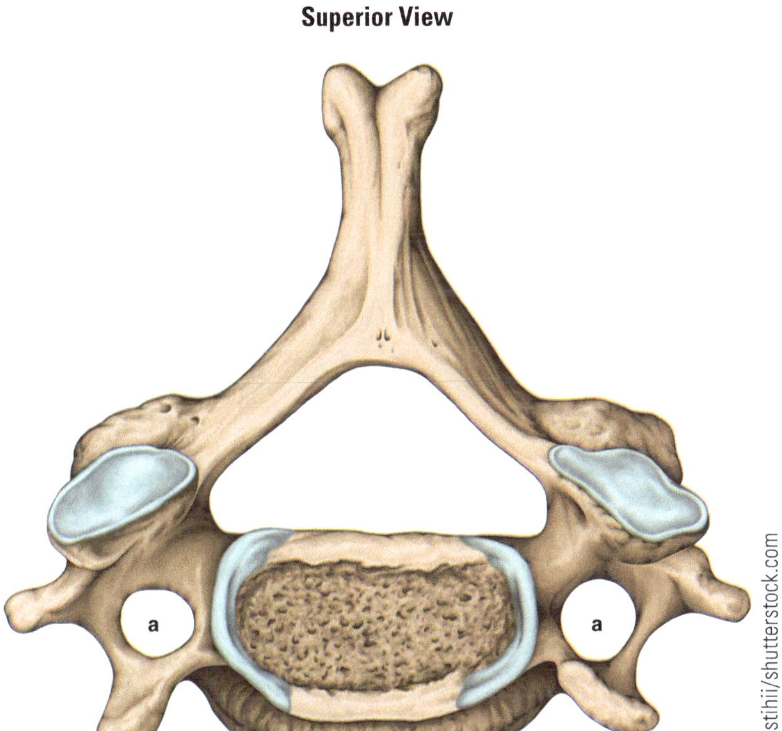

*Note that the remaining cervical vertebrae have the normal anatomy of a vertebrae discussed on previously. The exceptions are the **transverse foramen (a)** and the **bifid spinous process (except C7).** The bifid result from two separate ossification centers and they help with muscle and ligament attachment (especially ligamentum nuchae).*

Anterior view

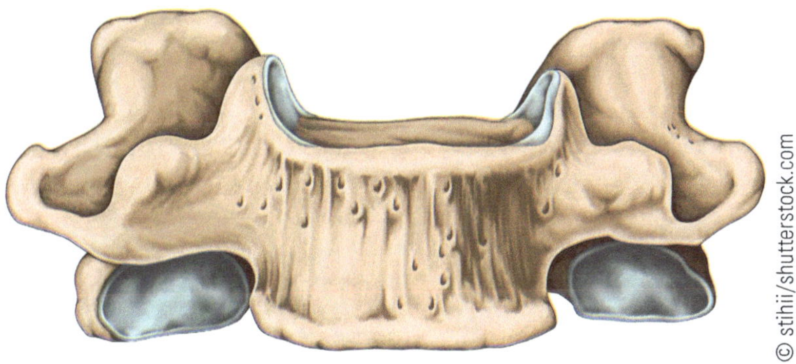

*Note the **uncinate processes (a)** that arise from the edges of the cervical bodies. These processes will allow for stability and mobility in the cervical region; allowing for neck flexion and extension.*

Cervical Vertebrae *continued...*

Cervical vertebrae articulated:

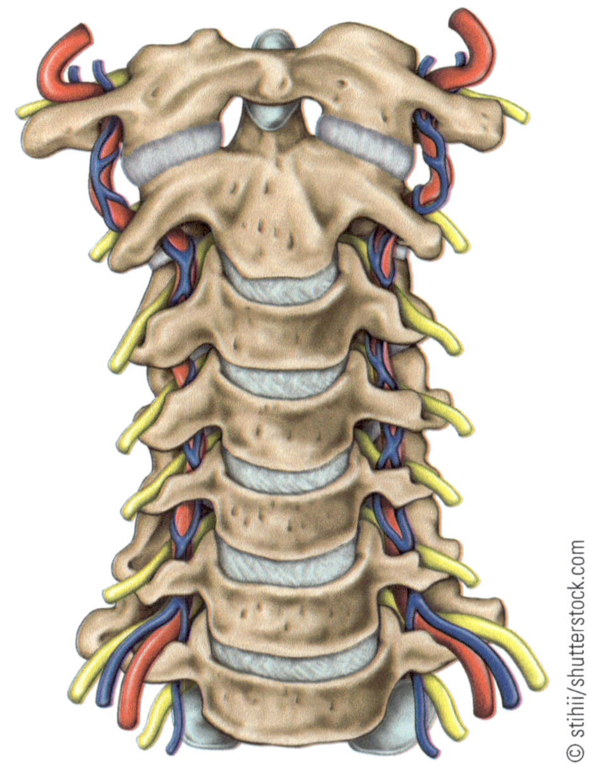

© stihii/shutterstock.com

Note the vertebral artery and vein as they run through the transverse foramen. The spinal nerve roots are running through the intervertebral foramen.

Thoracic Vertebrae

THE VERTEBRAL COLUMN

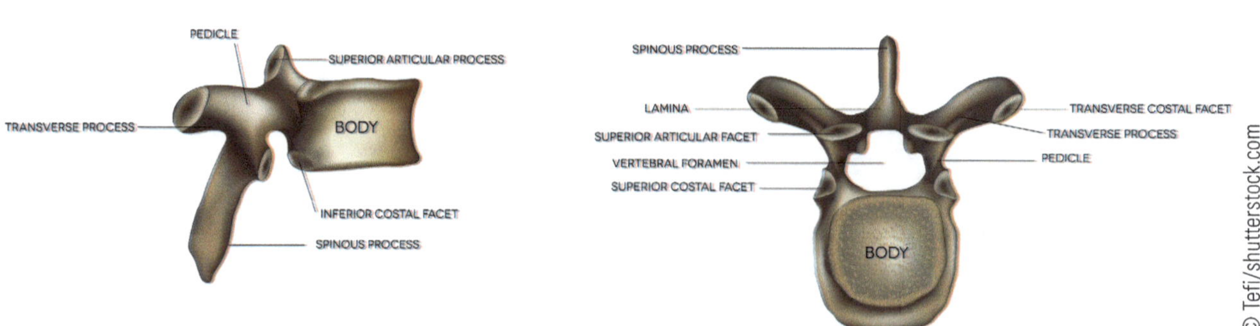

© Tefi/shutterstock.com

*Note the primary difference, the **transverse costal facet** on the transverse processes. These facets will articulate with the **tubercle of the rib.** There is also a **superior costal facet** and **inferior costal facet** on the posterior edges of the bodies. Note the downward pointing spinous process.*

Lumbar Vertebrae

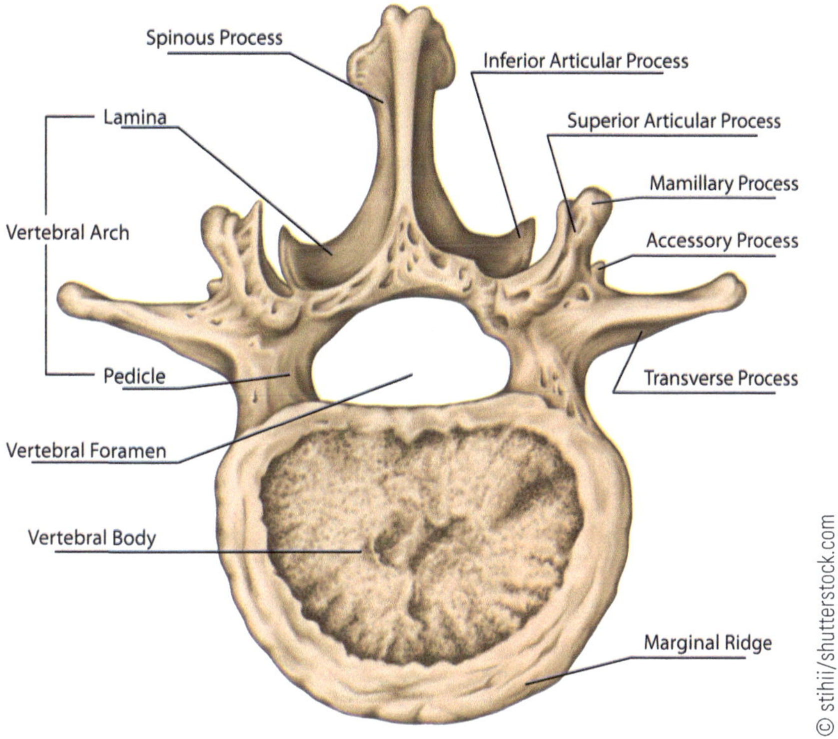

Note the larger bodies of the lumbar vertebrae, primarily for weight bearing purposes

Sacrum

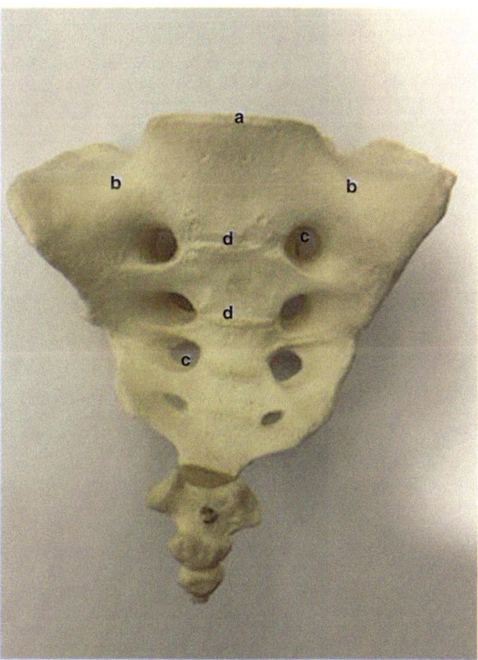

Source: Shireen Rahman

*Start at the top at the **promontory (a).** Moving out to the sides, you will find the **ala (wing of sacrum)(b).** The bilateral holes are the **anterior sacral foramina (c)** which enable ventral spinal nerves to pass posteriorly. The horizontal lines between the foramen are the **transverse lines(d),** and they are the fused bodies.*

Sacrum *continued...*

Posterior View

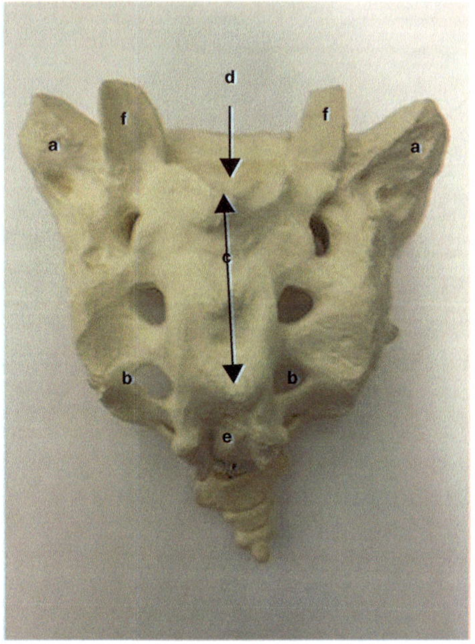

*Start at the **lateral part (a)** and move medially to the **posterior sacral foramen (b)**. Most medially, you will find the fused spinous processes, the **median sacral crest (c)**. At the superior aspect of the crest, you will find a hole entering the **sacral canal (d)**. At the inferior aspect of the crest, you will find the end of the sacral canal, the **sacral hiatus(e)**. This canal will house the sacral nerves via the cauda equina and filum terminale (discussed later). Note the **superior articulating facet (f)** for articulation with the L5 inferior articulating facet.*

Skull (as important for back muscle attachment)

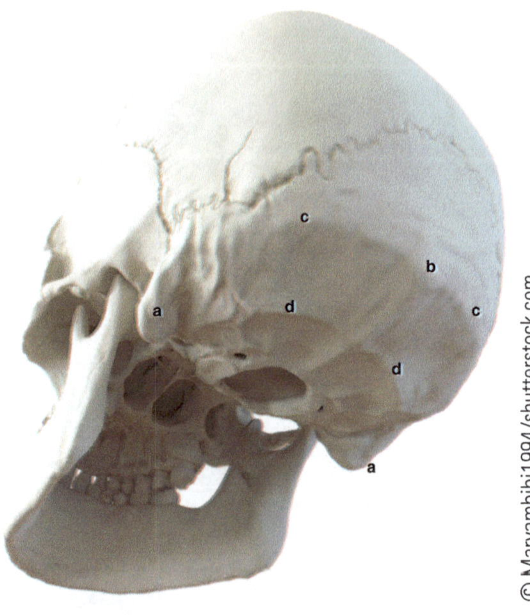

*Many of the back muscles/ligaments will attach to the following: **mastoid process (a), external occipital protuberance (b), superior nuchal line (c), inferior nuchal line(d).** The **occipital condyles (e)** will articulate with the superior articulating facet of the atlas.*

<u>Rib</u> *I hate the ribs so I put little effort forth in this next section ☺ My apologies in advance.*

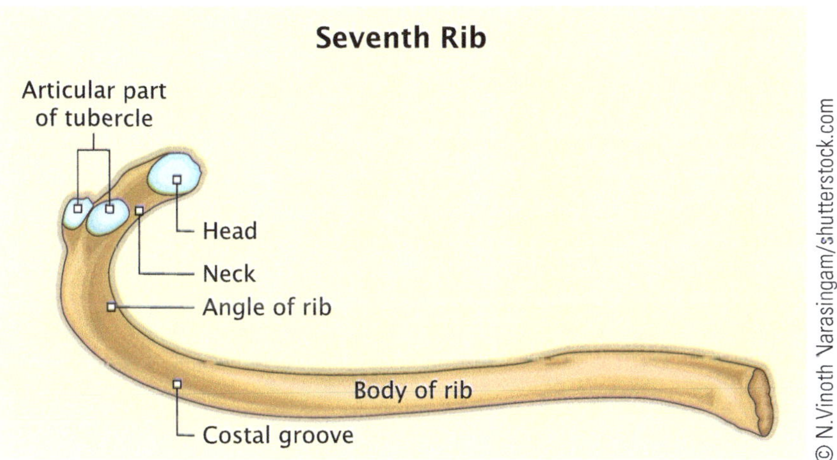

Joints and Ligaments of the Vertebral Column

1. <u>Atlanto-Occipital Joint</u>:
   a. Occipital condyles of skull with Superior articulating facet of atlas
   b. Condyloid synovial joint; thin joint capsules
   c. Reinforced with **anterior and posterior atlanto-occipital membranes**
   d. Flexion/Extension "yes" and Lateral Flexion "um I don't' know?"
2. <u>Atlanto-Axial Joints</u>:
   a. **2 gliding joints** between the articulating facets of the atlas and axis
   b. **1 pivot joint** between the articulating facet for the dens of the atlas with the dens of the axis; rotation ("no")

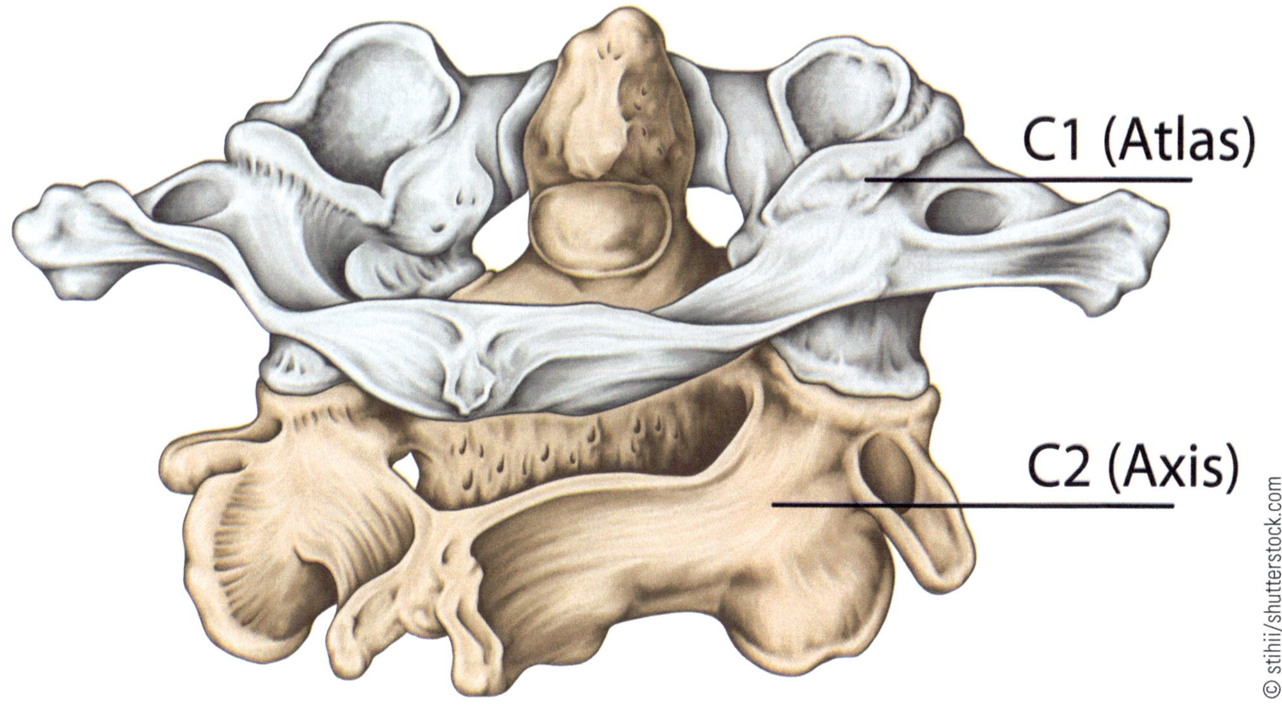

Ligaments of the Atlanto-Occipital and Atlanto-Axial Joints:

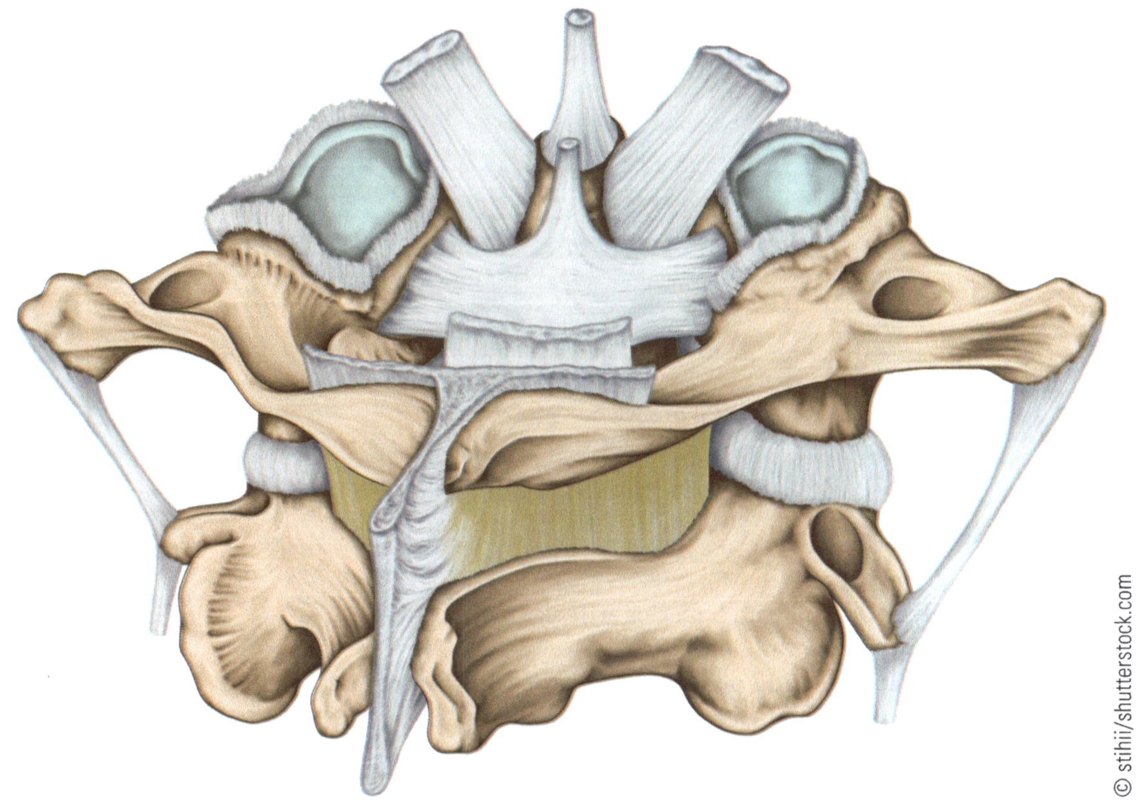

a. <u>Cruciform Ligament</u>
   Forming a cross-like pattern at the posterior dens to hold dens in place.
   i. Transverse Ligament moves across
   ii. Longitudinal Ligament moves up and down
b. <u>Apical Ligament</u> extends from the tip of the dens to the occipital bone (foramen magnum)
c. <u>Alar Ligament</u> extends bilaterally from the sides of the dens to occipital condyles

** **Atlantoaxial dislocation****
   1. Damage to the cruciform ligament (fracture to dens is typical)
   2. Dislocation can injure the spinal cord and medulla

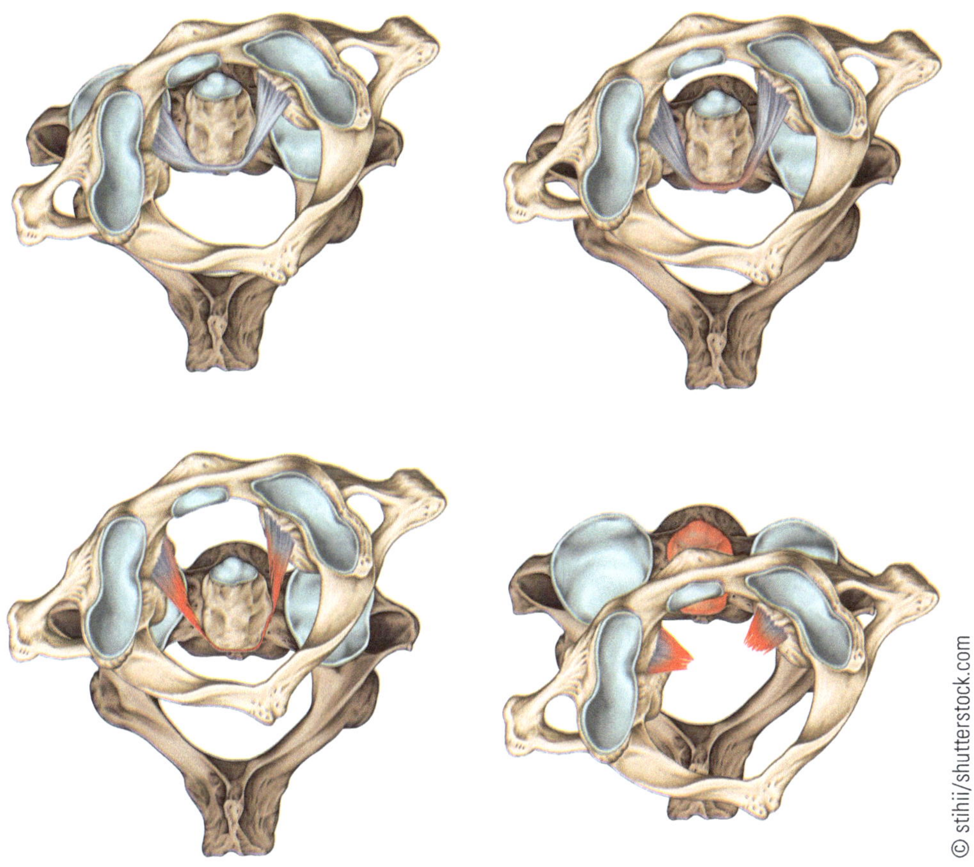

Joints continued . . .
1. <u>Intervertebral Joint</u>
   a. Body to adjacent Body
   b. **Intervertebral Disc** sits between the bodies; Fibrocartilaginous joint

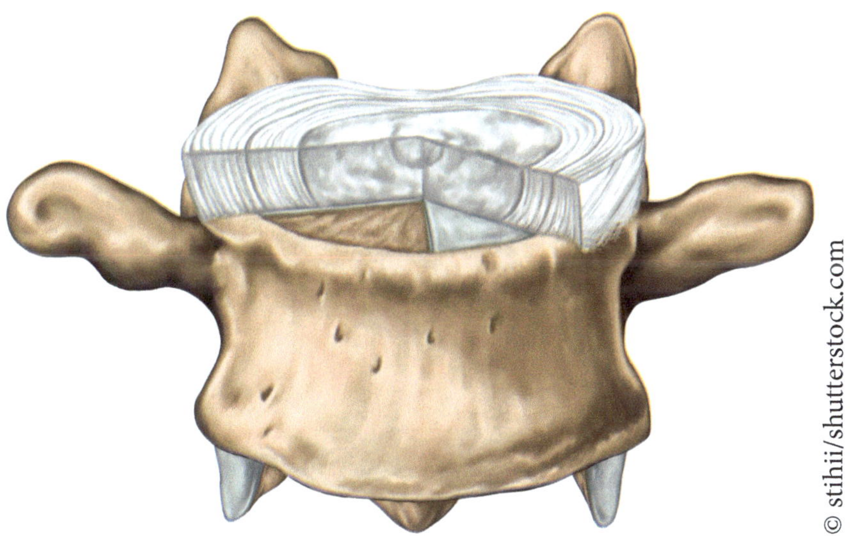

*The outer aspect of the intervertebral disc is the **annulus fibrosis** consisting of fibrocartilage. This is shock absorber that ties the adjacent vertebrae. The **nucleus pulposus** is the center "jelly" of the disc. It also helps with shock absorption. All aspects, except out aspects are avascular. The nucleus can become herniated from trauma or degeneration and press on the spinal nerves; **herniated disc.***

Ligaments of the Atlanto-Occipital and Atlanto-Axial Joints: *continued . . .*

Herniated Disc

1. Nucleus protrudes through the annulus fibrosis
2. Typically occurs at the posterior-lateral aspect where there is a lack of the ligament reinforcement surrounding the disc (posterior longitudinal ligament)

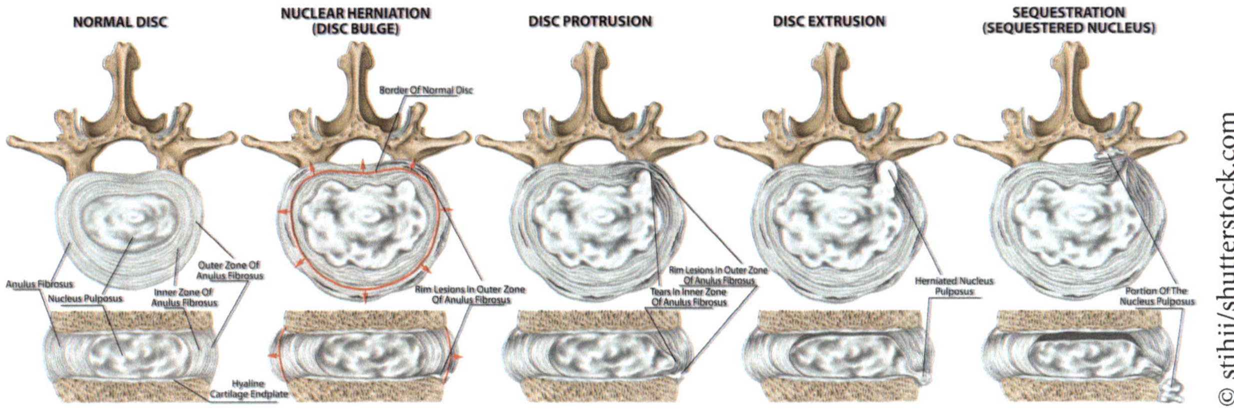

© stihii/shutterstock.com

Ligaments of the Vertebral Column

Draw/outline/color accordingly

1. <u>Anterior Longitudinal Ligament</u>
    a. Runs longitudinally down the anterior bodies
    b. Restricts:_________________________________

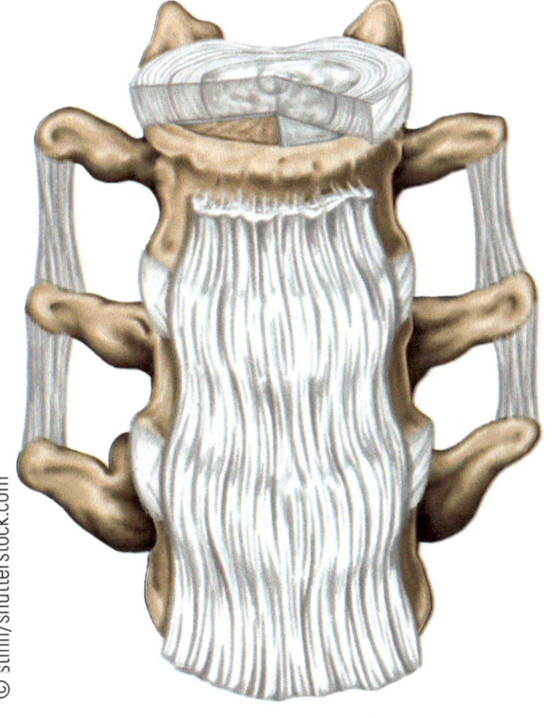
© stihii/shutterstock.com

Draw/outline/color accordingly

2. <u>Posterior Longitudinal Ligament</u>
    a. Runs longitudinally down the posterior bodies
    b. Anterior vertebral foramen
    c. Not as broad as the anterior long. ligament
    d. Leaves the posterior lateral aspect at risk for disc herniation
    e. Restricts:_________________________________

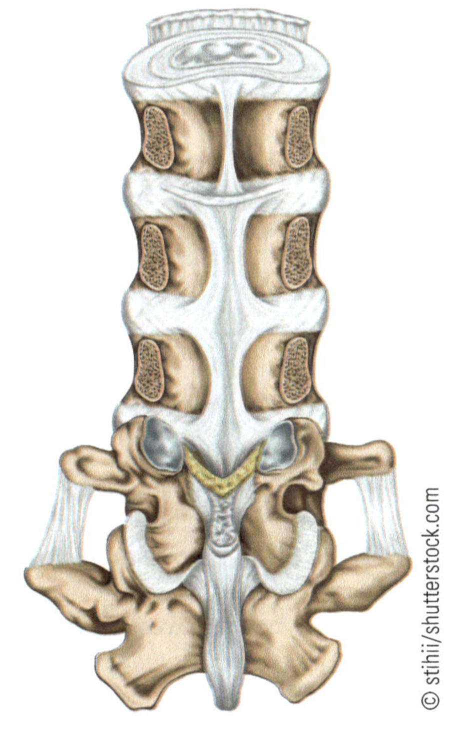
© stihii/shutterstock.com

3. Ligamentum Flavum
   a. Runs down the posterior wall of the intervertebral foramen;
   b. One aspect on each side of the lamina
   c. Maintains erect posture

4. Interspinous Ligament
   a. In between the bodies of the spinous processes
   b. Restricts: _______________________________

5. Supraspinous Ligament
   a. Runs from tip to tip of the spinous processes
   b. Restricts: _______________________________

6. Intertransverse Ligament
   a. Bilaterally, transverse process to transverse process
   b. Restricts: _______________________________

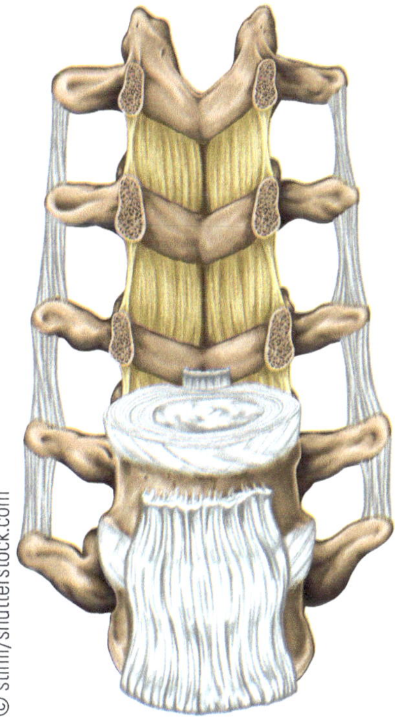

© stihii/shutterstock.com

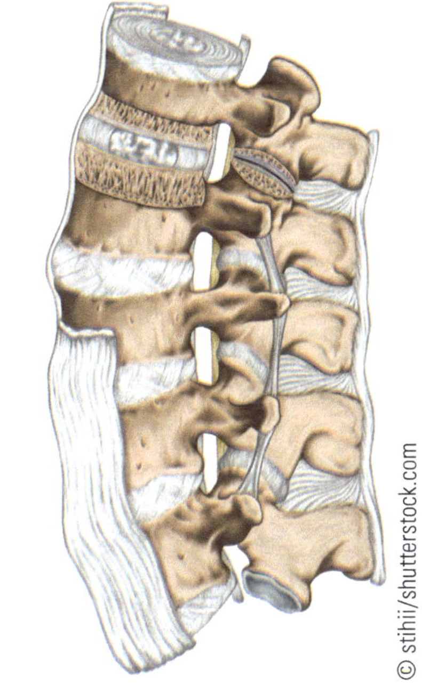

© stihii/shutterstock.com

7. Nuchal Ligament:
   a. Broad continuation of the supraspinous ligament
   b. C7 to external occipital protuberance

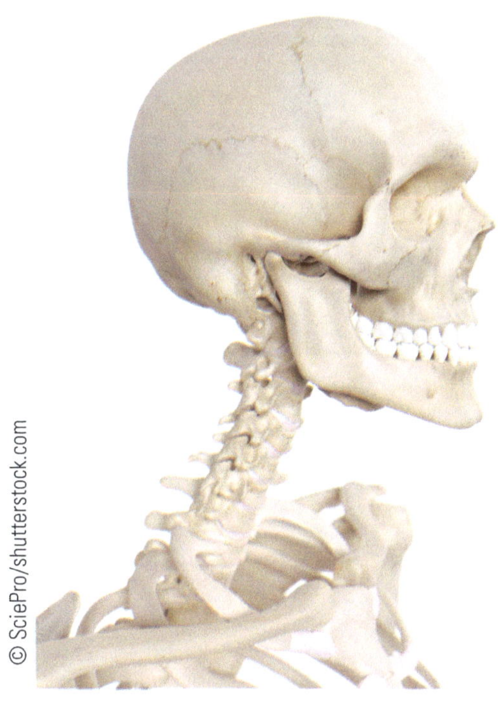

© SciePro/shutterstock.com

© Iryna Kuznetsova/shutterstock.com

Clinical Quick Hit

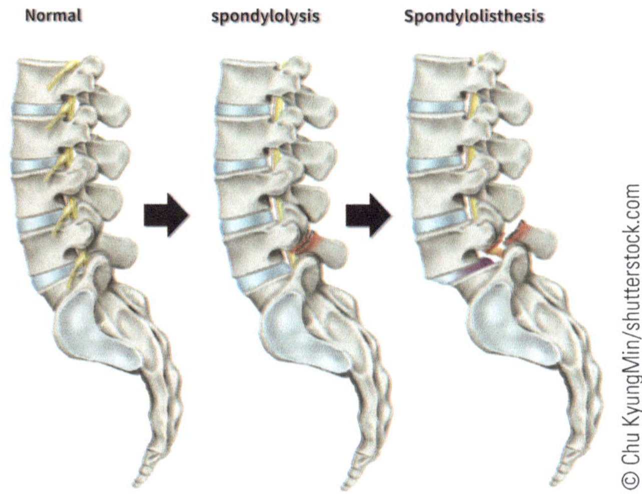

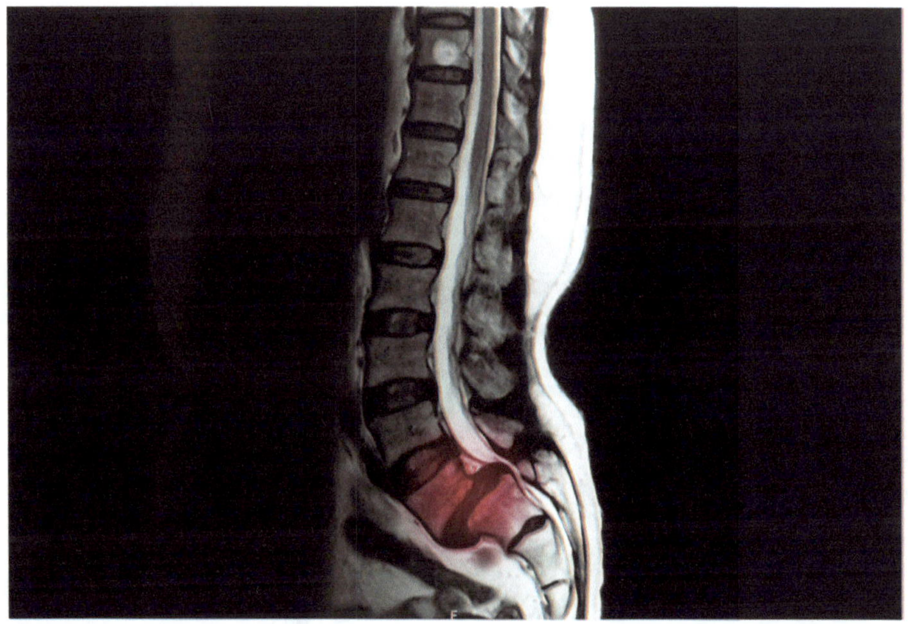

Muscles of the Back

Superficial Back

1. Trapezius
2. Latissimus dorsi
3. Levator scapulae
4. Rhomboid majo/minor

<u>Intermediate Back</u> *Color/Trace/Outline accordingly*

Serratus Poterior, Levator Costarum

Muscle	Insertion To	Origin	Function	Notes
Serratus Posterior Superior	Superior border ribs 2-4	Nuchal ligament Spinous process C7-T3	Elevates ribs	*Between Rhomboids and Erector Spinae*
Serratus Posterior Inferior	Inferior border ribs 8-12	Spinous process T11-L2	Depresses ribs	*Often gets lost in the lats*
Levator Costorum *Longus, Brevis*	Medial rib	Transverse process (thoracic)	Elevate ribs	*Draw on*

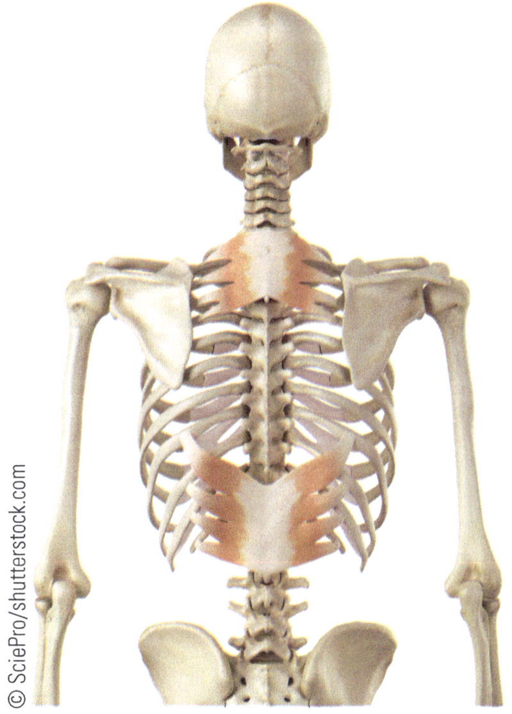

© SciePro/shutterstock.com

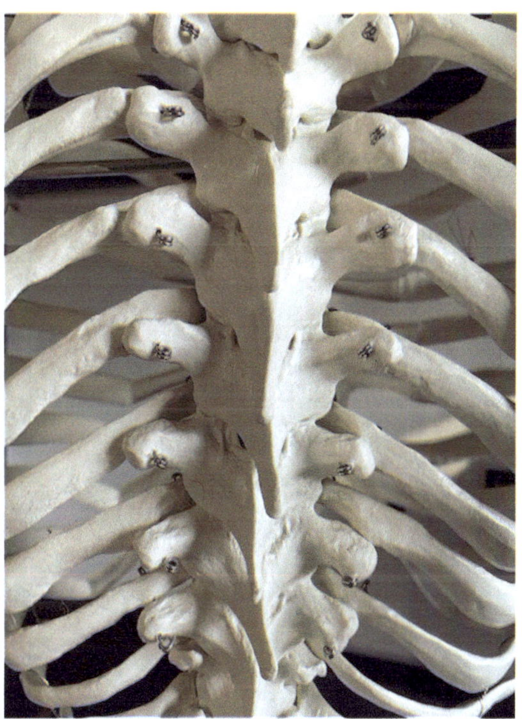

Source: Shireen Rahman

Muscles of the Back *continued . . .*

Deep Back

1. <u>Erector Spinae: Iliocostalis</u>

*This is a long muscle running from the sacrum-**ribs**- to transverse processes of cervical vertebrae; when compared to other erector spinae muscles, it is the most lateral.*

Bilateral: Extension
Unilateral: Lateral flexion and rotation

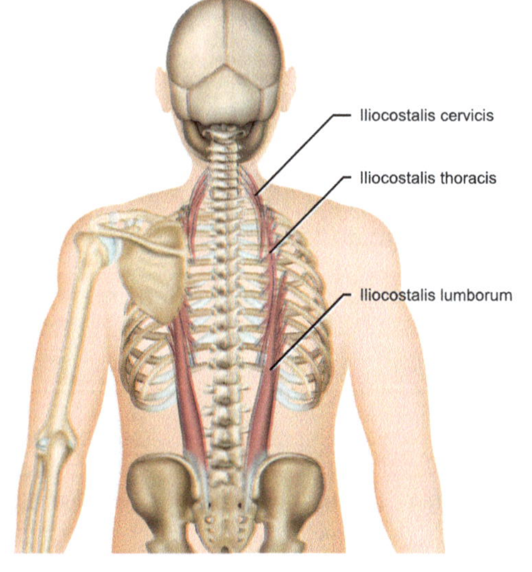

2. <u>Erector Spinae: **Longissimus**</u>

This is the longest *muscle of the erector spinae, as it runs from the sacrum all the way to the mastoid process of the skull. It is the middle erector spinae.*

Bilateral: Extension
Unilateral: Lateral flexion and rotation

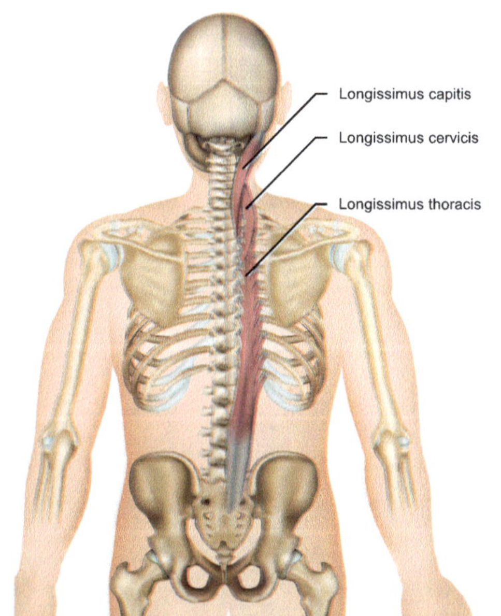

3. <u>Erector Spinae: **Spinalis**</u>

This is the most medial erector spinae and runs along the **spine.** *It will run from the spinous process to the spinous process.*

Bilateral: Extension
Unilateral: Lateral flexion

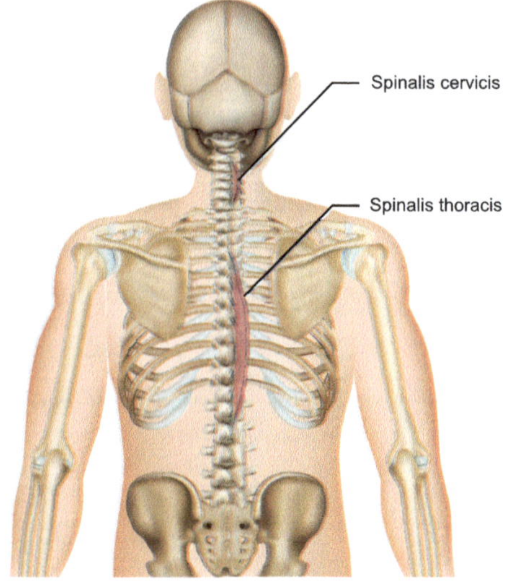

<u>Deep Muscles Continued:</u> *Color/Outline accordingly*

*Splenius refers to "bandage" as these muscles wrap around deeper muscles. The splenius muscles will be most recognizable where they originate, **running from the most inner aspect (spinous process) to a more lateral aspect (transverse process or mastoid process); in to out.***

4. Splenius Capitis (to head)

Bilateral: Extends
Unilateral: Lateral flexion, Lateral rotation
Note: *In to out*

5. Splenius Cervicis (to neck)

Bilateral: Extends
Unilateral: Lateral flexion, Lateral rotation
Note: *In to out*

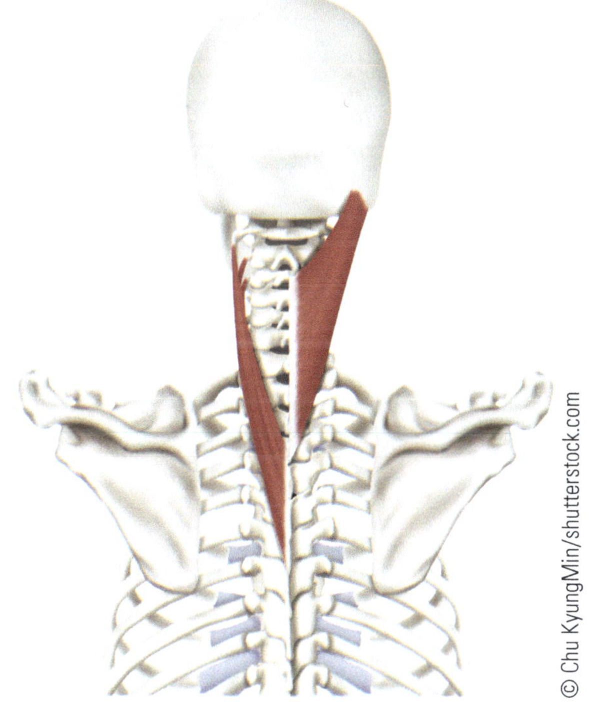

Since the Splenius muscles are bandages, there are **deeper** muscles to come . . . . exciting news!!!

Deeper Muscles

*The semispinalis muscles will be bandaged in by the splenius muscles. They will move in the opposite direction than the splenius muscles; **Out to In** as they move to their proximal attachment.*

Bilateral: Extension
Unilateral: Rotation
   1. Semispinalis Capitis (to head)
   2. Semispinalis Cervicis (to neck)
Note: Out (transverse process), In (spinous process/nuchal line); here the cervicis is just under the capitis

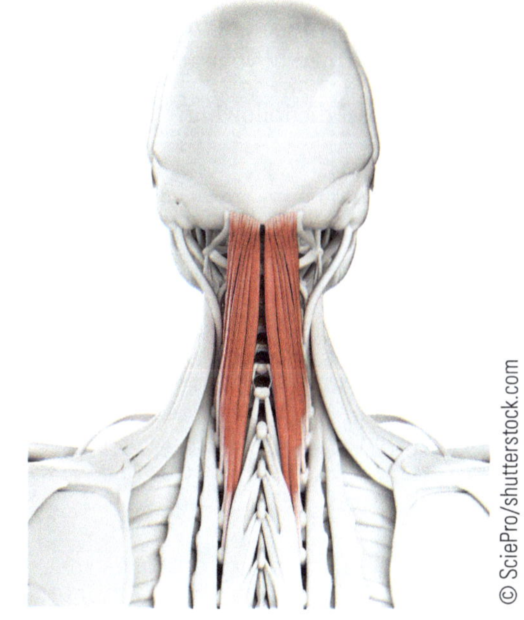

Muscles of the Back *continued* . . .

3. <u>Multifidus</u>

This muscle is deep to the semispinalis muscle and run the length of the vertebrae. Best located and tagged in the lumbar region.

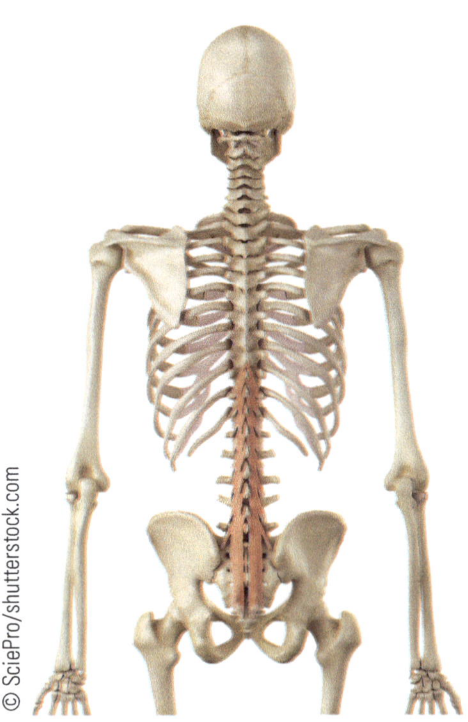

4. <u>Rotatores Longus and Brevis</u>

Deepest, along the length of the vertebrae. They will pass from the transverse process to the spinous process.

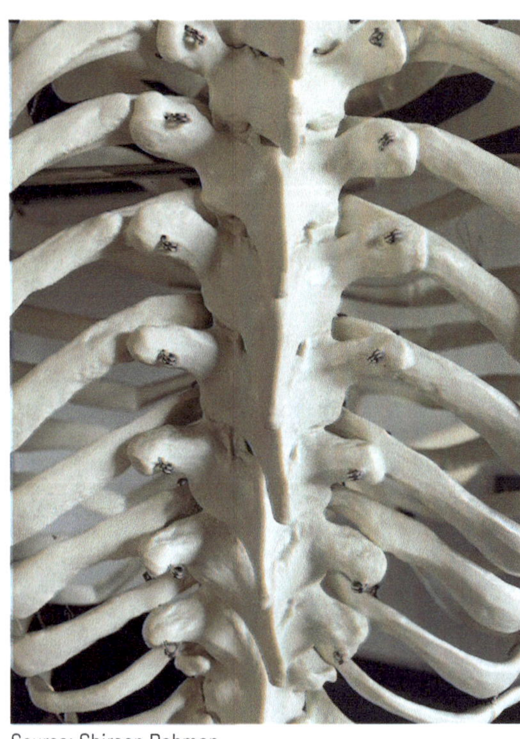

Source: Shireen Rahman

Remember..as you get more stressed and tired..how hard you worked to get here—and how excited you were when you learned of your acceptance. Celebrate for a moment..and then get ***back*** to work

Can you label the following?

. . . And Last but Coolest . . .

The Suboccipital Triangle

Muscle	Insertion TO	Origin	Function	Helpful
Rectus Capitis Mnior	Medial occipital *Nuchal line*	C1 posterior tubercle	Extension	**STAND**
Rectus Capitis Major	Medial occipital *Nuchal line*	C2 spinous process	Extension Rotation	**FALLING**
Obliquus Capitis Inferior	Transverse process of C1	Spinous process C2	Rotation	**FALLEN**
Obliquus Capitis Superior	Lateral occipital *Nuchal line*	Transverse process C1	Extension	**STAND**

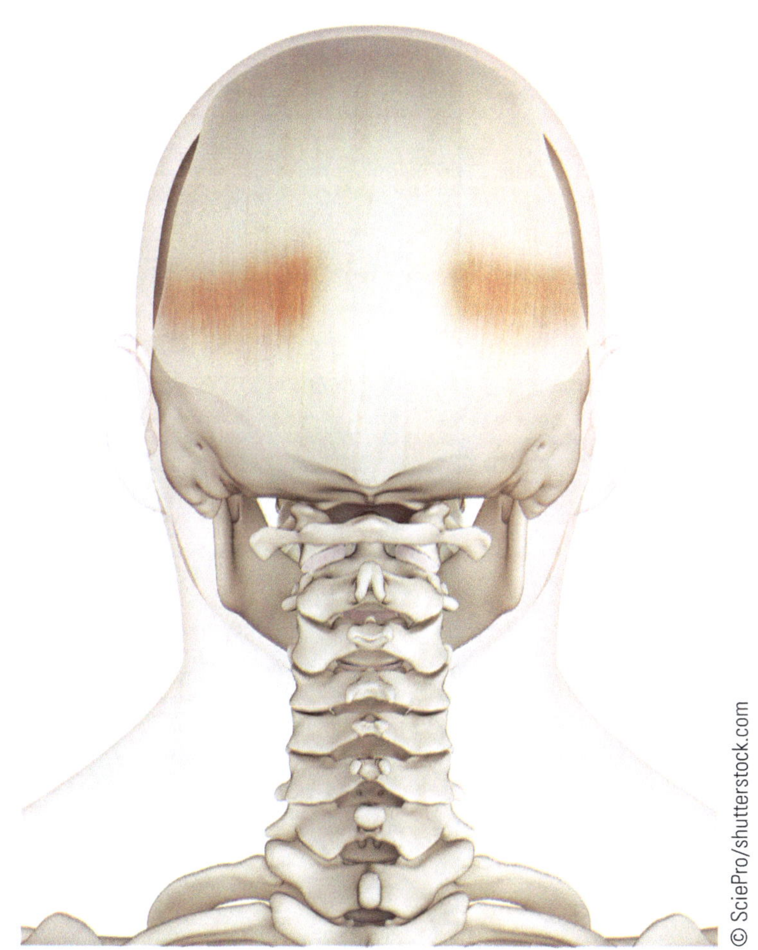

© SciePro/shutterstock.com

Spinal Cord Anatomy

Quick Hit:

1. Part of the **central nervous system**
2. Continuous with the **medulla oblongata of the brainstem.**
3. Surrounded by **meninges** (discussed later)
4. **Foramen magnum** to L2 at the **conus medullaris** (the end of the spinal cord; T12-L1), running through the **intervertebral foramen**
5. Continues as the
    a. **cauda equina (L1-L5):** motor and sensory for legs, genitourinary, and perineum
    b. **filum terminale;** pia mater continuation of the conus medullaris to coccyx
6. Two **enlargements:**
    a. Cervical enlargement (C5-T1; think brachial plexus); _note below_
    b. Lumbar enlargement (L1-S3; think lumbrosacral plexus); _note below_

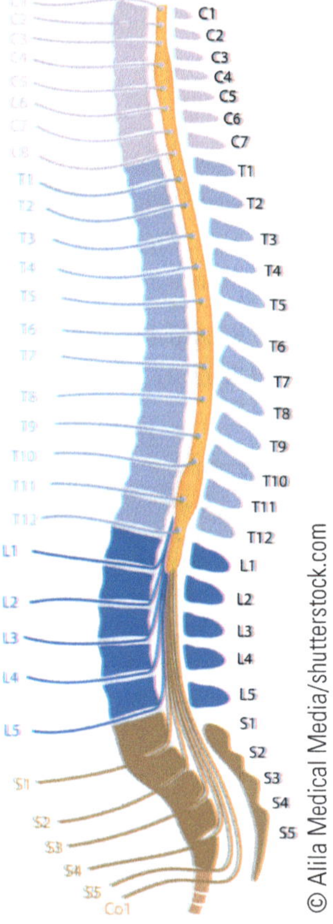

© Alila Medical Media/shutterstock.com

Spinal Cord Cross Section Anatomy

© Jose Luis Calvo/shutterstock.com

Meninges (DAP) and More

Dura Mater	Outermost
	Fibrous/protective
	Epidural Space:
	   • superficial to the dura mater
	   • fat
	   • venous plexus
	Subdural Space:
	   • between dura and arachnoid
	   • minimal fluid for protective lubrication
	Epidural Block

Arachnoid Mater	Web-like
	Connects to pia mater
	Avascular
	Ends at S2
	Subarachnoid Space:
	   • between arachnoid and pia mater
	   • houses cerebrospinal fluid (CSF)
	   • **Lumbar Cistern** is area of enlarged subarachnoid space at L1-L2; increased volume of CSF
	Lumbar Puncture/Spinal Tap

Pia Mater	Innermost Sits on spinal cord and cannot be separated Vasculature Extends laterally as the **denticulate ligament** to provide stability Extends inferiorly as the **filum terminale** from L2-S2 *Meningitis*

Blood Supply	**Anterior spinal artery (1) and its branches:**
	Branches from the **vertebral artery**
	Vertebral artery stems from the subclavian artery
	Continues to run inferiorly on the anterior surface of the spinal cord along the **anterior median fissure of the spinal cord.**
	
	Posterior spinal artery(2) and its branches:
	Primarily branches from:
	   • vertebral artery
	   • posterior inferior cerebellar arteries
	AND branches from the vertebral artery, intercostal arteries, and lumbar arteries

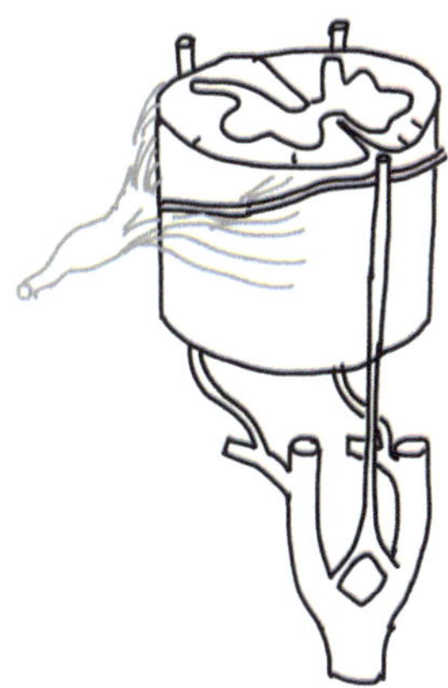

Source: Shireen Rahman

Spinal Cord within the Vertebral Foramen
Posterior (Lamina)

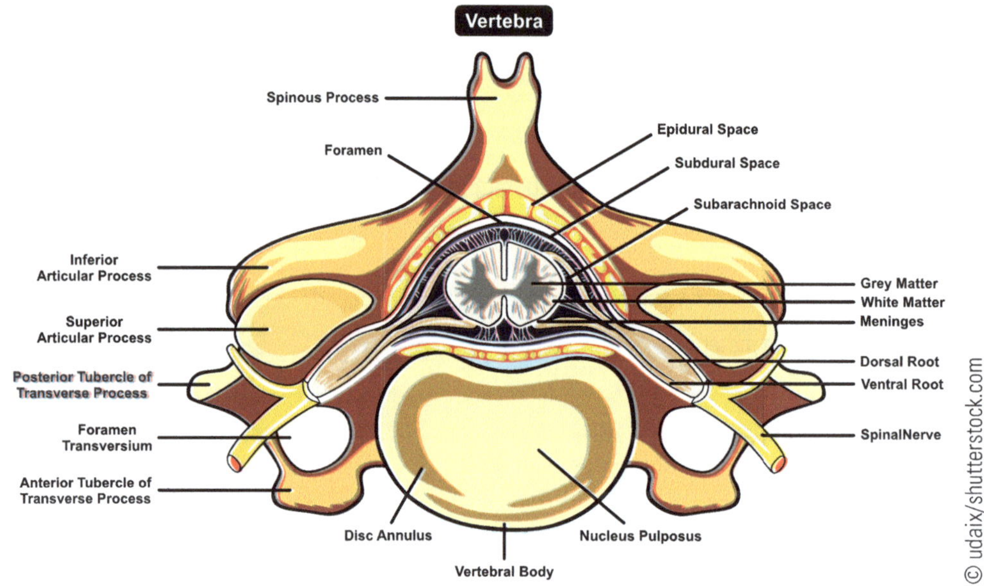

ANTERIOR (POSTERIOR ASPECT OF BODY)

Putting it All Together

Muscle	Insertion TO	Origin	Function	N Think Region	A Think Region(S)
Trapezius				Spinal Accessory	Transverse Cervical
Latissimus Dorsi				Thoracodorsal	Thoracodorsal
Rhomboids				Dorsal Scapular	Dorsal Scapular
Posterior Serratus Superior	Upper ribs	Lower C, Upper T Spinous Processes	Elevates ribs	Intercostal N	Intercostal (post)
Posterior Serratus Inferior	Lower ribs	Lower T, Upper L Spinous Processes	Depresses ribs	Ventral Rami Spinal n T9-T12	Intercostal (post)
Erector Spinae					
Iliocostalis	Ribs Cervical Transverse P	Sacrum, Ilium, Spinous Processes L and T	Extension Lateral Flexion Rotation	Posterior Rami *(C8-L1)*	Branches of Vertebral Intercostal (post) Lumbar
Longissimus	To Mastoid Process of skull	Sacrum, Ilium, Spinous Processes L and T	Extension Lateral Flexion Rotation	Posterior Rami *(C1- L5)*	Branches of Vertebral Intercostal (post) Lumbar
Spinalis	To Cervical Spinous Processes	Sacrum, Ilium, Spinous Processes T	Extension Lateral Flexion Rotation	Posterior Rami *(C-T)*	Branches of Vertebral Intercostal (post)
OTHER					
Levator Costarum	Rib	Thoracic transverse processes	Elevate ribs	Posterior Rami *(Lower cervical)*	Branches of Intercostal (post)

(continued)

Muscle	Insertion TO	Origin	Function	N Think Region	A Think Region(S)
Splenius Capitis	Occipital	Cervical Spinous Processes	Extension Lateral Flexion Rotation	Posterior Rami	Branches of Occipital Transverse cervical
Splenius Cervicis	Cervical Transverse Processes	Thoracic Spinous Processes	Extension Lateral Flexion Rotation	Posterior Rami	Branches of Occipital Transverse cervical
Semispinalis *(general)*	Occipital	Thoracic Spinous Processes (and cervical)	Extension Lateral Flexion Rotation	Posterior Rami	Branches of Occipital Post Intercostal
Multifidus *(general)*				Posterior Rami	
Rotatores	Thoracic Transverse Processes	Thoracic Spinous Processes	Extension Rotation	Posterior Rami	Branches of Post Intercostal
SUBOCCIPITAL					
Rectus capitis posterior minor	Nuchal line (occipital)	Posterior Tubercle (C1)	Extension	Suboccipital N (C1 Post Rami)	Vertebral Occipital
Rectus capitis posterior major	Nuchal line (occipital)	Spinous Process (C2)	Extension Rotation	Suboccipital N (C1 Post Rami)	Vertebral Occipital
Obliquus capitis inferior	Transverse Process of C1	Spinous Process of C2	Rotation	Suboccipital N (C1 Post Rami)	Vertebral Occipital
Obliquus capitis superior	Nuchal line (occipital)	Transverse Process of C1	Extension	Suboccipital N (C1 Post Rami)	Vertebral Occipital

****In case you were wondering****

1. Posterior Rami serve all deep back muscles; deep to posterior serratus muscles
2. Posterior Intercostal and Lumbar Arteries are branches of the Thoracic Aorta
3. Occipital Artery branches from the External Carotid Artery
4. **Greater Occipital N (C2)** is a <u>cutaneous</u> nerve for the skin of upper neck/posterior scalp (passes inferiorly to obliquus inferior and then up towards occipital)
5. **Lesser Occipital N (C3)** is a <u>cutaneous</u> nerve for the skin of the lateral head-posterior to the ear (passes laterally to the greater)

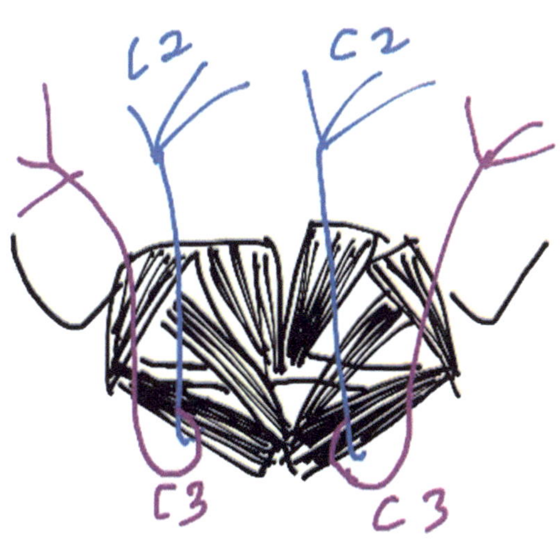

Source: Shireen Rahman

Putting It All Together

<u>Putting It All Together</u>

Putting It All Together

Putting It All Together

Skull, Brain, Cranial Nerves

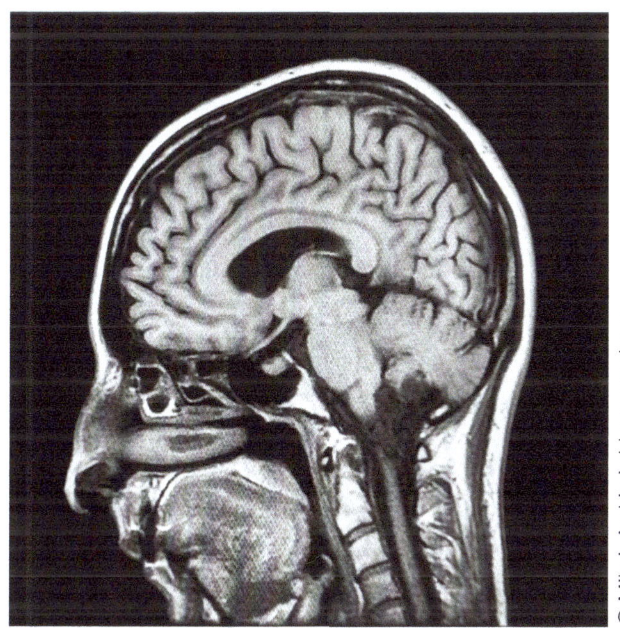

CHAPTER 5

© Mihai_Andritoiu/shutterstock.com

So Much Fun Stuff!!!

Bony Landmarks of the Skull

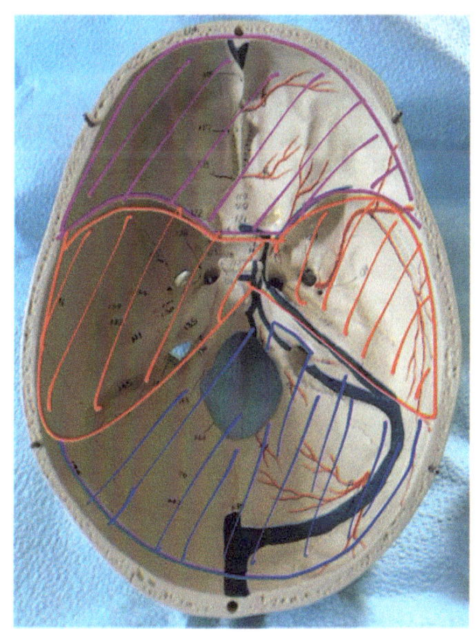

Source: Shireen Rahman

Cranial Fossa:

Anterior Cranial Fossa (Purple)
Frontal, Ethmoid, Sphenoid

Middle Cranial Fossa (Red)
Sphenoid, Temporal, Parietal

Posterior Cranial Fossa (Blue)
Spenoid, Parietal, Temporal, Occipital

Primary Bones of the Skull

Frontal Bone	
Frontal Sinus	
Parietal Bone	Superior/lateral
Temporal Bone	
Squamous part	Largest, fossa aspect Superior to zygomatic arch Muscle attachment
Petrous part	Houses middle and inner ear *fracture- CSF draining from ear
Zygomatic process	Muscle attachment As it forms the arch, protection to the eye
Mastoid process	Muscle attachment
Styloid process	Muscle attachment
Occipital Bone	
External occipital protuberance	Muscle attachment
Internal occipital protuberance	
Superior and inferior nuchal lines	Muscle attachment
Occipital condyles	Articulation with the atlas
Sphenoid Bone	
Greater and lesser wings	Protective of brain Contain many foramen for passage of nerves and arteries (provided later)
Medial and lateral pterygoid plates	Muscles of mastication attachment
Sella turcica	U-shaped structure containing: Tuberculum sellae: anterior wall of sella Hypophyseal fossa: u aspect; houses pituitary gland Dorsum sellae: posterior wall of sella
Anterior/posterior clinoid process	Anterior/posterior sella turcica Provide attachment sites for the tentorium cerebelli
Ethmoid Bone	Superior to the nasal cavity and between the orbits
Cribiform plate	Has several little holes for olfactory nerve receptors Olfactory nerve *fracture: CSF from nose
Crista galli	*The dens of the ethmoid;* attachment for the falx cerebri
Perpendicular plate	Help form the nasal septum (upper)
Nasal	
Vomer	Splits the nasal cavity; helps form the nasal septum (lower)
Nasal conchae	Warming of air, airflow direction, humidification
Maxillary Bone	*Maxillary teeth*
Zygomatic Bone	Cheekbones *common fracture zone*
Mandible Bone	*Mandibular teeth*
Occipital Bone	

<u>Ready, Set, Go . . . . !!!!</u>

LATERAL VIEW

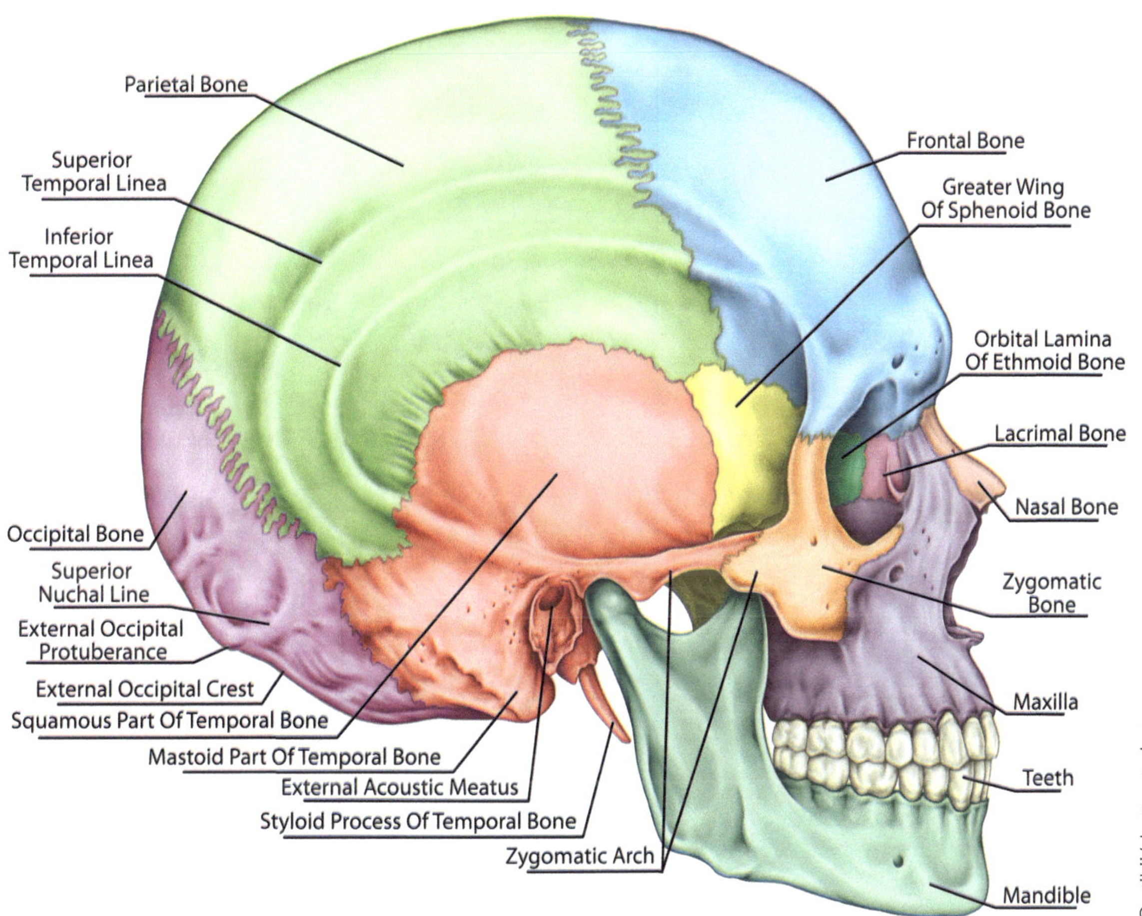

FRONTAL VIEW

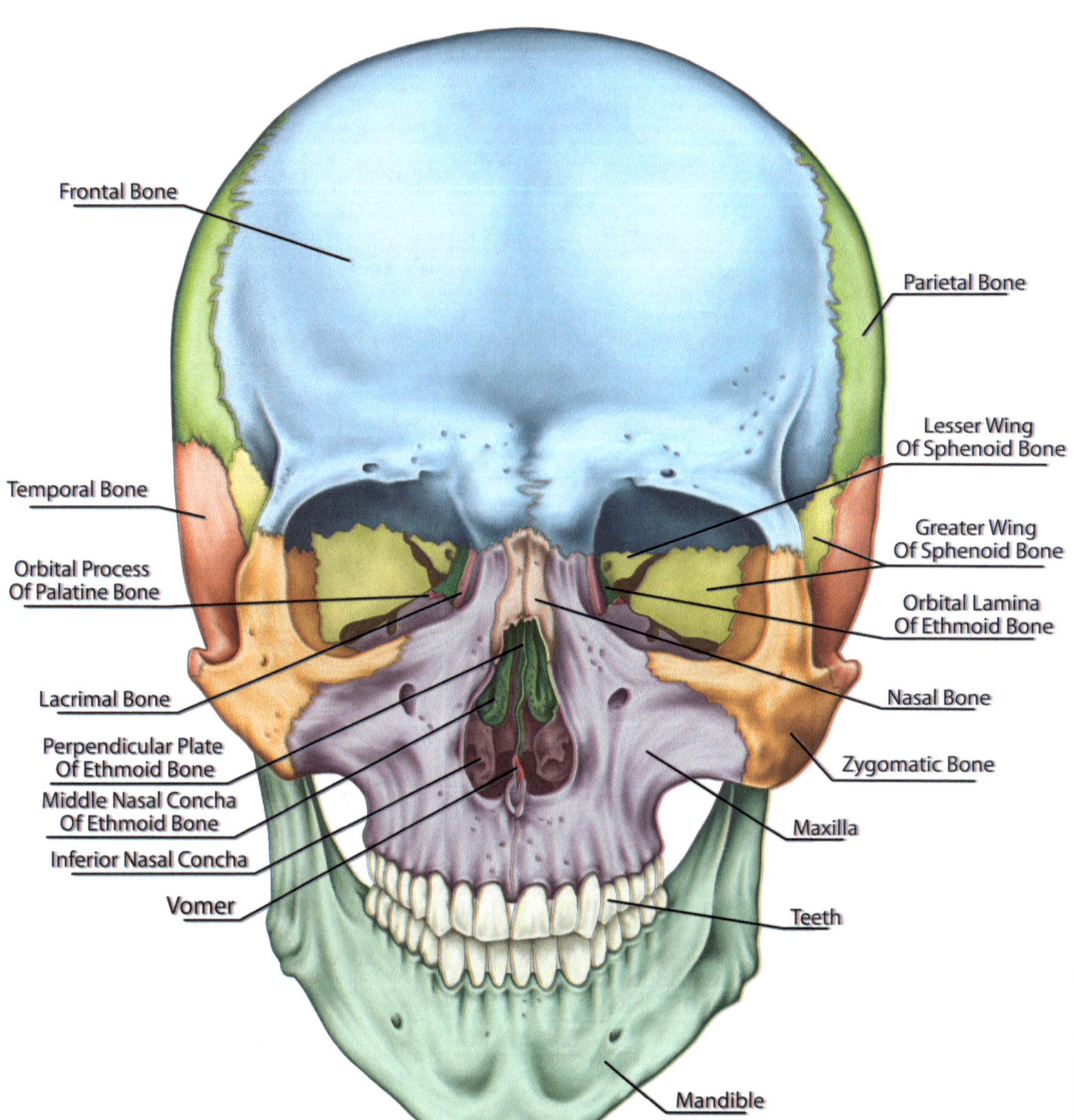

POSTERIOR VIEW

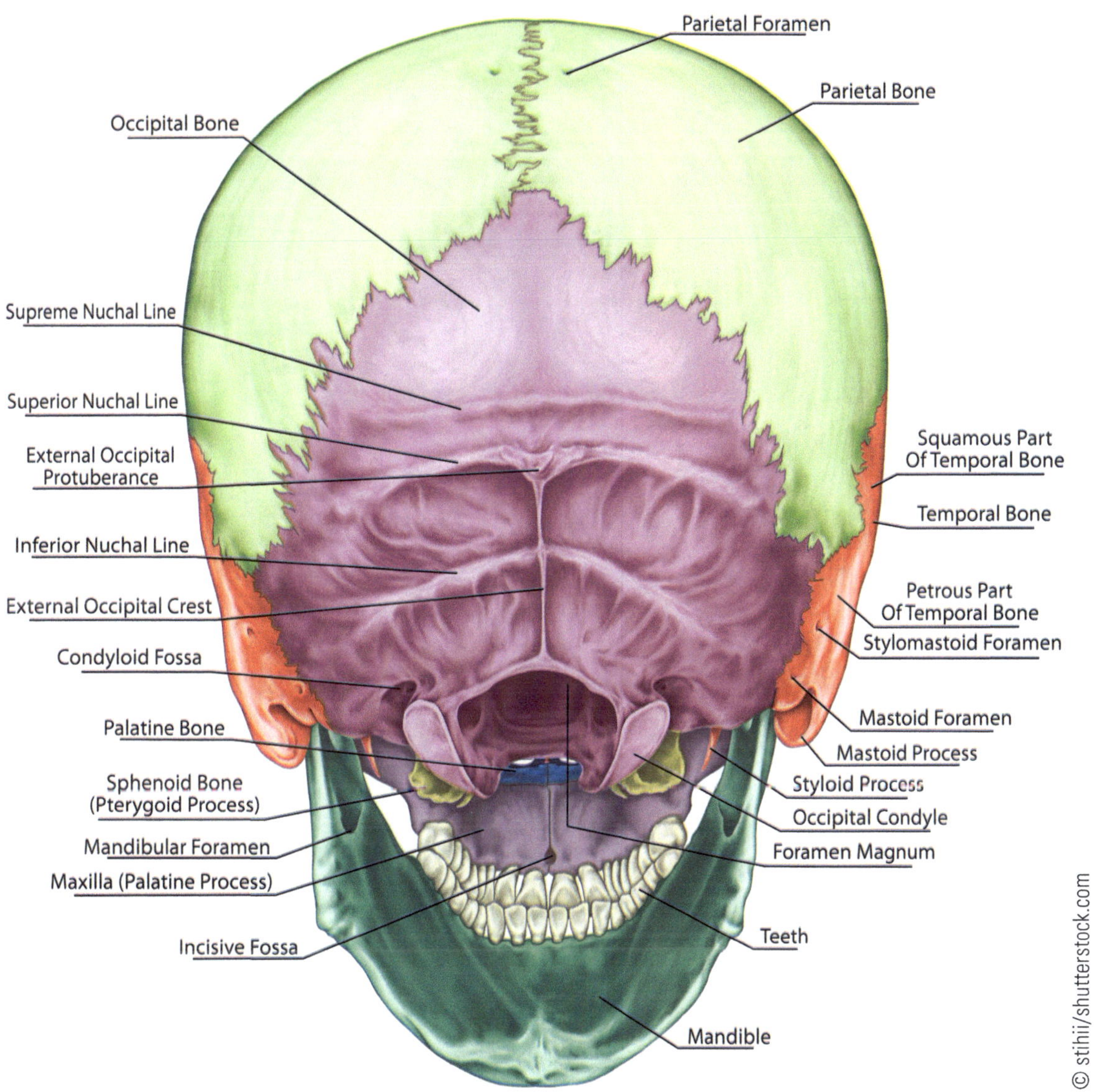

INFERIOR VIEW

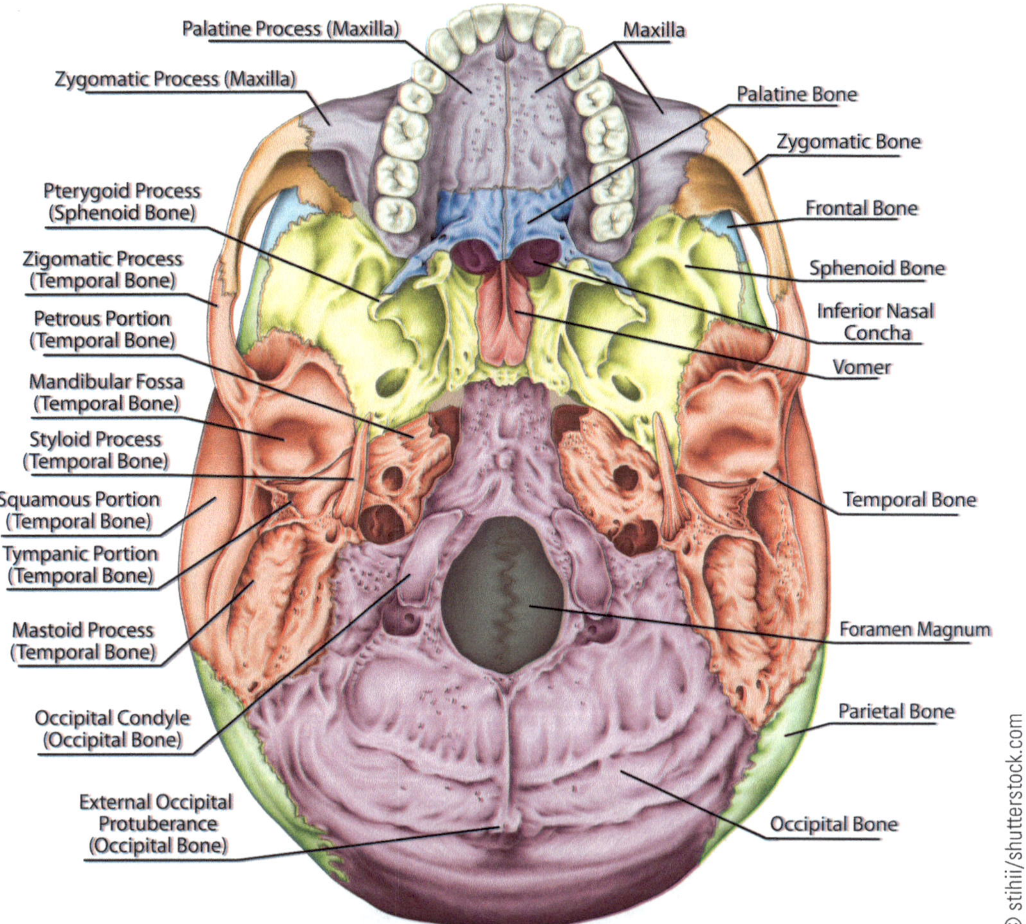

Sphenoid Bone

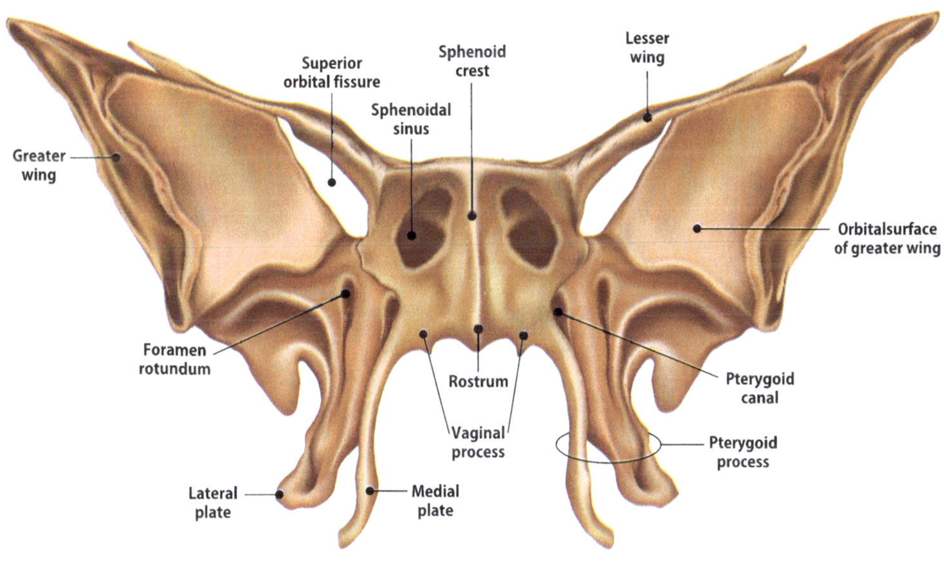

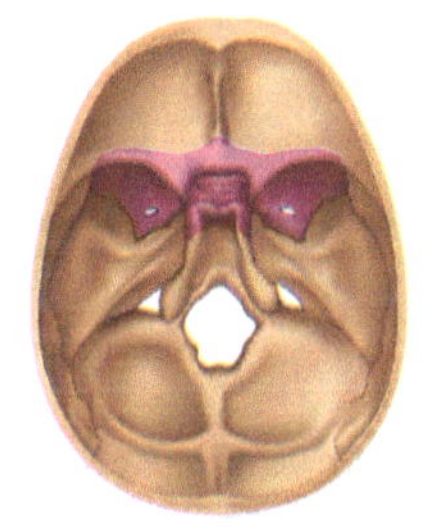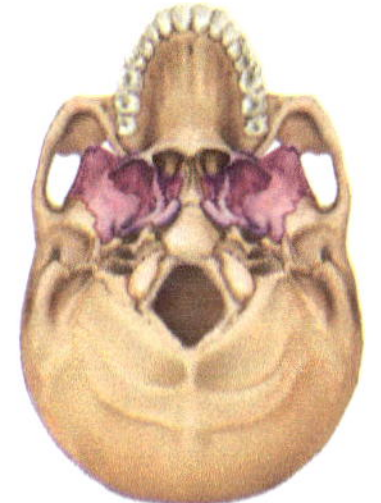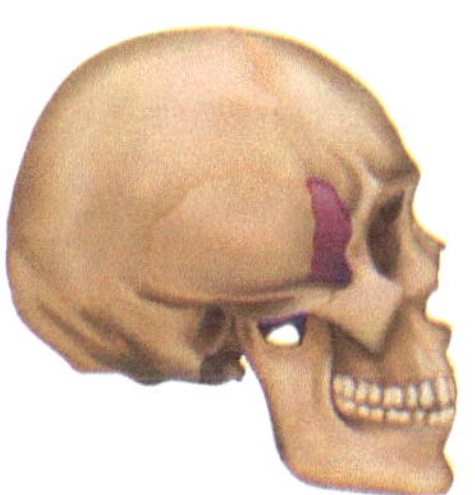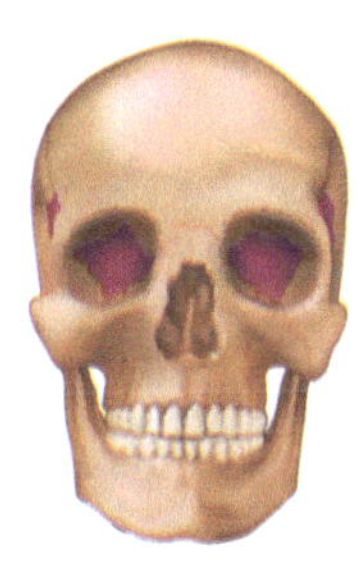

© studiovin/shutterstock.com

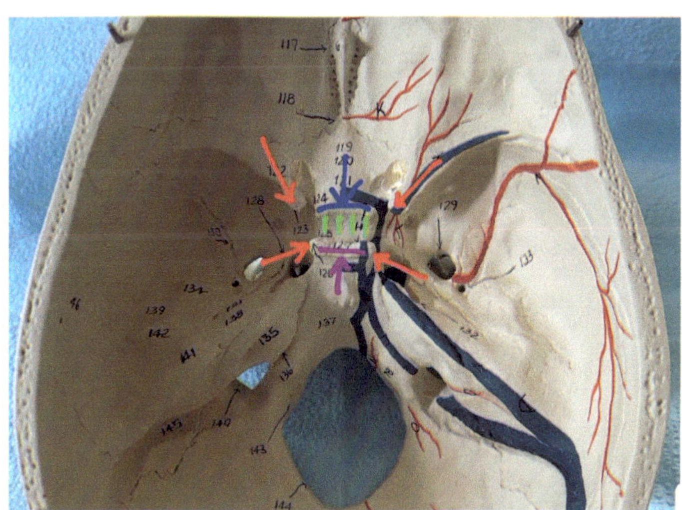

Anterior and Posterior Clinoid Processes (Red arrows)
Tuberculum Sellae (Blue)
Hypophyseal Fossa (Green)
Dorsum Sellae (Purple)

Ethmoid

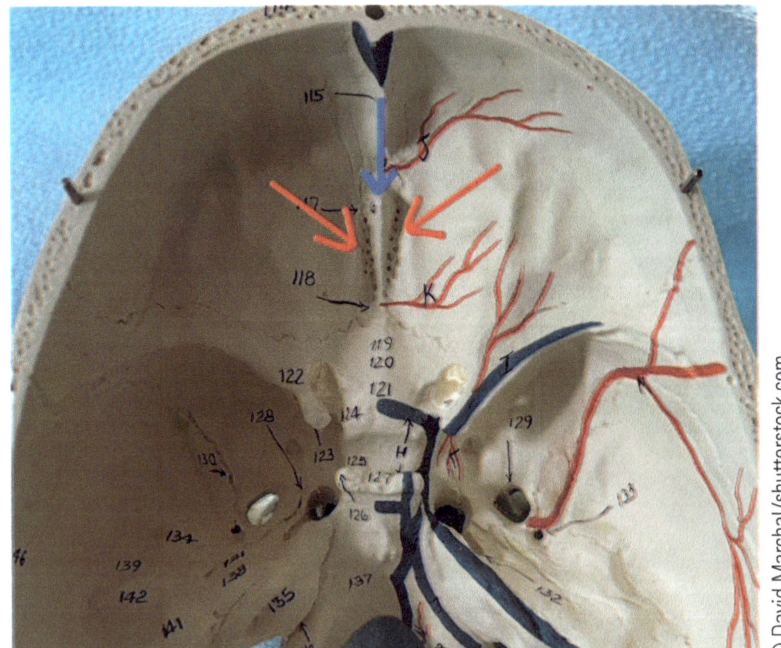

Cribiform plate (Red)
Crista galli (Blue)

© David Marchal/shutterstock.com

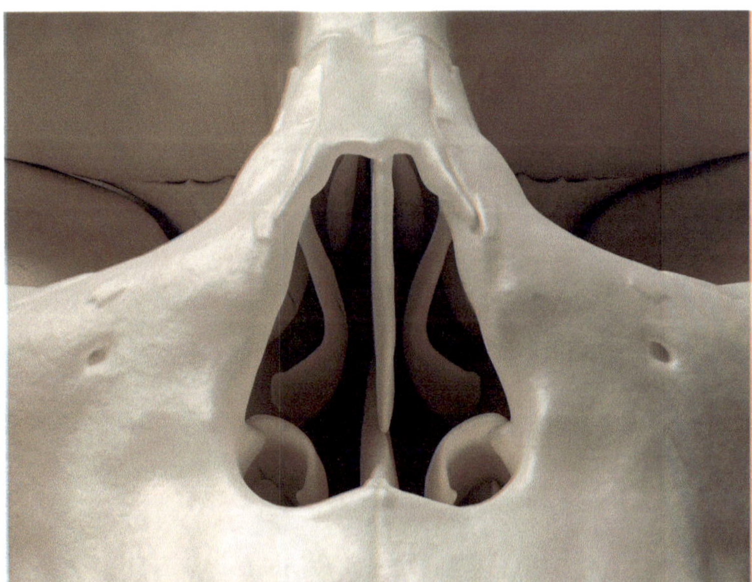

Source: Shireen Rahman

The perpendicular plate of ethmoid and vomer run vertically to make the nasal septum. Bilateral to the septum are the nasal conchae.

Some Skull Stress Relief

Sutures and Junctions

Sutures are immoveable fibrous joints; Junctions are found where bones/sutures meet
Draw and label the sutures/junctions accordingly

Sutures	
Coronal Suture	Between the frontal and 2 parietal bones
Sagittal Suture	Between 2 parietal bones (labeled)
Squamous Suture	Between parietal bones and temporal bone
Lamboidal Suture	Between 2 parietal bones and occipital bone
Sphenofrontal Suture	Between sphenoid bone and frontal bone
Sphenosquamous Suture	Between sphenoid bone and temporal bone

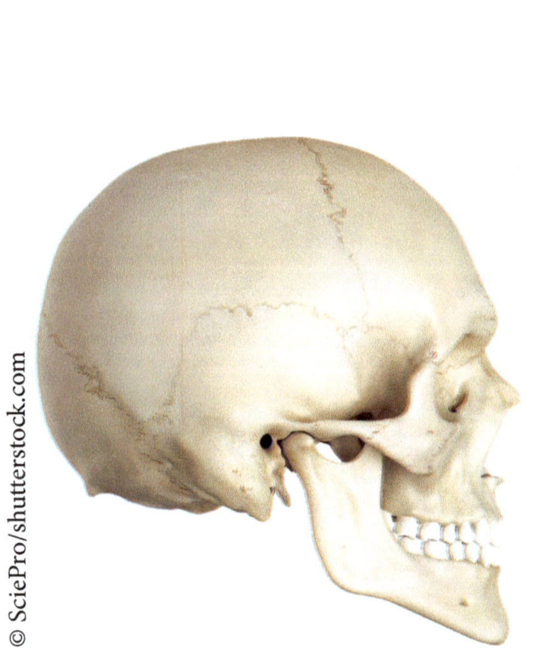

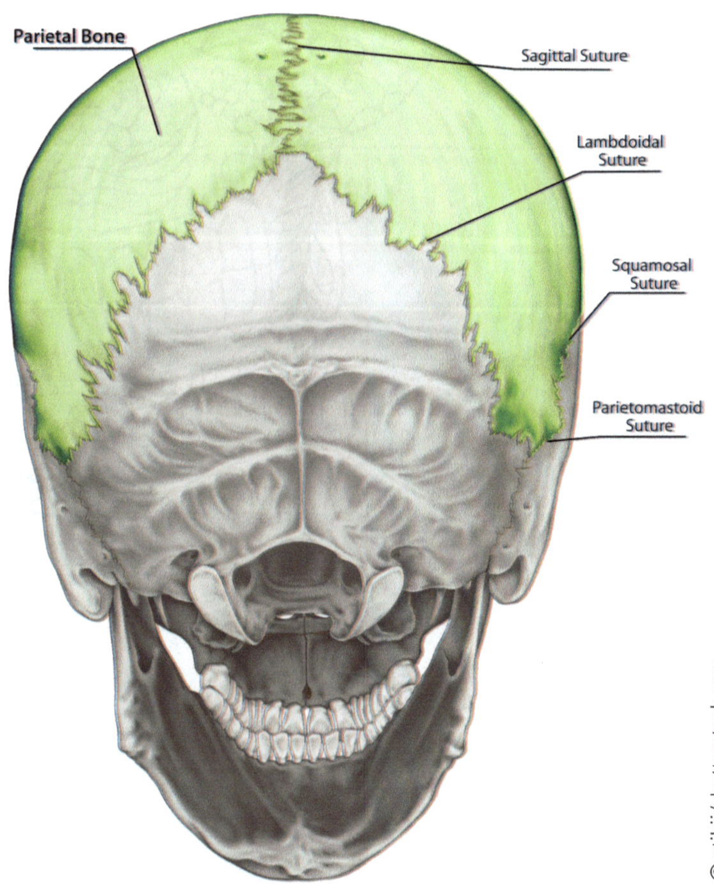

Junctions	
Bregma	Intersection between coronal and sagittal sutures
Lambda	Intersection between lamboid and sagittal sutures

Pterion	Intersection between frontal, sphenoid, temporal, and parietal bones
	• location: width of two fingers superior to zygomatic arch
	• clinical importance:
	  ○ location of the **middle meningeal artery**
	  ○ injury may lead to epidural hemorrhage

Nasion	Intersection between frontal and nasal bones
Asterion	Intersection between parietal, temporal, and occipital bones
Inion	Prominence at the external occipital protuberance

Holes (Foramina) of the Head (italics is FYI)

From anterior to posterior

Hole	Passage	Hole	Passage
Foramen cecum	*Veins from nasal mucosa*	Foramen magnum	Spinal cord Spinal accessory N Vertebral A Venous plexus Ant/Post spinal A
Cribiform plate	Olfactory N	**FRONT OF SKULL**	
Optic canal	Optic N Opthalmic artery A/V to retina	Supraorbital foramen Infraorbital foramen	Supraorbital N/A/V Infraorbital N/A/V
Superior orbital fissure	Oculomotor Trochlear Abducens Trigeminal I (opthal)	Mental foramen	Mental N/A/V
Foramen rotundum	Trigeminal II (max)	**OTHERS**	
Foramen ovale	Trigeminal III (mand)	Stylomastoid foramen	Facial N
Foramen spinosum	Middle meningeal A		
Foramen lacerum	*Internal carotid A crosses*		
Carotid canal	Internal carotid A Carotid plexus (SNS)		
Internal acoustic meatus	Facial N Vestibulocochlear N Labyrinthine A		
Jugular Foramen	Glossopharyngeal N Vagus N Spinal accessory N		
Hypoglossal canal	Hypoglossal N Meningeal A		

Foramen for Cranial Nerves

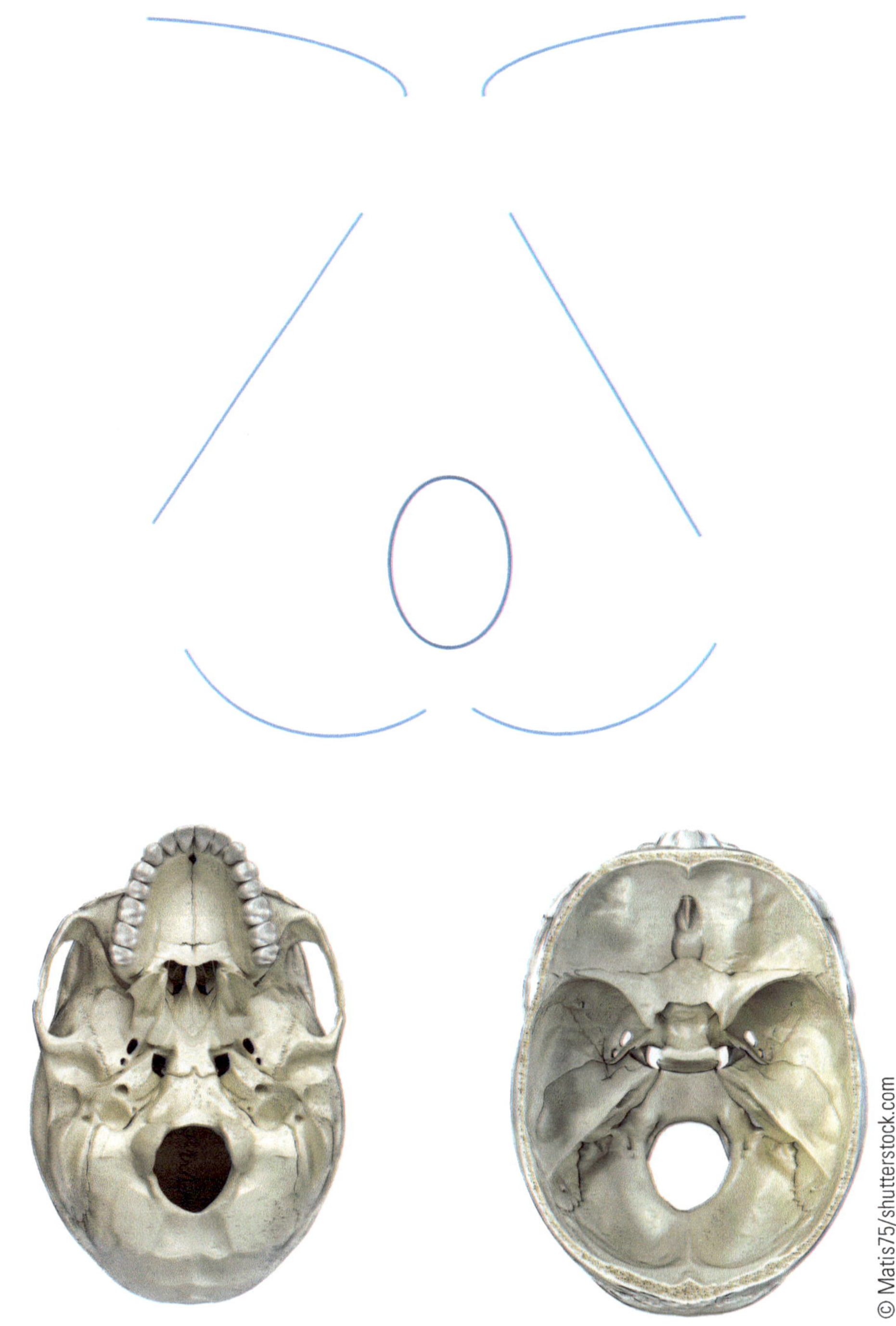

Before leaving the skull . . . .

Paranasal Sinuses:
1. Extension of nasal cavity (air-filled)
2. Lined with mucous secreting cells
3. Innervated by Trigeminal N
4. Exact function unknown
5. *Sinusitis*

Brain

Meninges of the Brain
1. Protect brain
2. Support framework for arteries, veins, venous sinuses
3. Innervated by the Trigeminal Nerve
4. 3 layers: (DAP)

Dura Mater:
1. Outermost, fibrous layer
2. Sensory nerve fibers
3. Epidural superficial to the dura mater
4. Subdural deep to dura mater

Brain *continued* . . .

**Clinical Quick Hit:

EPIDURAL HEMATOMA VS SUBDURAL HEMATOMA

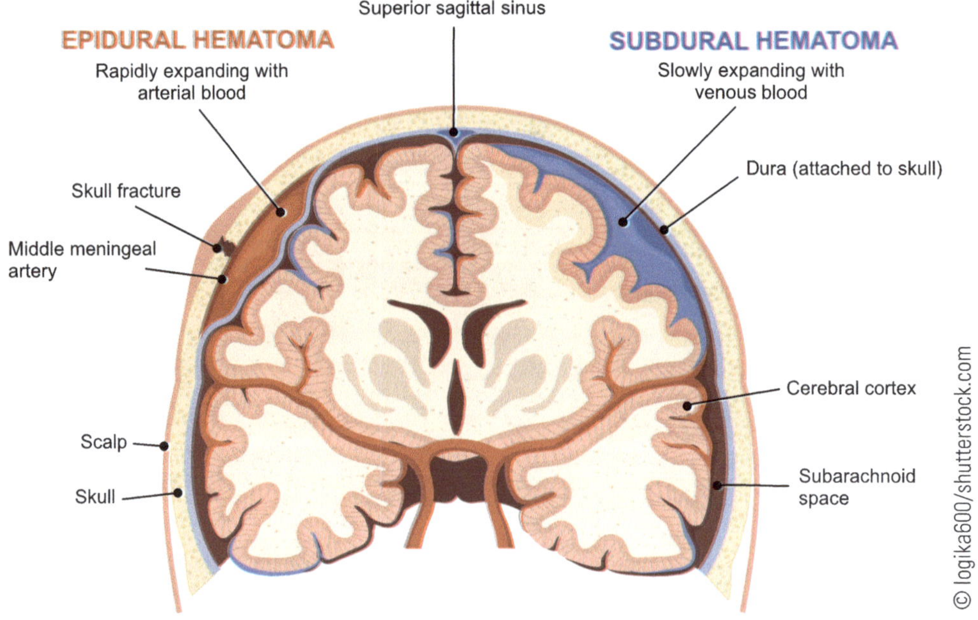

5. **Dura Folds:**
   a. Folds in the dura mater that divides cranial cavity into compartments
   b. Cerebral falx, Tentorium cerebelli, Cerebellar falx, Sellar diaphragm

Cerebral Falx

   i. Largest fold
   ii. Contains venous sinus (in blue below)
   iii. Lies within the longitudinal fissure (between right and left cerebral hemispheres)

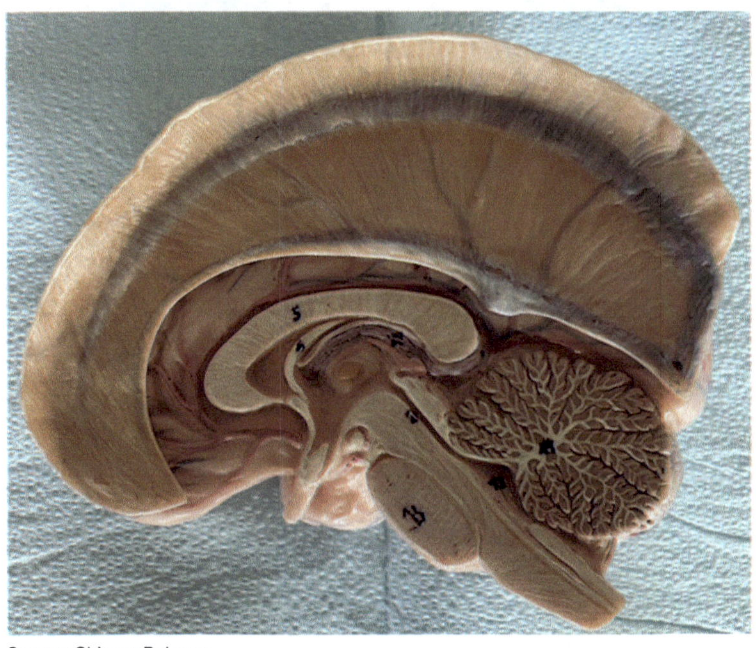

Source: Shireen Rahman

Tentorium Cerebelli
  iv. Divided the occipital lobe from the cerebellum
  v. Perpendicular to the cerebral falx

Cerebellar Falx
  vi. Just posterior and inferior to tentorium cerebelli; small triangular region
  vii. Attached to the occipital bone

Sellar Diaphragm
  viii. Smallest of the dural fold
  ix. Covers the pituitary gland

Arachnoid Mater

1. Thin avascular middle layer
2. Subarachnoid Space
   a. Between arachnoid mater and pia mater
   b. Contain cerebrospinal fluid
      i. CSF produces in the choroid plexus of the ventricles
      ii. CSF moves from the ventricular systems to the subarachnoid space
      iii. Nourishes and protects the brain

Pia Mater

1. Internal vascular layer
2. Directly surrounds the brain and spinal cord

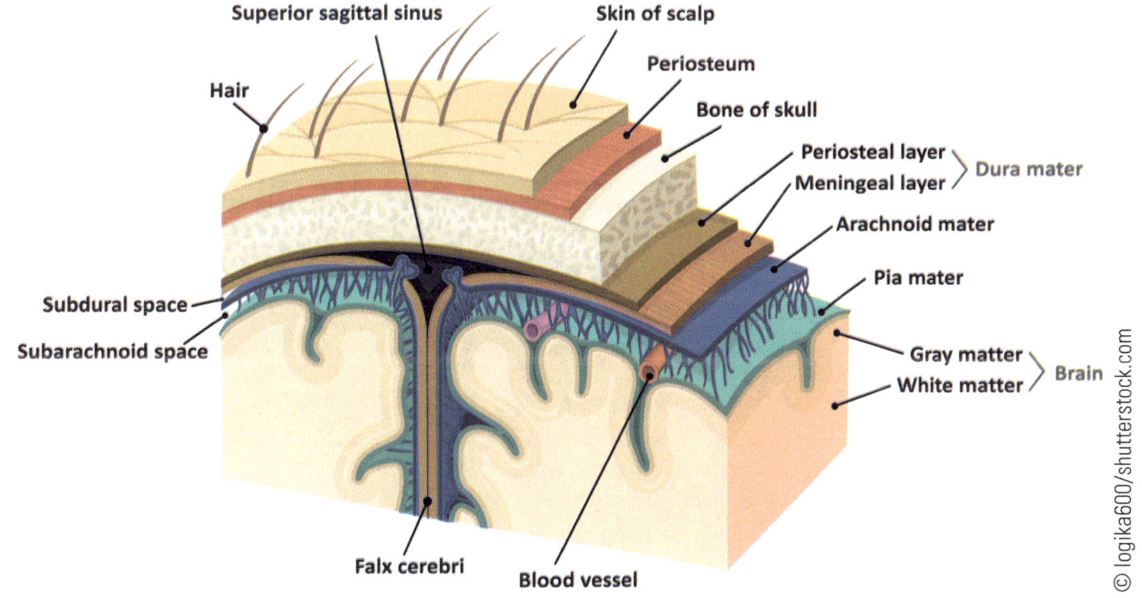

Aspects of the Brain

Telencephalon	Cerebrum Lobes, Sulci, and Gyri
Diencephalon	Thalamus Hypothalamus Pineal gland *And related structures*
Mesencephalon	Midbrain Superior brainstem Cerebral peduncles, Colliculi *and related structures*
Metencephalon	Middle brainstem Pons, Cerebellum *Tracts, Cranial nerve nuclei*
Mylencephalon	Inferior brainstem Medulla oblongata Ends at the foramen magnum *Tracts, Cranial nerve nuclei*

Cerebrum Plus

Longitudinal Fissure

Separates the right and left cerebral hemispheres; location for the **cerebral falx.**

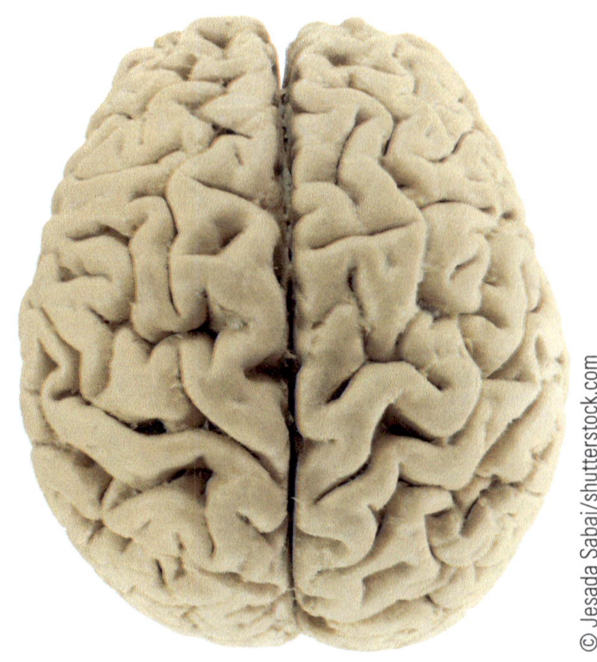

© Jesada Sabai/shutterstock.com

Lobes (in general first)

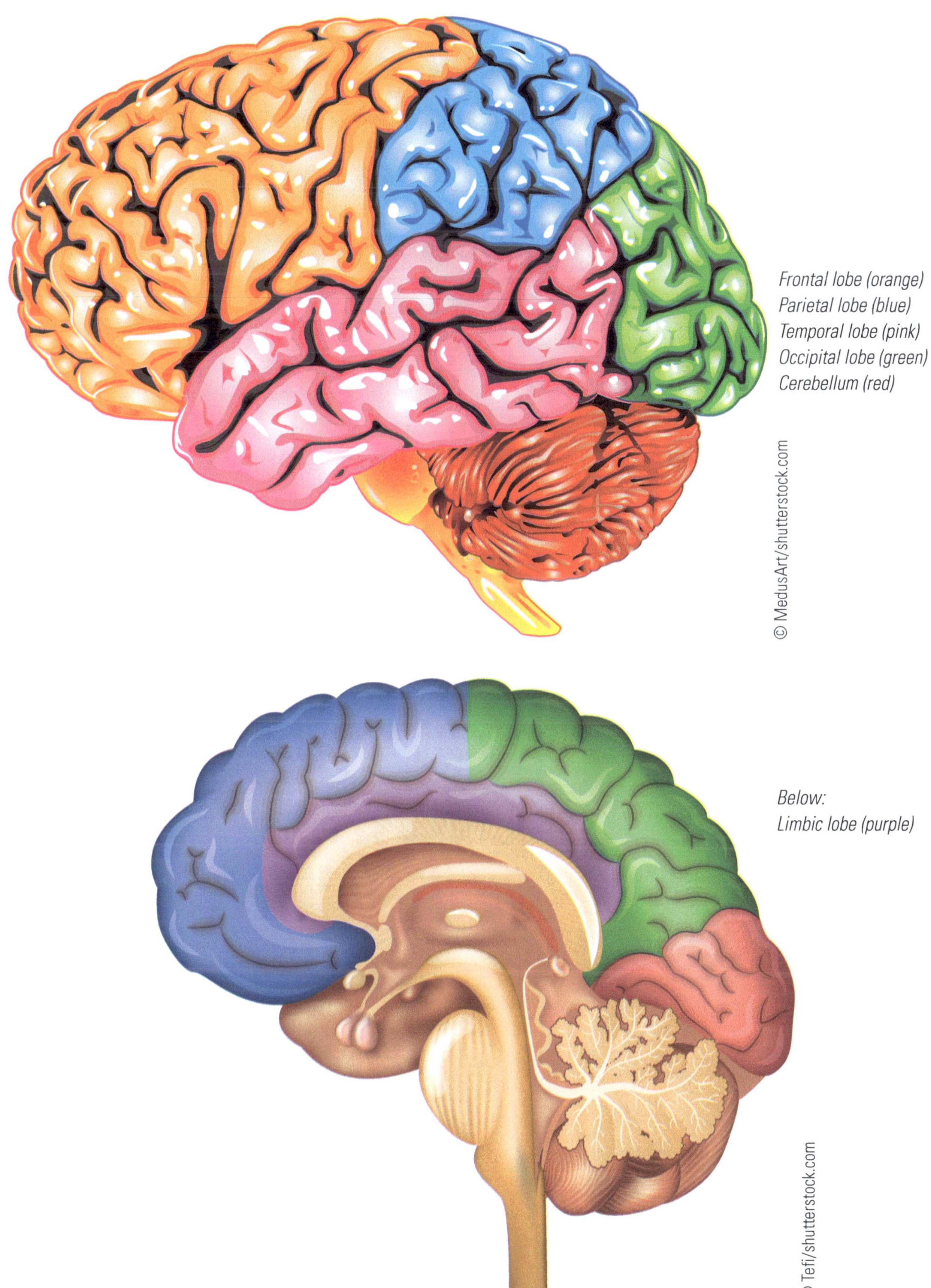

Frontal lobe (orange)
Parietal lobe (blue)
Temporal lobe (pink)
Occipital lobe (green)
Cerebellum (red)

© MedusArt/shutterstock.com

Below:
Limbic lobe (purple)

© Tefi/shutterstock.com

Sulci

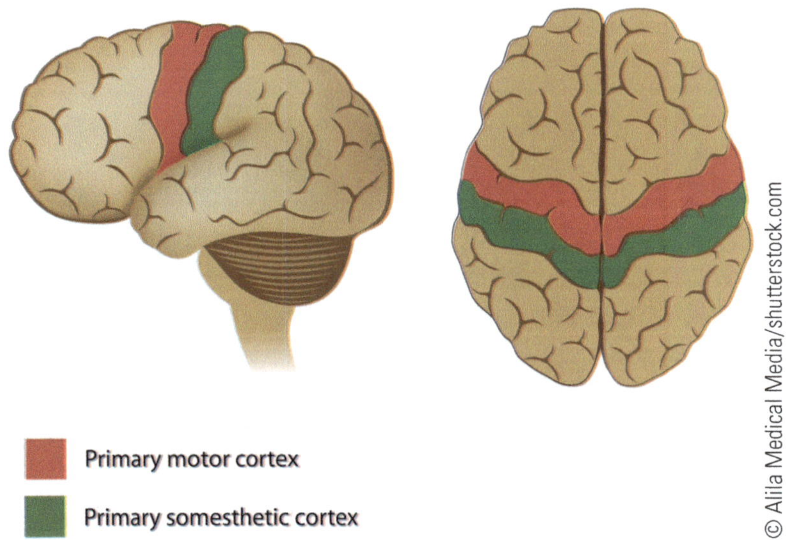

The **Central Sulcus** is seen as the groove between the green and red above. The central sulcus will separate the motor and the sensory cortices. Divides the frontal and parietal lobes.

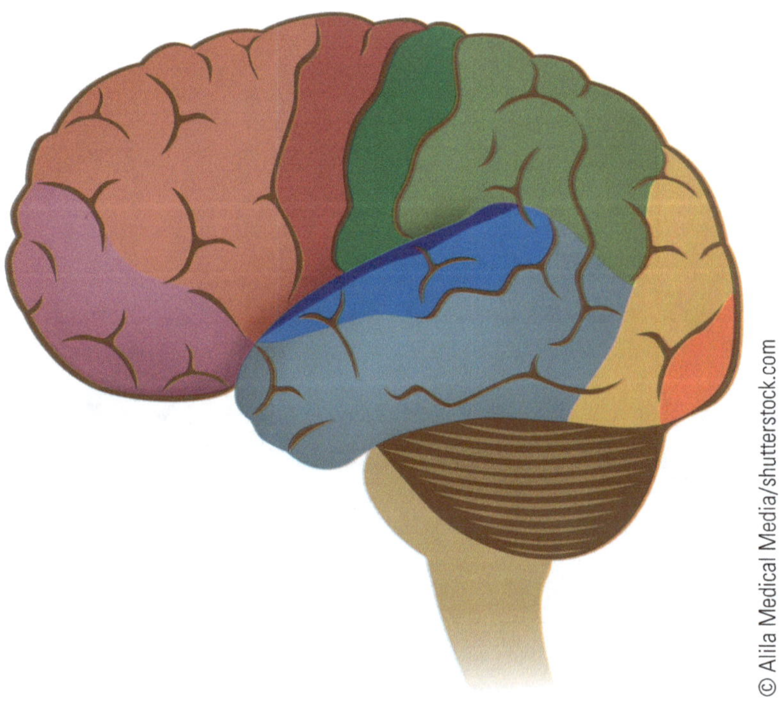

The **Lateral Sulcus** is seen above in deep blue. This sulcus separates the frontal and the temporal lobes.

Frontal Lobe

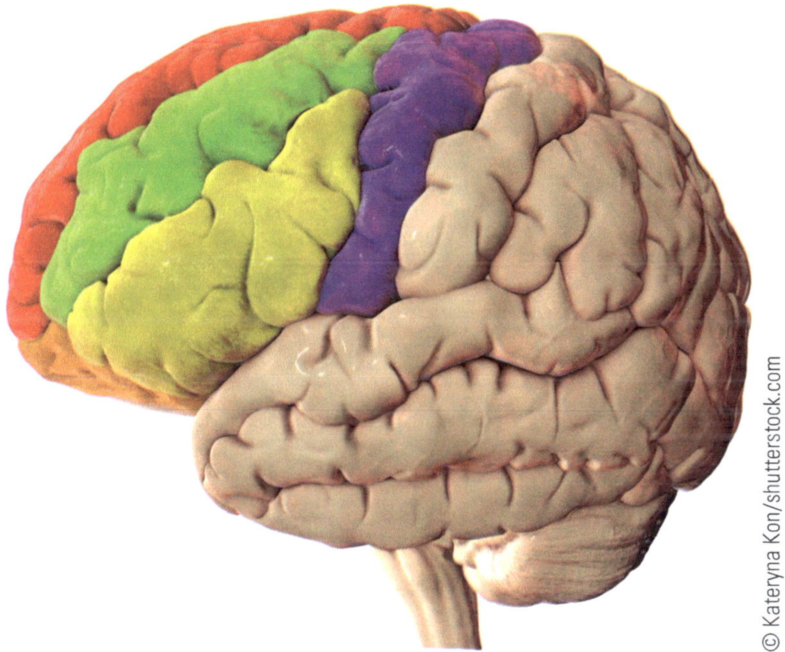

© Kateryna Kon/shutterstock.com

The **Central Sulcus** is the area of separation between the functional motor and sensory cortices. Above it is the sulcus just posterior to the purple gyrus.

The **Precentral Gyrus (purple)** is just anterior to the central sulcus. It is the area of the **primary motor cortex,** an area for voluntary movement.

The **Superior Frontal Gyrus (red), Middle Frontal Gyrus (red), and Inferior Frontal Gyrus (yellow)** are areas of cognition, memory, problem solving, etc. Some aspects of these gyri will also serve as the **pre-motor cortex** which aid in voluntary movement. The **Broca's area** will be here, for production of speech.

Parietal Lobe

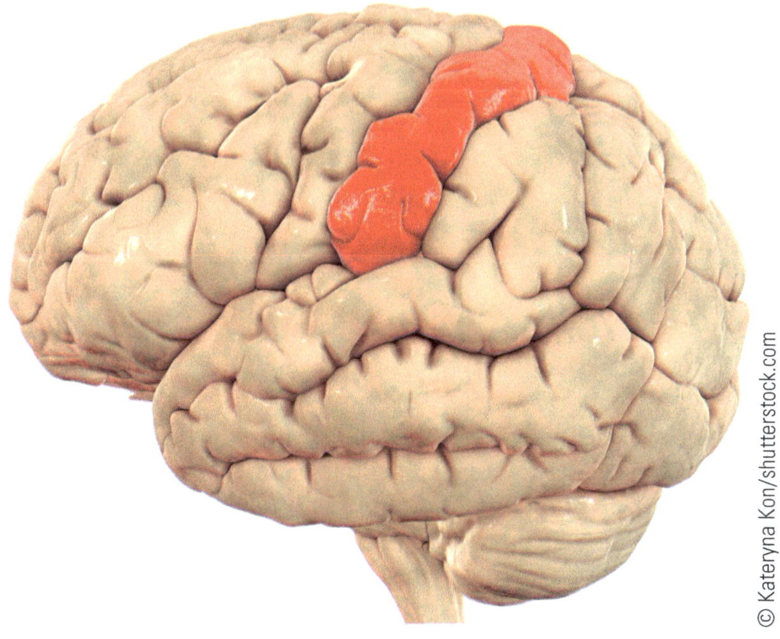

© Kateryna Kon/shutterstock.com

The **Postcentral Gyrus** (red) is just posterior to the central sulcus. It is the area of the **primary somatosensory cortex,** an area to feel sensations.

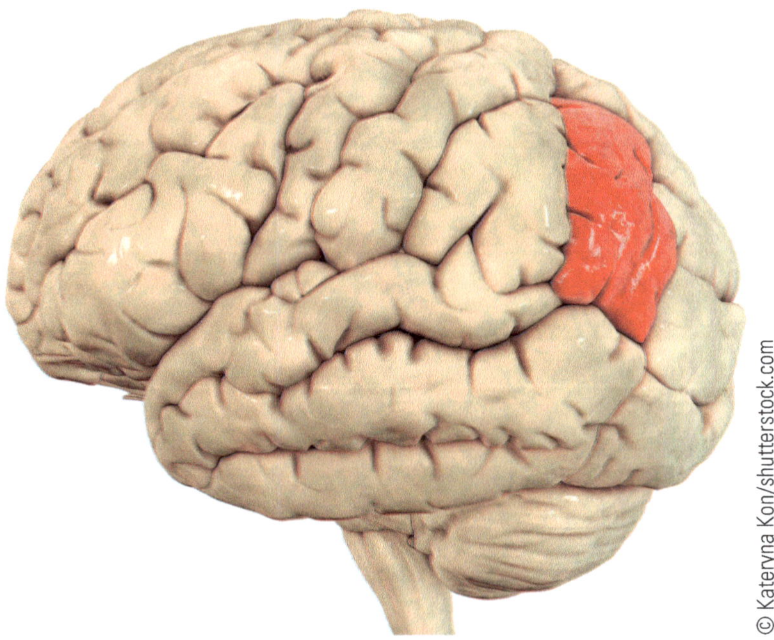

The parietal lobe also houses the **angular gyrus** (red) which houses the areas of language, processing numbers, spatial cognition, memory, attention, etc.

Parieto-occipital Sulcus

The groove between the parietal lobe and occipital lobe

Temporal Lobe

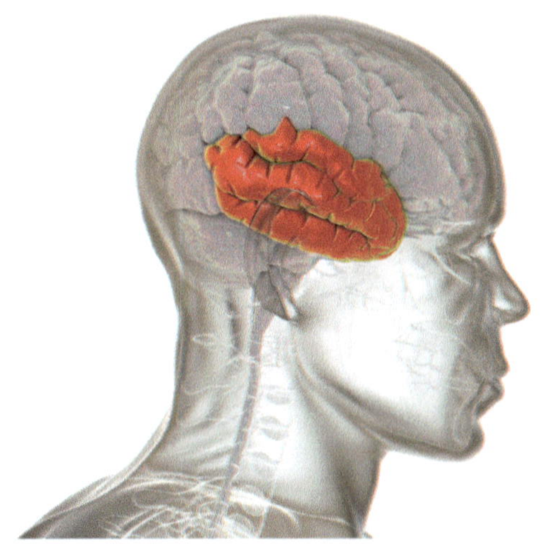

The gyri of the temporal lobe are just inferior to the **lateral sulcus**, which divides the frontal and temporal lobes. The **superior temporal gyrus** (yellow) is the **primary auditory cortex.** It also houses the **Wernicke's area,** comprehension of speech The **middle temporal gyrus (orange)** will help with sound recognition, language memory, language processing. The **inferior temporal gyrus (red)** will aid in facial recognition –though specific functions are not 100% clarified for any of these areas.

Occipital Lobe

The **parieto-occipital sulcus** will separate the parietal and occipital lobe.

The **calcarine sulcus** will separate the superior and inferior occipital lobe.

The occipital lobe will serve as the **primary visual cortex.**

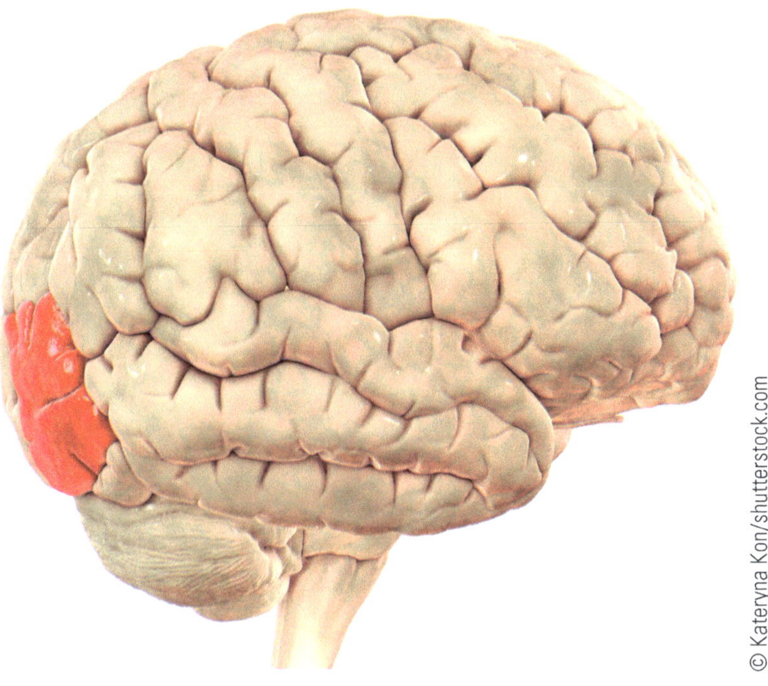

© Kateryna Kon/shutterstock.com

Limbic Lobe (color accordingly)

The **corpus collosum (a)** will separate the internal right and left hemispheres. Above the corpus collosum is the **limbic lobe within the cingulate gyrus (b),** for emotion and memory, etc.

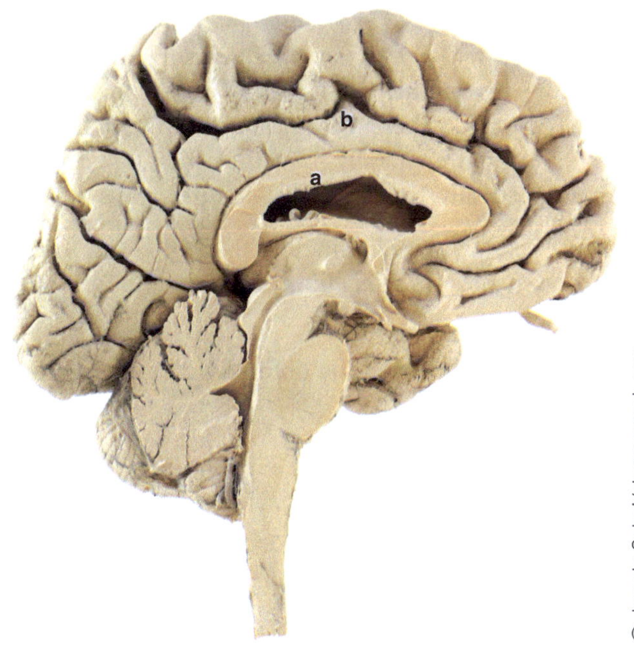

© Jesada Sabai/shutterstock.com

Insula Lobe

Found deep within the lateral sulcus, the insula is the **primary gustatory center.** It is also involved in language, sympathetic tone, and ties to the limbic lobe.

Inferior Surface of the Brain

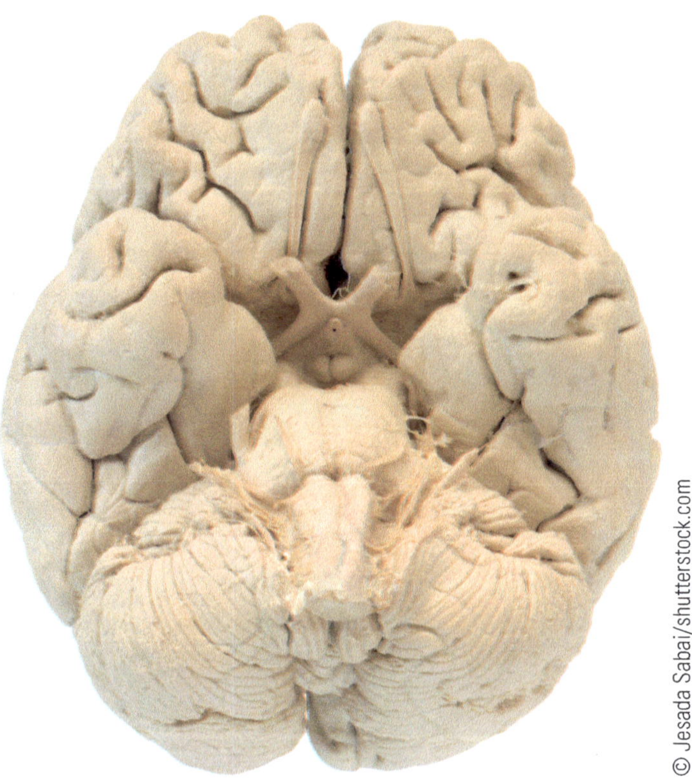

© Jesada Sabai/shutterstock.com

Start with the **olfactory bulb (CN I),** which extends posteriorly via the **olfactory tract.** Soon you see the **optic nerve (CN II),** which come together from its bilateral aspects to meet at the **optic chiasm.** The chiasm is where optic nerve fibers may decussate. From the chiasm, or the bottom aspects of the "x" as the **optic tracts,** which allow the visual message to eventually reach the occipital lobe.

The **pituitary gland** sits just inferior to the "x". Moving more inferiorly and just under the pituitary gland, you will find the **mammillary bodies,** which will help with memory formation. To both sides of the mammillary bodies, the **cerebral peduncles** are located. These are an aspect of the midbrain that house nerve tracts (esp descending motor tracts). If you move more laterally from the cerebral peduncles, you will find the **uncus,** a primary **olfactory center,** and more laterally, the **parahippocampal gyrus,** which will contribute to memory.

Now come back to the center, just below the midbrain, and find the **pons.** The pons will serve as a nerve fiber connection between the cerebrum and the cerebellum, serve as a home plate for cranial nerve nuclei, and where tracts will pass.

Moving inferior to the pons, you will find the **medulla oblongata**, which houses the pyramids and olives (for tracts), and cranial nerve nuclei.

The **brainstem,** or the midbrain, pons, and medulla all together, in addition to their own functions, combine to function many critical involuntary functions (*respiratory, heart rate, BP, temperature, wake and sleep cycles, digestion, sneezing, coughing, vomiting, and swallowing.etc*).

Medial View of Brain

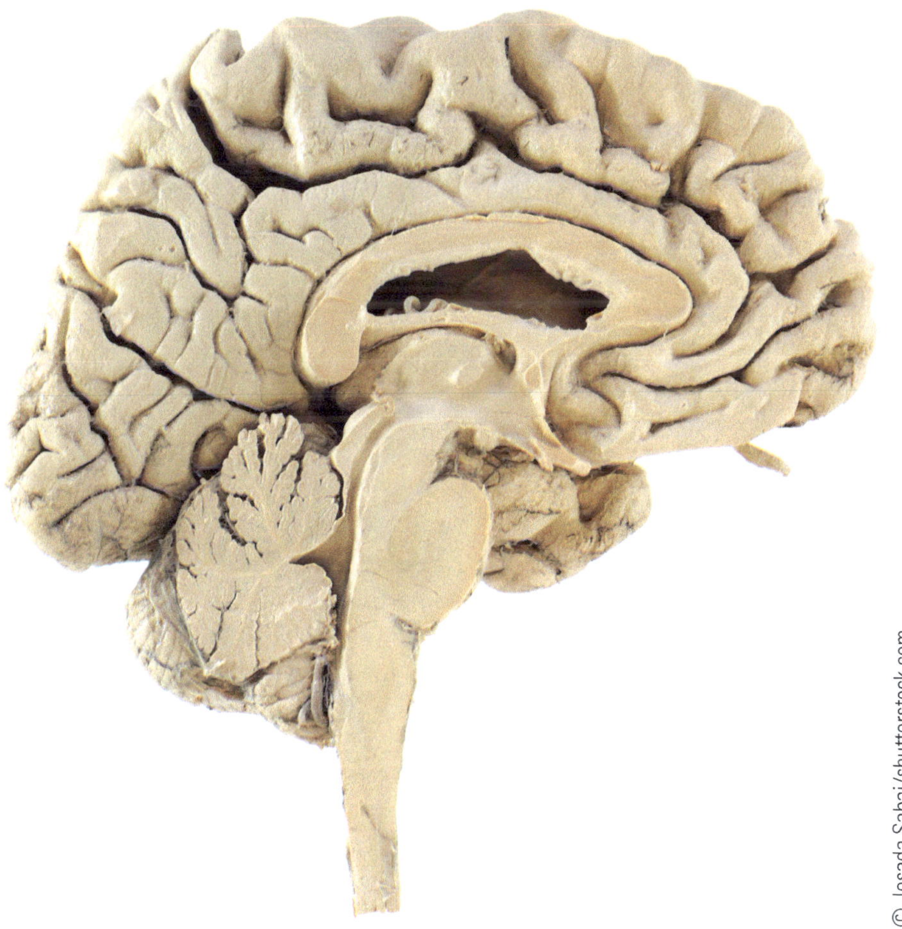

First, find the **corpus callosum,** which separates the right and left cerebral hemispheres internally. Just below, you find the one side of the bilateral **lateral ventricles**, which houses the cerebrospinal fluid. Separating the lateral ventricles is the **septum pellucidum**, which is not shown here. The **fornix** is the inferior aspect of the lateral ventricle.

Dropping below the fornix, you find the **thalamus** which will serve as a control center for many functions. The area for which the thalamus sits is also the **third ventricle.** Moving posteriorly, you will find a small bulb, the **pineal gland,** which secretes melatonin. Just inferior to the gland is the **corpora quadrigemina, aka the superior and inferior colliculi** (superior-vision; inferior-auditory). Just anterior to the colliculi, you find the **cerebral aqueduct** which brings CSF from the 3rd ventricle to the **4th ventricle**, which is just anterior to the **cerebellum** (balance, coordination). **Cerebellar peduncles** will bridge the brainstem to the cerebellum (superior, middle, and inferior)

On this medial view, if you go back to the thalamus, you can find the **hypothalamus** (anterior and inferior to thalamus); just under, you find the **mammillary bodies,** to the **cerebral peduncles, pons, and medulla.**

Of Clinical Significance

Basal Ganglia

1. Subcortical (deep cortex) mass of gray matter within the lower aspects of each cerebral hemisphere;
2. Motor integration to enable SMOOTH movements; output from and input to the motor cortex
3. Damage to this area will lead to dyskinesia

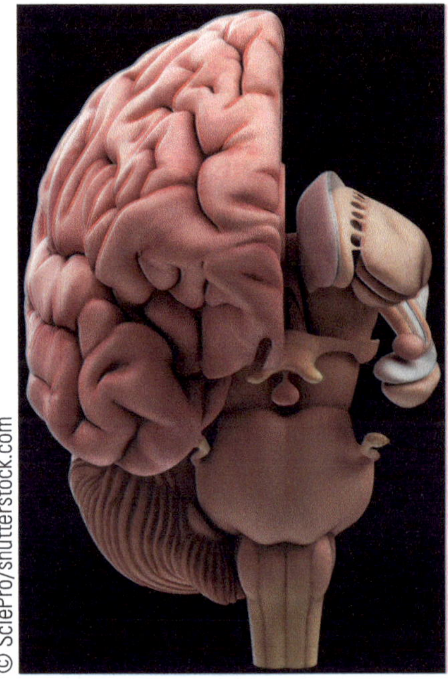

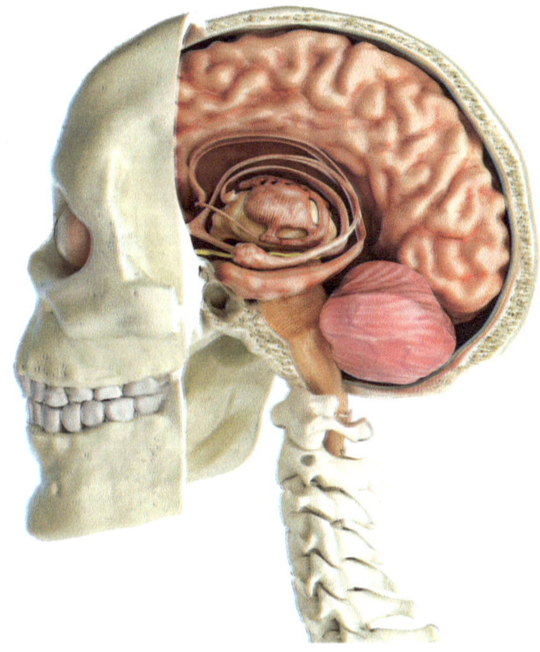

Anatomy of the Basal Ganglia:

Caudate Nucleus	Large mass of cells Head, Body, Tail Receive fibers from the motor cortex and substantia nigra Send to Globus Pallidus
Putamen	Largest aspect of the basal ganglia Receive fibers form the motor cortex and substantia nigra Send to Globus Pallidus
Striatum	Caudate nucleus and Putamen Secrete GABA (inhibitory) Direct and Indirect pathways
Globus Pallidus	Internal and External Secrete GABA Direct (internal) and Indirect (internal, external) pathways
Substantia Nigra Pars Reticularis	Direct pathway Secrete GABA
Subthalamus	Indirect pathway Secrete Glutamate
Substantia Nigra Pars Compacta	Dopamine pathway Acts on the Striatum

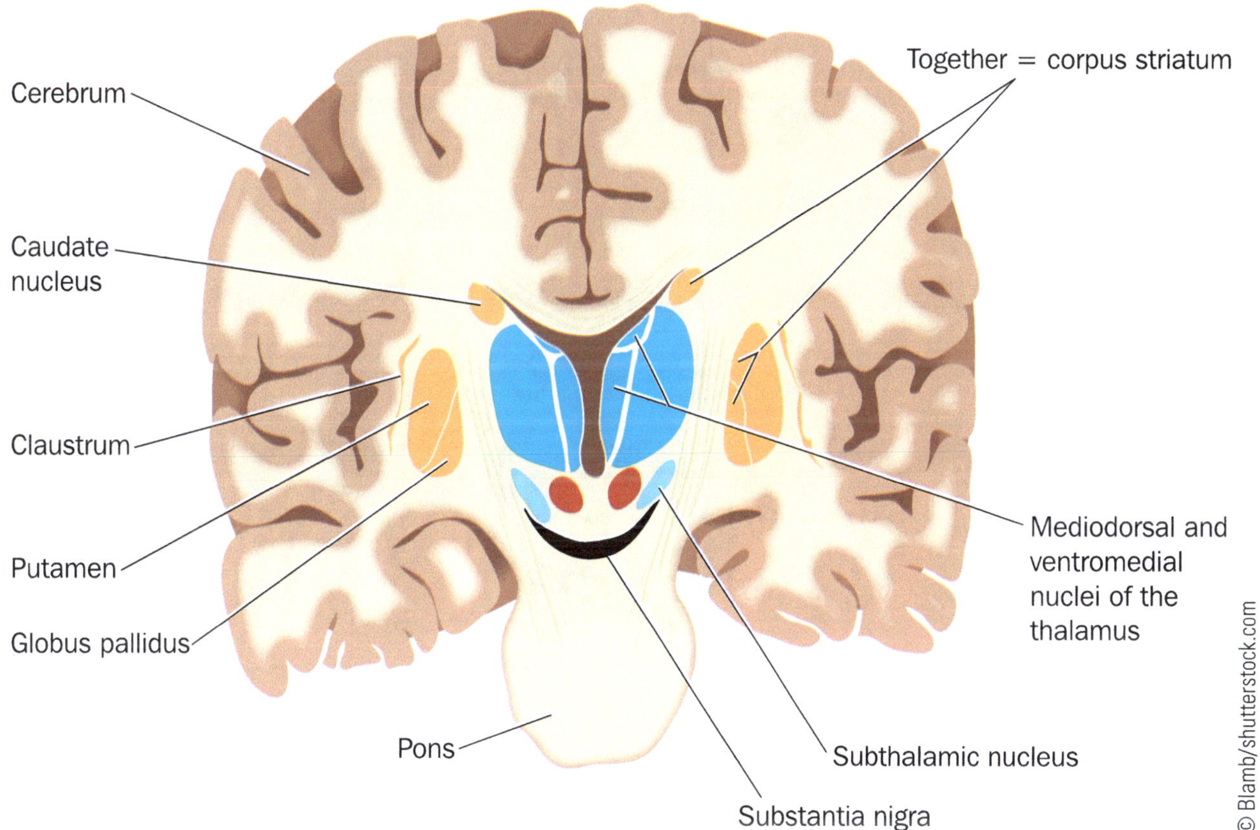
Cerebrum
Caudate
nucleus
Claustrum
Putamen
Globus pallidus
Pons
Together = corpus striatum
Mediodorsal and
ventromedial
nuclei of the
thalamus
Subthalamic nucleus
Substantia nigra
© Blamb/shutterstock.com

Direct Pathway of the Basal Ganglia: Facilitation of movement

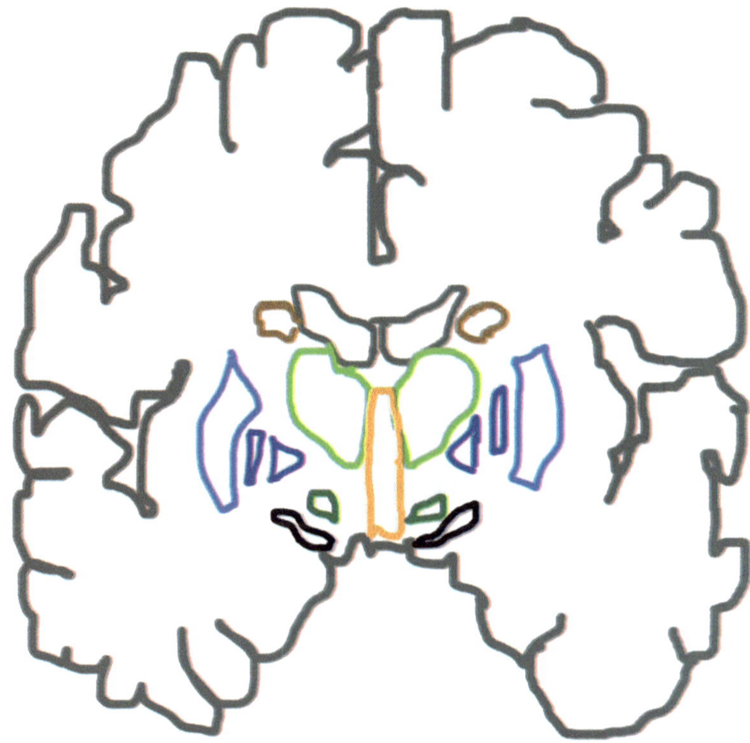

Indirect Pathway of the Basal Ganglia: Decrease unwanted movement

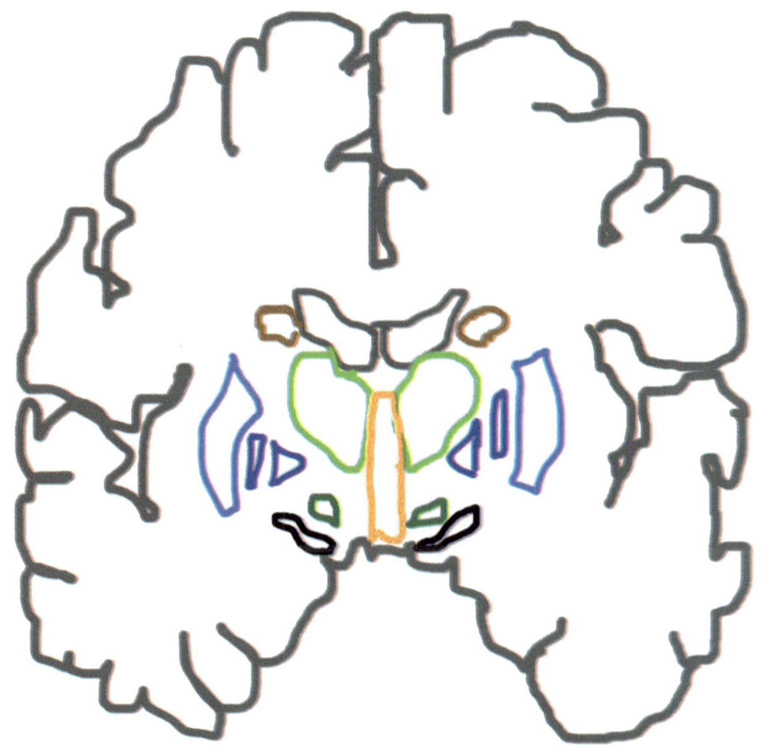

Nigrostriatal Pathway via Dopamine 1

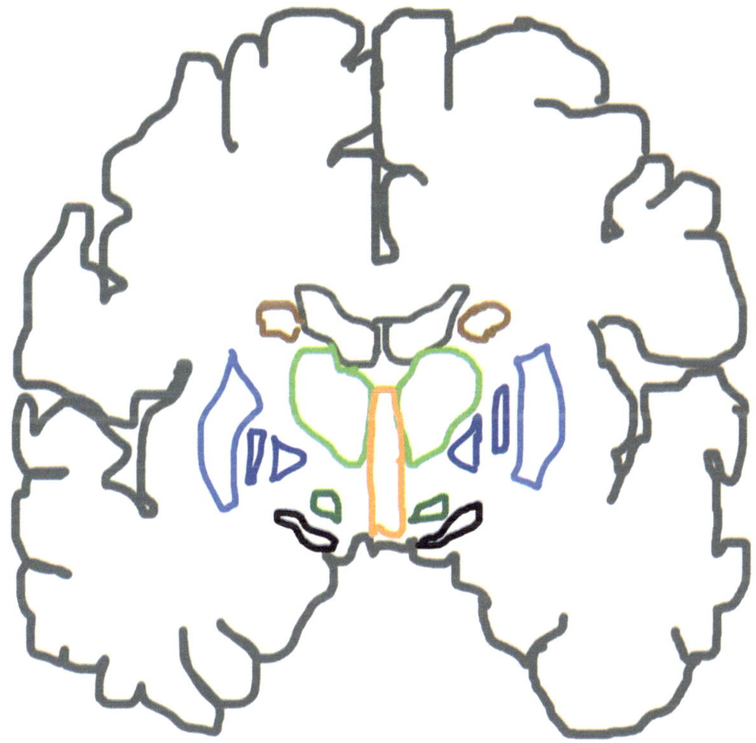

Nigrostriatal Pathway via Dopamine 2

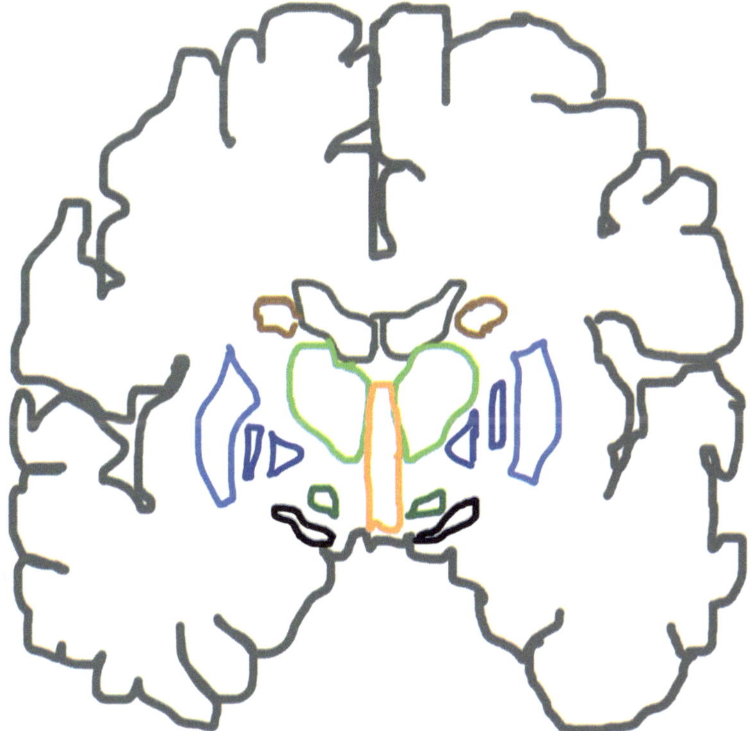

Of Clinical Significance

Of Clinical Significance

The Limbic System

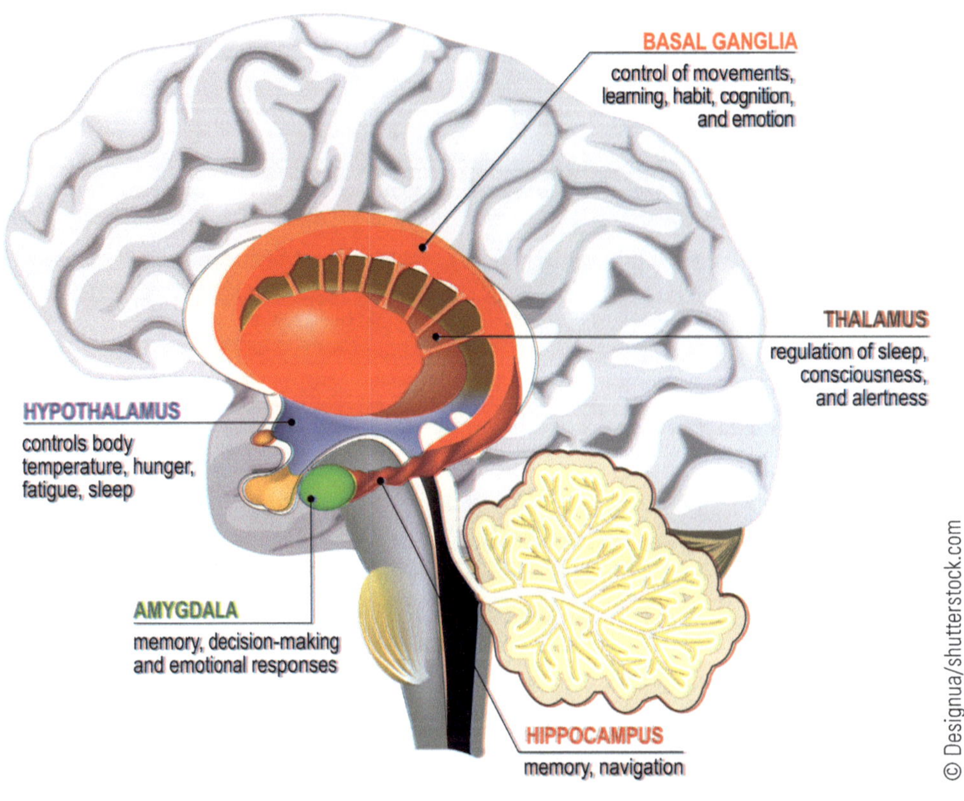

The limbic system (as we know it this second in time)
1. Emotional brain
2. Receives sensory information and has output to the endocrine, visceral motor, and somatic motor effectors

Structures of the limbic system (in general terms)

Hippocampus	Deep medial temporal lobe Memory; short-term memory processing Helps with hypothalamus function
Amygdala	Tail of caudate nucleus; uncus region Emotion, fear, regulation of aggression and rage, feeding beahviors, autonomic/endocrine links
Fornix	Relays messages from the hippocampus to the hyppthalamus
Mammillary bodies	Relay message to the thalamus; **mammillothalamic tract**
Thalamus	Integration center
Cingulate gyrus	Above the corpus callosum; receives information from the thalamus to the limbic/parahippocampal cortex Memory consolidation; perception of pain

Circuit of Papez
1. *Circular pathway that interconnects the major limbic structures*
2. *Emotional response-memory*

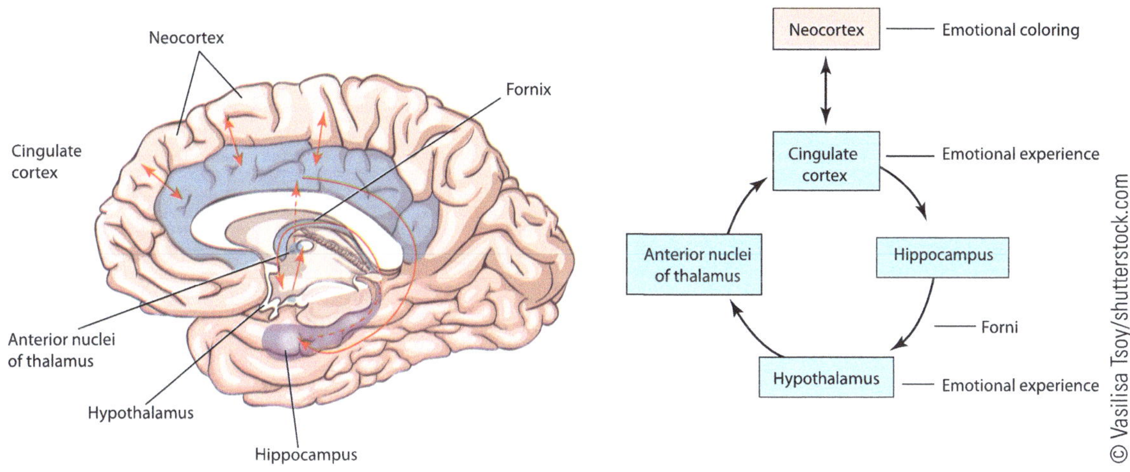

Cerebral Blood Supply
(Circle of Willis)

The cerebrum is supplied by branches of 2 primary arteries:
1. **Vertebral A**
   * Branches from the Subclavian A
   * Ascends through the transverse foramen of cervical vertebrae
   * Enters the foramen magnum
   * Branches into Anterior/Posterior Spinal arteries and PICA (below)
   * Serves posterior aspect
2. **Internal Carotid A**
   * Branches from Common Carotid A
   * Enter carotid canal
   * Serves the anterior aspect

Checklist for Circle of Willis:
Pay attention to names for region supplied

3 Cerebellar Arteries
   * Posterior inferior cerebellar (PICA)
   * Anterior inferior cerebellar (AICA)
   * Superior cerebellar

3 Cerebral Arteries
   * Posterior cerebral
   * Middle cerebral
   * Anterior cerebral

3 bridge arteries
   * Basilar
   * Posterior communicating
   * Anterior communicating

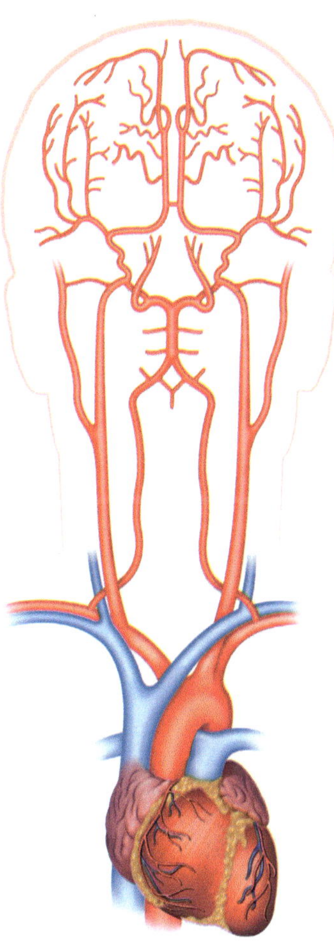

Circle of Willis

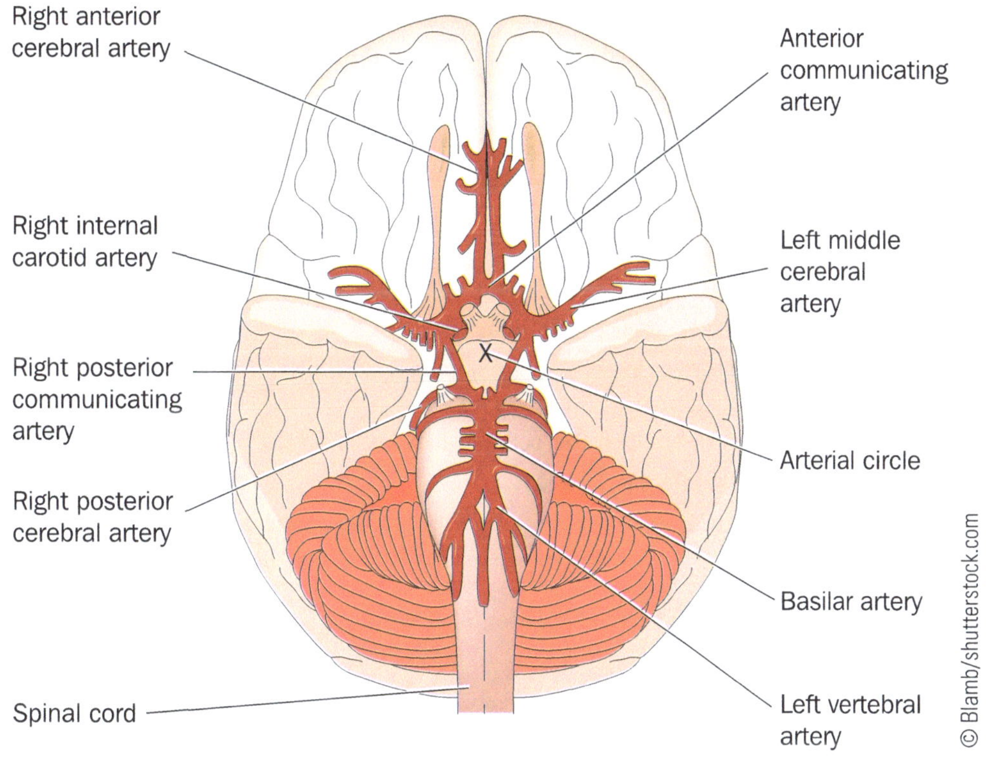

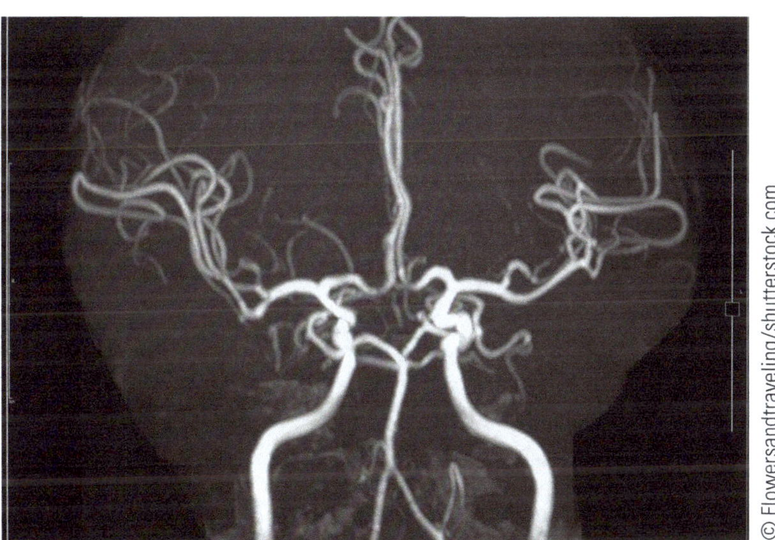

Cerebral Artery Supplies (in general)

Remember that arterial supply is not exact due to overlapping and branching. The best way to think about arterial supply is to think regionally—if it is close enough—it probably helps supply the area. The key is to know the brain anatomy so you know what falls in the region. **If you know where the distribution, you also know the potential signs of cerebral hemorrhage, tumor, and/or stroke/ischemia/infarction.**

From Vertebral A	
Posterior Inferior Cerebellar	Posterior-inferior cerebellum and some aspects of the medulla *Lateral medullary Syndrome; Wallenberg*
Basilar	2 vertebral arteries form together at lower pons; Branches into: **Pontine A:** Pons **Labyrinthine A:** Cochlea and vestibular aspects (auditory/balance) **Anterior Inferior Cerebellar** **Superior Cerebellar** **Posterior Cerebral**
Anterior Inferior Cerebellar	Supplies anterior inferior surface of cerebellum
Superior Cerebellar	Supplies superior cerebellum
Posterior Cerebral	Supplies midbrain, temporal lobes, occipital lobes
Posterior Communicating	Bridges to the Internal Carotid A Supplies optic chiasm, optic radiation, hypothalamus
	From Internal Carotid A
Ophthalmic	Supplies aspects of the orbit via the optic canal; runs with the optic nerve
Middle Cerebral	Moves laterally to supply lateral cerebrum (frontal, temporal, parietal)
Anterior Cerebral	Runs along the longitudinal fissure; medial/anterior aspects of frontal and parietal lobes

Distribution of cerebral arteries

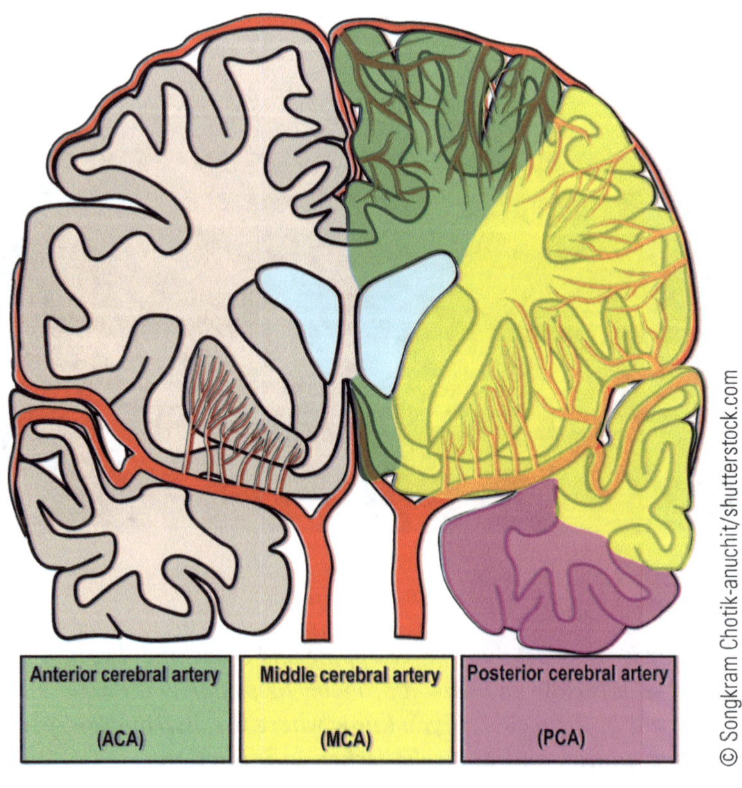

M1 of middle cerebral artery (MCA) occlusion

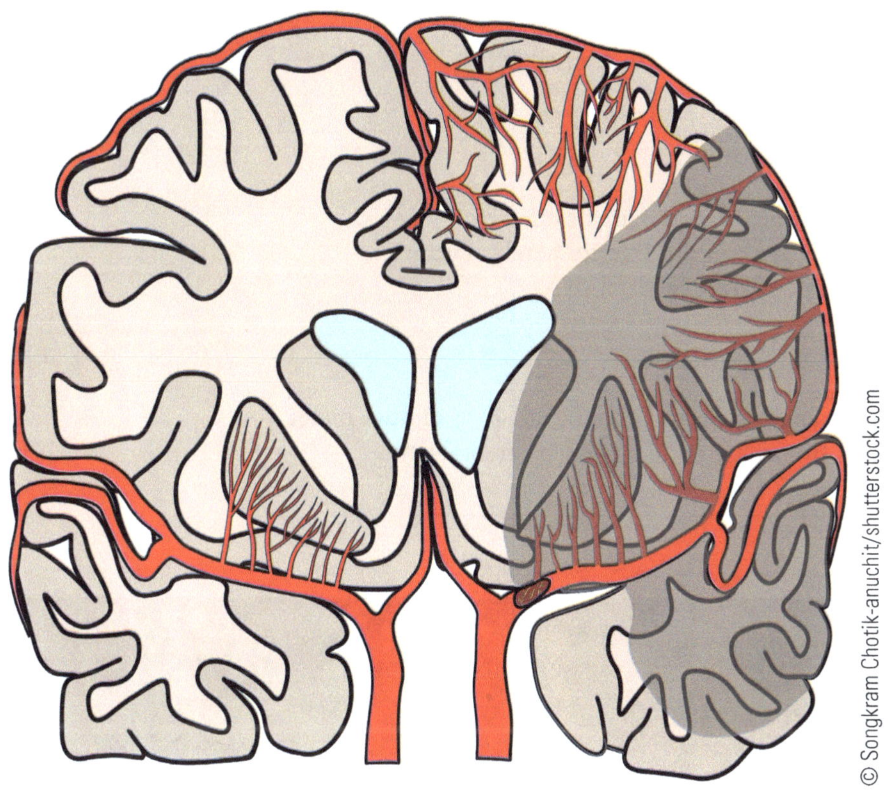

Inferior M2 segment occlusion

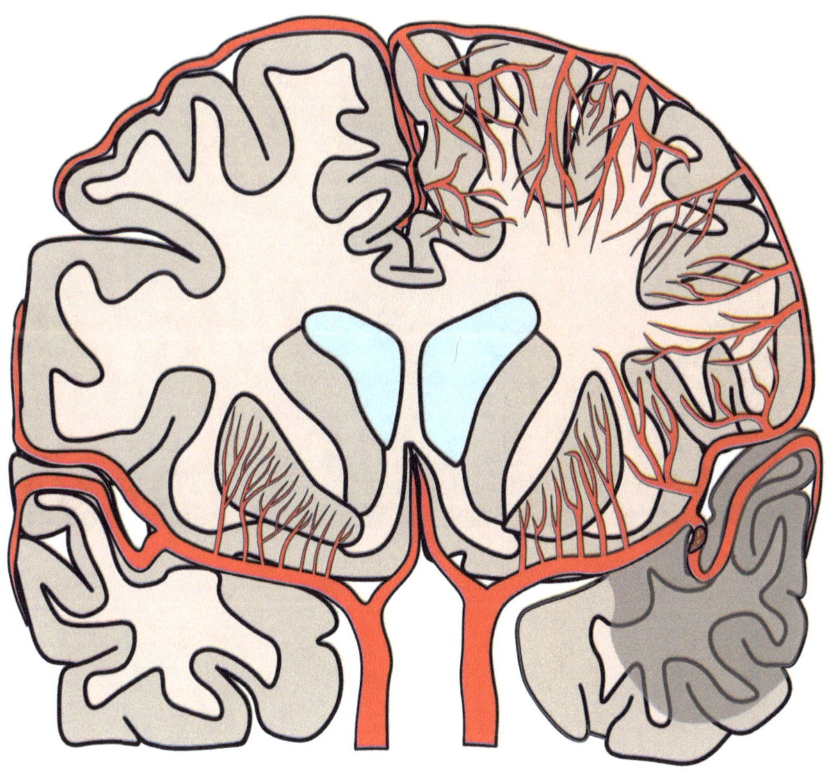

Venous Drainage of the Cerebrum

Drainage starts deep in the cerebrum via small veins that lead to larger veins, to the **dural venous sinuses.** The dural venous sinuses will empty into the **internal jugular vein.**

In general, the superficial cerebrum is drained by:
1. Superior cerebral vein
2. Middle cerebral vein
3. Inferior cerebral vein

The deep structures of the cerebrum are drained by:
1. Branches of superior, middle, inferior cerebral veins
2. Great vein of Galen (great cerebral vein)

Dural venous sinuses

1. Spaces between periosteal and meningeal layers of dura mater
2. Collect blood from superficial and deep cerebral veins
3. Absorb CSF

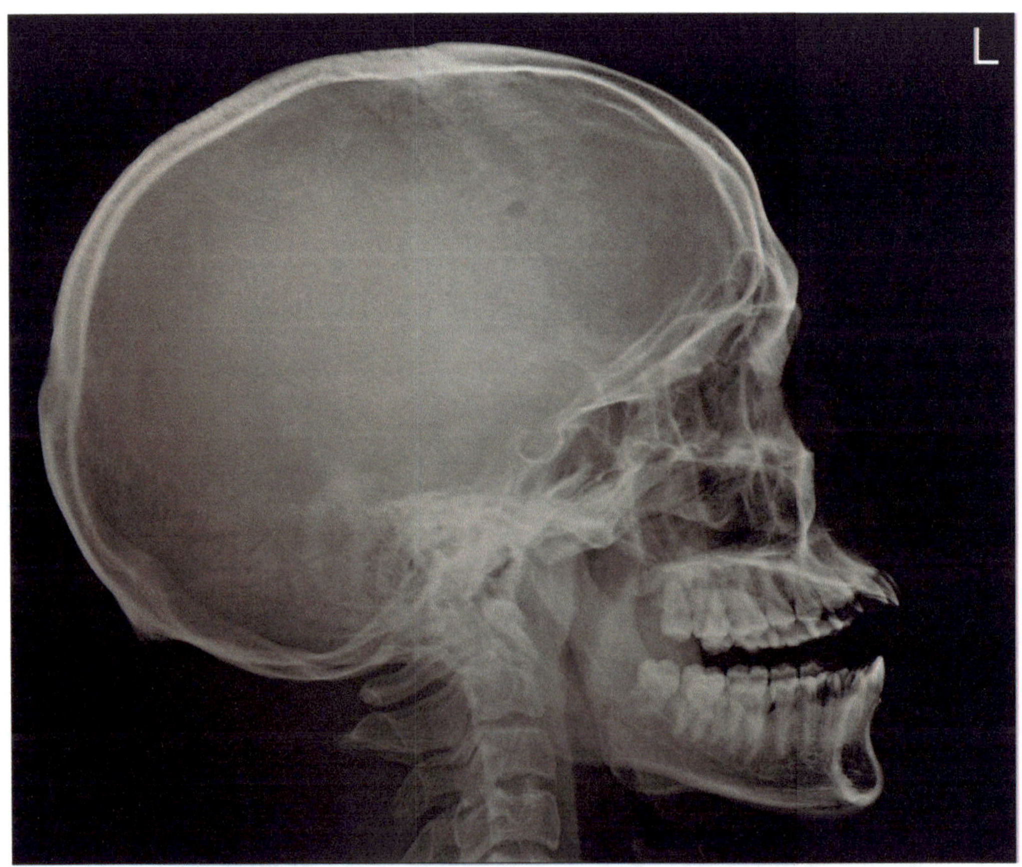

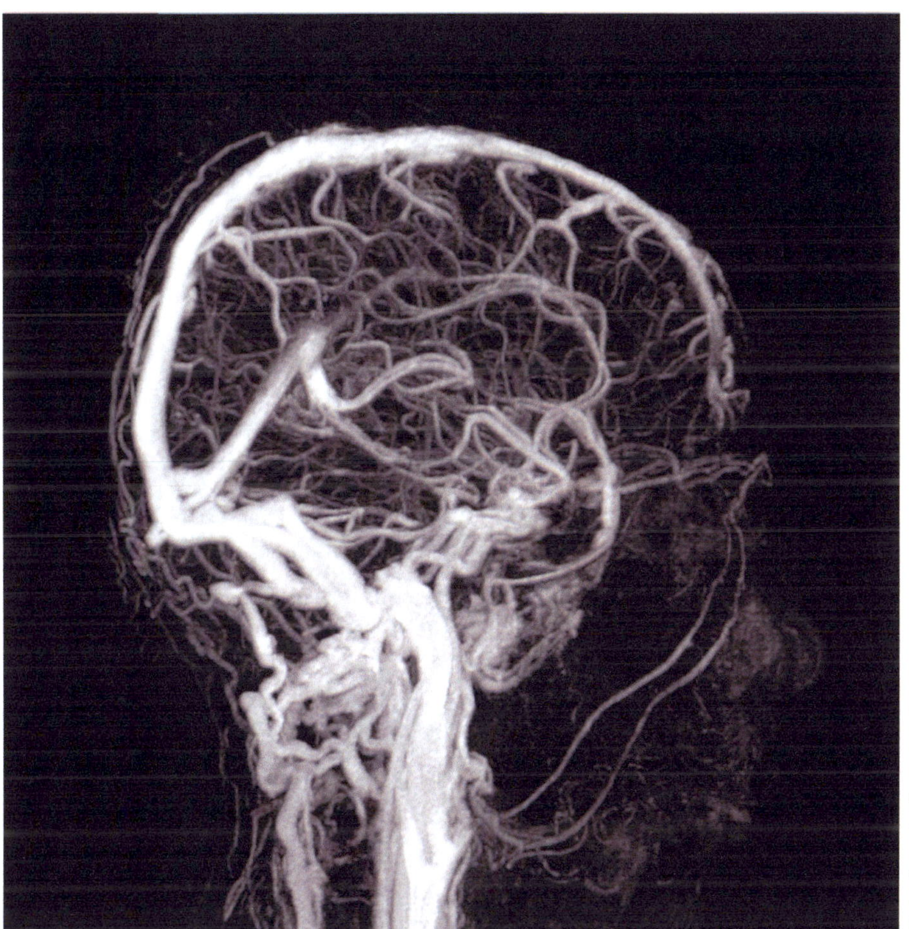

Superior Sagittal Suture	Associated with superior aspect of cerebral falx Begins at crista galli moving posteriorly like the longitudinal fissure	Receives from superior cerebral vein Drains CSF from subarachnoid space Will head to the confluence of sinuses
Inferior Sagittal Suture	Associated with inferior aspect of cerebral falx; Superior to the corpus callosum	Receives from some cerebral veins Merges into the great cerebral vein (of galen) to straight sinus
Straight Sinus	As cerebral falx meets with tentorium cerebelli Union of inferior sagittal sinus and great cerebral vein (of of galen)	Receives from inferior sagittal sinus, posterior cerebral veins, great cerebral vein, cerebellar veins, cerebral falx veins Will head to the confluence of sinuses
Confluence of Sinuses	Located at internal occipital protuberance	Receives from superior sagittal sinus, straight sinus, occipital sinus
Transverse Sinus	Right and left Moves across from the confluence of sinuses Associated with tentorium cerebelli	Receives from confluence of sinuses
Sigmoid Sinus	Right and left Continuation of the transverse sinuses	Receives from the transverse sinus and leads to the internal jugular vein

Other		
Cavernous Sinus	Bilateral at sella turcica Passing through this sinus area: Internal carotid A Abducens N Oculomotor N Trochlear N Ophthalmic N Maxillary N	Receives from great cerebral vein *Thrombosis*
Superior/Inferior Petrosal Sinuses	At the petrous aspect of the temporal bone (general area)	Superior receives from cerebellar, inferior cerebral, tympanic veins Inferior receives from veins of pons/medulla, auditory veins Receives from cavernous sinus Superior drains to transverse sinus Inferior drains to sigmoid sinus
Sphenoparietal Sinus	Along the lesser wing of the sphenoid bone	Receives from middle cerebral vein Drains into the cavernous sinus

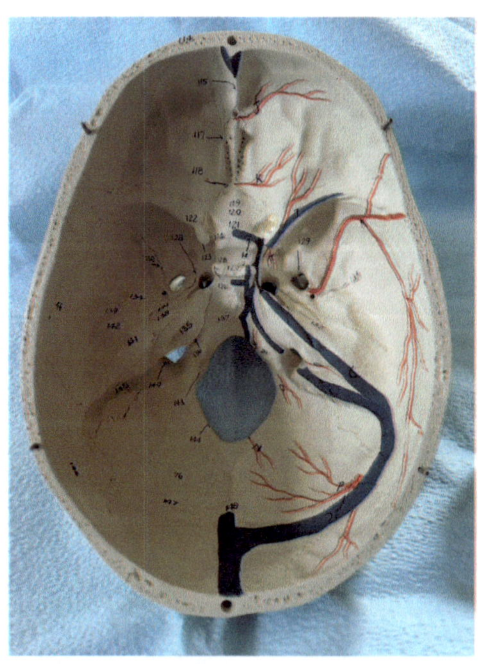
Source: Shireen Rahman

ARE YOU *DRAINED* YET???

Sample Drainage:

Superior cerebral vein drains into the <u>Superior Sagittal Sinus</u>→ <u>Confluence of Sinuses</u>→ <u>Transverse Sinus</u> → <u>Sigmoid Sinus</u>→ <u>Internal Jugular Vein</u>

<u>MEANWHILE</u>

The <u>Inferior Sagittal Sinus</u>→ <u>Straight Sinus</u> → <u>Confluence of Sinuses</u>→<u>Transverse Sinus</u>→ <u>Sigmoid Sinus</u>→<u>Internal Jugular Vein</u>

Ventricular system

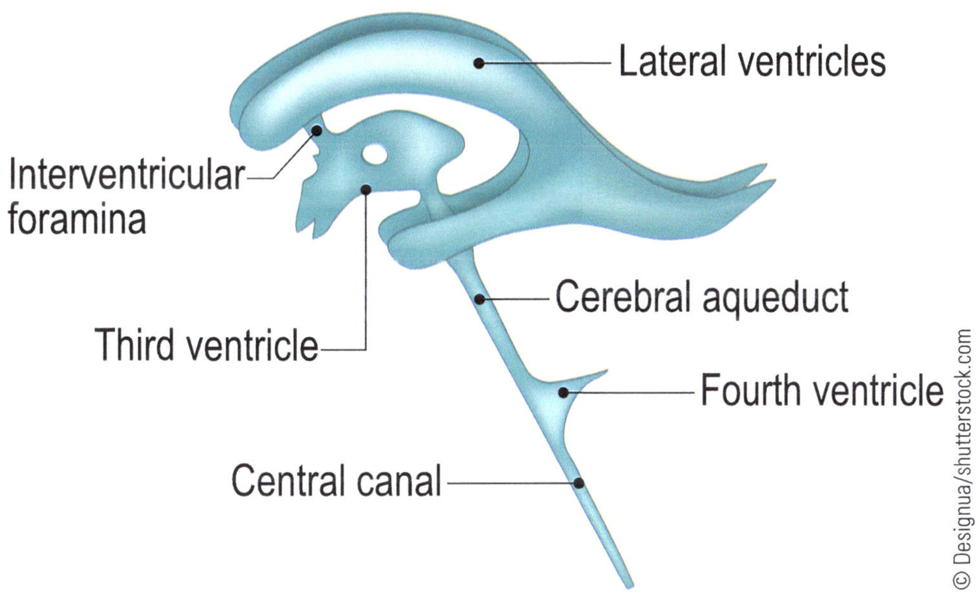

Ventricular system

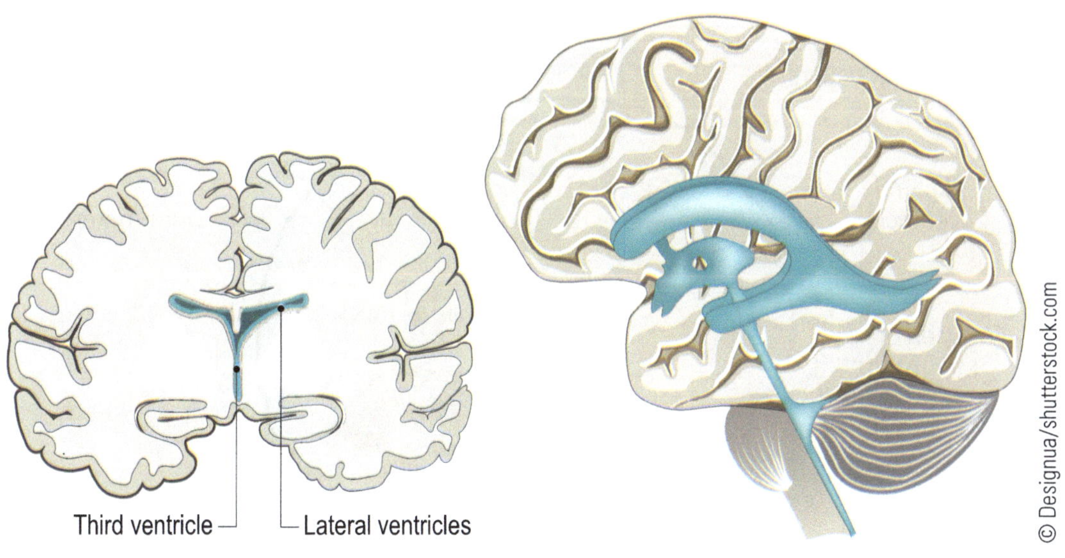

Ventricular System

<u>Cerebrospinal Fluid</u> supports and cushions the CNS, transports hormones, and removes metabolic wastes by absorbing them into the subarachnoid space → dural venous sinuses.

It is also important in maintain balanced intercranial pressure.

Choroid Plexus: will produce the CSF

Lateral Ventricles:
1. Within the cerebral cortex
2. Empties to 3rd ventricle via the <u>interventricular foramen</u>

Third Ventricle:
1. Within the diencephalon
2. Empties into the <u>cerebral aqueduct</u> which sends to the 4th ventricle

Fourth Ventricle:
1. Brainstem
2. Posterior to Pons and Medulla
3. Has <u>median and lateral aperatures</u> which empty into the subarachnoid space
4. The fourth ventricle empties into the <u>central canal</u> of the spinal cord

Aperatures:
CSF leave ventricle system to the subarachnoid space

Arachnoid Granulation:
Allows CSF to leave subarachnoid space and be absorbed into the dural venous sinuses

Cisterns:
Widening in subarachnoid space with large pools of CSF

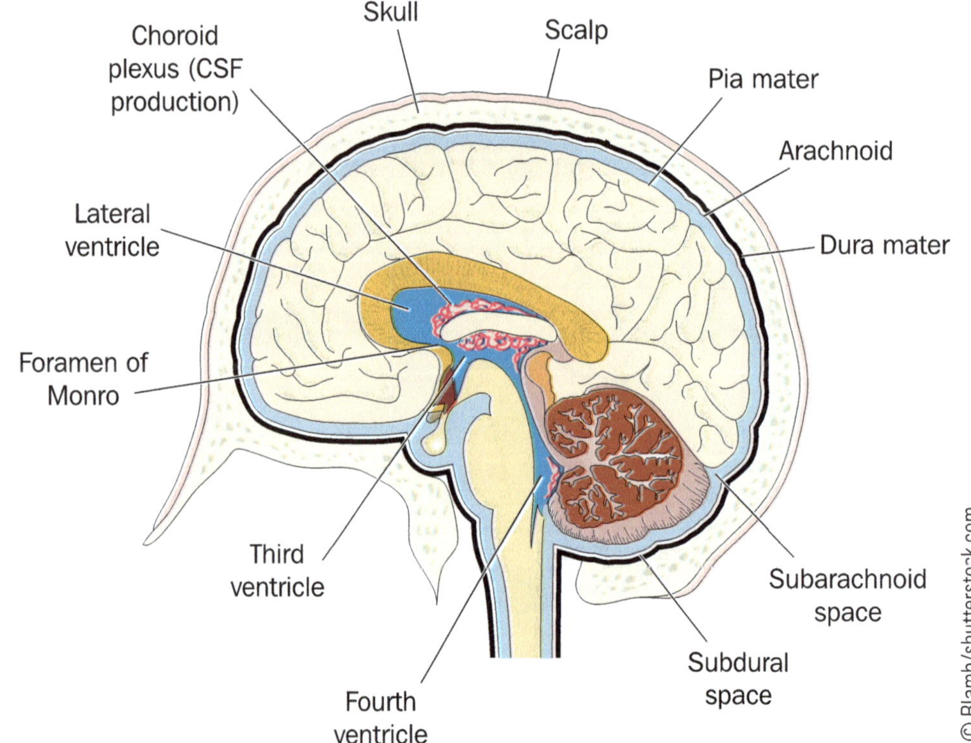

<u>**Of Clinical Significance**</u>
Intercranial Pressure

<u>Putting It All Together</u>

Putting It All Together

Putting It All Together

Putting It All Together

Cranial Nerves

1. <u>Afferent Fibers (to the Brain)</u>
2. <u>Effferent Fibers (from the CNS)</u>

Component		General Function	Cranial Nerve
General Somatic Afferent	**GSA**	Cell body is outside the CNS Innervates skin, skeletal muscles, tendons, and joints Pain, temperature, proprioception, touch, and pressure	**V, VII, IX, X Trigeminal* (primary) Facial Glossopharyngeal Vagus**
General Visceral Afferent	**GVA**	Receptors in the serous linings and smooth muscle of the viscera Sensory input from viscera; signals of thirst, hunger, visceral pain, and general unpleasant feelings Changes in BP	**IX, X Glossopharyngeal Vagus**
Special Visceral Afferent	**SVA**	Special senses utilizing **chemoreceptors** Olfaction, Taste	**I, VII, IX, X Olfactory Facial Glossopharyngeal Vagus**
Special Sensory Afferents	**SSA**	Special Senses Vision, Hearing	**II, VIII Optic Vestibulocochlear**
General Somatic Efferent	**GSE**	Cell Bodies in the Brainstem or spinal cord Motor innervations to voluntary muscle	**III, IV, VI, XII Oculomotor Trochlear Abducens Hypoglossal**
General Visceral Efferent *Via preganglionic parasym-pathetic axons*	**GVE**	Cell bodies within the CNS Motor innervations to smooth muscle, cardiac muscle, and glands	**III, VII, IX, X Oculomotor Facial Glossopharyngeal Vagus**
Branchial Efferent (Special Visceral Efferent)	**BE (SVE)**	Motor innervations to pharyngeal, face, palate, laryngeal skeletal muscles	**V, VII, IX, X, XI Trigeminal Facial Glossopharyngeal Vagus Spinal accessory**

Cranial Nerve Location

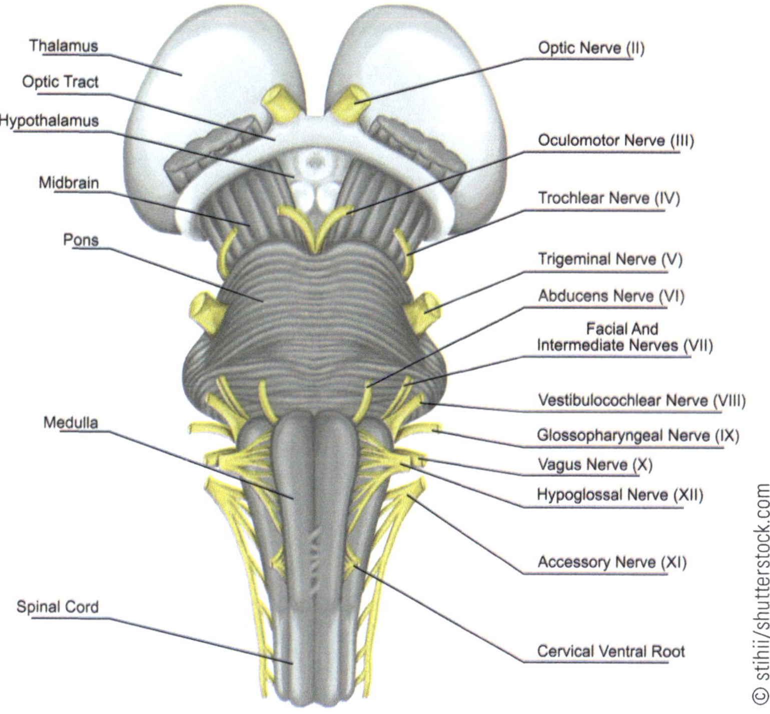

Understanding Cranial Nerve Nuclei

- A cranial nerve nucleus is a collection of neurons in the brain stem associated with one more cranial nerves
- Axons carrying information to and from the cranial nerves form a synapse first at these nuclei

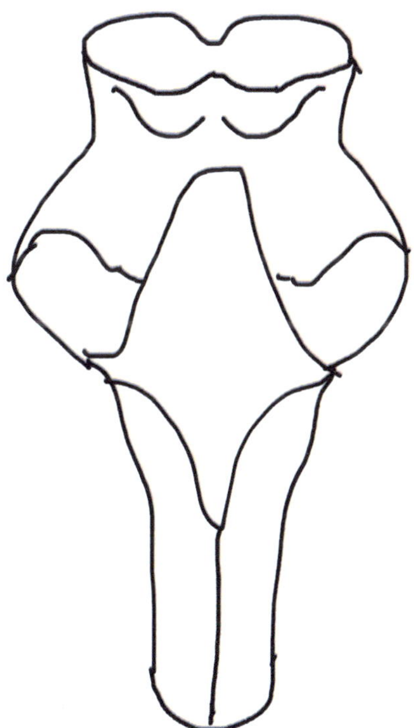

Source: Shireen Rahman

Nerve	Nucleus Location	Afferent Function	Efferent Function	Lesions
Olfactory I Cribiform plate of ethmoid	Olfactory Epithelium	**SVA**- SMELL	**NONE**	Anosium
Optic II Optic canal	Retinal Cells (Ganglion)	**SVA**-VISION	**NONE**	Vision loss or impairment
Oculomotor III Superior orbital fissure	Oculomotor Nucleus *Just medial to Superior Colliculus*	**NONE**	**GSE**- Via Axon • To Rectus muscles of eye (superior, inferior, medial) Moves eyes medially and up and down • To levator palpebrae superior; raises superior eyelid	Loss of medial and up and down movement of the eye; double vision; eye droop (ptosis)
	Edinger-Westphal Nucleus *Immediately inferior to Oculomotor Nucleus*		**GVE**- Through Ciliary Ganglion (Pre/Post Fibers) Parasympathetic innervations to • Sphincter papillae • Ciliary Muscle *Pupil constriction* *Accommodation of the lens*	Loss of pupil reflex to light (remains dilated) and poor lens accommodation;
Trochlear IV Superior Orbital Fissure	Trochlear Nucleus *Medial to the Inferior Colliculis*	**NONE**	**GSE**- Via Axon • Superior Oblique muscle *Down and out*	Excessive outward rotation of the eye; usually result in a head tilt to accommodate vision
Trigeminal V	Mesencephalic Nucleus of Trigeminal Nerve Main Sensory Nucleus of Trigeminal Ganglion Spinal Trigeminal Nucleus	**GSA**- • Ophthalmic (V1)- ***Superior Orbital Fissure To:*** *Orbital Region* *Cornea sensation; skin of forehead, scalp, eyelids, nose; mucosa of the nose* • Maxillary (V2) ***Foramen Rotundum To:*** *Maxillary region* *Sensation from skin of face over maxilla; upper lip; maxillary teeth; nose mucosa; maxillary sinus; palate* • Mandibular (V3) ***Foramen Ovale To:*** *Mandibular Region* *Sensation of skin over mandible; lower lip, side of head, mandibular teeth; temporomandibular joint; mouth mucosa; anterior tongue*		Loss of sensation to areas noted
	Motor Nucleus of Trigeminal Ganglion; *Medial to Cerebellar Ped. On Pons;*		**SVE (BE)**-Via Axon ***Foramert Ovale To:*** • Muscles of mastication • Temporalis Muscle	Weakness or paralysis of mastication muscles

Nerve	Nucleus Location	Afferent Function	Efferent Function	Lesions
Abducen VI Superior Orbital Fissure	<u>Nucleus of Abducens</u> *Posterior Medial Pons* *In between the Motor Trigeminal and Facial Nucleus*	**NONE**	**GSE-** Via Axon • Lateral rectus muscle *Turn eye laterally*	Paralysis of Lateral Rectus
Facial VII	Geniculate Ganglion to <u>Spinal Trigeminal Nucleus</u> to Thalamus	**GSA-** Via Afferent Neurons *From* • External acoustic meatus; posterior ear skin • *Info of touch, pain, temp.*		Loss of sensation from the noted area
	Submandibular Ganglion to Geniculate Ganglion to <u>Nuclei of Solitary Tract</u> to Thalamus	**SVA-** Via Afferent Neurons *From* • Taste (anterior tongue)		Loss of taste sensation from anterior tongue
	<u>Superior Salivatory Nuclei</u> *Just inferior to Facial Nucleus; Lower Pons*		**GVE-** Parasympathetic innervations to Pre through Geniculate Ganglion synapses with Post at Pterygopalantine Ganglion • Lacrimal Gland • Nasal Mucosa • Oral Cavity Mucosa Submandibular Ganglion • Submandibular Gland • Sublinguil Gland	Decrease in secretions
	<u>Facial Motor Nuclei</u> *Pons; Inferior to Abducens Nuclei*		**SVE-(BE)** *Axon through Internal Acoustic Meatus to Geniculate Ganglion to Stylomastoid Foramen to* • Muscles of facial expression	
				Loss of facial expression
Vestibulocochlear *Internal Acoustic Meatus*	<u>Vestibular Nuclei</u> *Pons*	**SSA:** • Vestibular: Balance	**NONE**	Balance and Hearing Dysfunctions
	<u>Cochlear Nuclei</u> *Pons Lateral to Vestibular Nuclei*	• Cochlear: Hearing		

Nerve	Nucleus Location	Afferent Function	Efferent Function	Lesions
Glossopharyngeal *Jugular Foramen*	Afferent Neurons to Inferior Ganglion to <u>Nuclei of Solitary Tract</u> *Pons under vestibular nuclei*	**GVA:** Via Afferent Neurons *From* • Carotid body and Sinus • Palate/Pharynx		Disruption to Carotid reflex
	Afferent neurons to Inferior Ganglion to <u>Nuclei of Solitary Tract</u>	**SVA:** Via Afferent Neurons *From* • Taste (posterior tongue)		Loss of taste sensation from posterior tongue
	Afferent Neurons to Superior Ganglion to <u>Spinal Trigeminal Nuclei</u>	**GSA:** Via Afferent Neurons *From* • Sensation to external ear; • Sensation to posterior tongue		Loss of sensation from noted areas
	<u>Inferior Salivatory Nucleus</u> *Just inferior to the superior salivatory nuclei*		**GVE:** Parasympathetic innervations To Pre through Foramen Ovale to synapse with Post in Otic Ganglion to Parotid salivary gland	Loss of parotid gland secretion; reduction of salivary
	<u>Nucleus Ambiguus</u> *Lateral Medulla just inferior to Salivatory Nuclei*		**SVE (BE):** Efferent neuron to Stylopharyngeus muscle *Elevates pharynx for speech and swallowing*	Impairment of gag reflex
Vagus *Jugular Foramen*	Afferent Neurons to Superior Ganglion to <u>Spinal Trigeminal Nucleus</u> to Thalamus	**GSA:** Via Afferent Neurons *From* • Skin posterior ear; external acoustic meatus; dura of posterior cranial fossa		Loss of sensation from noted areas Loss of sensation from reported areas Disruption to respiration; minimal loss of taste
	Afferent Neurons to Inferior Ganglion to <u>Nuclei of Solitary Tract</u>	**GVA:** Via Afferent Neurons *From* • Sensory abdominal and thoracic viscera • Baroreceptors of Aortic Arch		
	Afferent Neurons to Inferior Ganglion to <u>Nuclei of Solitary Tract</u>	**SSA:** Via Afferent Neurons *From* • Taste from epiglottis and palate • Mediates respiratory function via aortic body		
	<u>Dorsal Motor Nucleus of Vagus</u> *Just medial to Nucleus Ambiguus*		**GVE:** Parasympathetic innervations to Pre through Jugular foramen to Post at Viscera Smooth muscle and glands in pharynx, larynx, thoracic and abdominal viscera; peristalsis	Hyperactivity of vagus nerve
	<u>Nucleus Ambiguus</u>		**SVE(BE):** Via Efferent Fiber Through Jugular Foramen to One tongue muscle; soft palate muscles; pharynx muscles, larynx muscles *Sounds in speech*	Difficulty swallowing and hoarseness

Nerve	Nucleus Location	Afferent Function	Efferent Function	Lesions
Accessory *Spinal cord*	<u>Anterior Horn Nucleus of Spinal Accessory</u> *Spinal cord level*	**NONE**	**SVE (BE):** Via Efferent Fiber Through Jugular Foramen to Spinal Root to • Sternocliedomastoid; Trapezius	Weakness in noted muscle functions
Hypoglossal *Hypoglossal Canal*	<u>Hypoglossal Nucleus</u> *Medulla; most medial to Dorsal Motor N of Vagus*	**NONE**	**GSE:** Via Efferent Fiber Through Hypoglossal Canal to • Hypoglossus, genioglossus, styloglossus; **intrinsic tongue muscles** *Protrusion of tongue*	Deviation of the tongue to the side

Efferent Pathways

Oculomotor N (III) GSE   CNS→Effector

Oculomotor Nucleus of midbrain → GSE→Superior Orbital Fissure→Superior Rectus, Inferior Rectus, Medial Rectus, Inferior Oblique, Levator Papilbrae

---

Oculomotor N (III) GVE   CNS—Ganglionic Fibers→Effector

Edinger-Westphal Nucleus of midbrain →GVE (preganglionic fiber) →Superior Orbital Fissure→ Cilary Ganglion (synaptic)→GVE (postganglionic fiber) →
Sphincter Pupillae   (constriction)
Ciliary Muscle   (accommodation via suspensory ligaments)

Trochlear N (IV) GSE   CNS→Effector

Trochlear Nucleus of midbrain → GSE → Superior Orbital Fissure →Superior Oblique Muscle

---

Trigeminal N (V) BE   CNS→Effector

Motor Nucleus of Trigeminal N of Pons → BE → Foramen Ovale → Muscles of Mastication, Temporalis, Tensor Tympani Muscle

---

Abducen N (VI) GSE   CNS→ Effector

Nucleus of Abducens of Pons →GSE → Superior Orbital Fissure→ Lateral Rectus

---

Facial N (VII) GVE   CNS –Ganglionic Fibers→ Effector

Geniculate Ganglion collects branches of Facial N (color pencil box)

Superior Salivatory Nucleus of Pons → GVE (preganglionic fiber)→ Internal Acoustic Meatus→Geniculate Ganglion (non-synaptic) → GVE preganglionic fiber →2 Ganglion

1. Ptyergopalatine Ganglion→ GVE (postganglionic fiber) → Lacrimal Gland, Nasal Mucosal Glands, Oral Mucosal Glands (secretions)
2. Submandibular Ganglion→ GVE (postganglionic fiber)→ Submandibular and Sublinguil Salivary Glands

---

Facial N (VII) BE   CNS→Effector

Facial Motor Nuclei of Pons →BE → Internal Acoustic Meatus→ Geniculate Ganglion (non-synaptic)→ Stylomastoid Foramen→ Muscles of Facial Expression,

The Stapedius Muscle is also innervated: Facial Motor Nuclei of Pons →BE →Internal acoustic meatus → Geniculate Ganglion →Stapedius

---

Glossopharyngeal N (IX) GVE   CNS--Ganglionic Fibers—Effector

Inferior Salivatory Nucleus of Medulla→GVE(preganglionic fiber)→Jugular Foramen →GVE (preganglionic fiber)→Foramen Ovale→Otic Ganglion (synaptic)→ GVE (postganglionic fiber)→ Parotid Gland (salivatory)

Glossopharyngeal N (IX) BE   CNS→ Effector

Nucleus Ambiguus of Medulla → BE → Jugular Foramen → Stylopharyngeus muscle (elevation of pharynx)

Vagus N (X) GVE   CNS-Ganglionic Fibers-Effector

Dorsal Motor Nucleus of Medulla → GVE (preganglionic fiber)→Jugular Foramen →GVE (preganglionic fiber)→synapse directly with postganglionic fiber at smooth muscle and glands in thoracic and abdominal viscera

Vagus N (X) BE   CNS→Effector

Nucleus Ambiguus of Medulla →BE→ Jugular Foramen→ Muscles of pharynx, larynx, and palate (swallowing, phonation)

Spinal Accessory (XI) BE   CNS →Effector

Anterior Horn Nucleus of Spinal Cord→ BE→(up through) Foramen Magnum → (out through) Jugular Foramen →Trapezius and Sternocliedomastoid

Hypoglossal (XII) GSE   CNS→Effector

Hypoglossal Nucleus of Medulla → GSE →Hypoglossal Canal→ Muscles of the tongue (tongue protrusion)

Afferent Pathways

Olfactory (I) SVA   Effector→Cortex directly or Thalamus→Cortex

Odor→Nasal Mucosa via chemoreceptors →SVA→ Cribiform Plate → Olfactory Bulb→SVA (mitral cells)→Olfactory Tract → Olfactory cortex (as discussed with olfaction)

Optic (II) SSA   Effector →Thalamus→Cortex

Light→Pigmented Epithelium→ Photoreceptor→ Bipolar Retinal Neuron→ Retinal Ganglion Cells→Optic N (SSA)→Optic Canal→ Optic Chiasm→ Optic Tract→Thalamus→Geniculocalcarine Tract → Visual Cortex of Occipital Lobe

Oculomotor (III) NONE

Trochlear (IV) NONE

Trigeminal (V) GSA   Effector→Thalamus→ Sensory Cortex   (Ophthalmic, Maxillary, Mandibular)

Ophthalmic (V1):
Sensation from orbital area, cornea, forehead, nose→GSA→ Superior Orbital Fissure→ Trigeminal Ganglion (non-synaptic) →
Mescencephalic Nucleus of Trigeminal N
Main Sensory Nucleus of Trigeminal N
Spinal Trigeminal Nucleus

<u>Maxillary (V2):</u>
Sensation from Maxillary region, maxillary teeth, node, sinus, palate, etc →GSA→Foramen Rotundum→
Trigeminal Ganglion (non-synaptic)→
Mescencephalic Nucleus of Trig
Main Sensory Nucleus of Trig
Spinal Trigeminal Nucleus

<u>Mandibular (V3):</u>
Sensation from Mandibular region, mandibular teeth, TMJ, mouth, anterior tongue, etc→GSA→Foramen
Ovale→Trigeminal Ganglion (non-synaptic)→
Mescencephalic Nucleus,
Main Sensory Nucleus of Tri,
Spinal Trigeminal Nucleus   (Nuclei found in Medulla)

Abducens (VI) NONE

Facial (VII) GSA   Effector→Thalamus→Sensory Cortex

Sensation from Posterior Ear, Auricle, Meatus Region→ GSA→Stylomastoid Foramen→ Geniculate Ganglion
(non-synaptic)→Internal Acoustic Meatus→ Spinal Trigeminal Nucleus_of Medulla→Thalamus →Sensory
Cortex

Facial (VII) SVA   Effector→Thalamus→Cortex

Taste via chemoreceptors in taste pores→SVA→Submandibular Ganglion→Geniculate Ganglion (non-
synaptic)→Internal Acoustic Meatus→ Nuclei of Solitary Tract of Medulla→Thalamus→ Gustatory cortex
(insula)

Vestibulocochlear (VIII) SSA   Effector→Thalamus→Cortex
this nerve never exits the skull but travels through the internal acoustic meatus

Sound→External Acoustic Meatus→Tympanic Membrane→Ossicles to Stapes to Oval Window→Vibration
wave to Scala Vestibule→Vibration wave to Scala Media through Vestibular Membrane→ Causing Basilar
Membrane to wave→ Forces hair cells to bend on Tectorial Membrane→Hair cells to SSA Spiral Ganglion
(non-synaptic)→Cochlear Nerve →Merges with Vestibular N (coming from Semicircular Canal) to form
Vestibulocochlear N→Cochlear Nucleus of Medulla→ Thalamus→ Auditory Cortex of Temporal Lobe

We are not doing the Vestibular aspect at this time . . .

Glossopharyngeal (IX) GVA   Effector→Thalamus→Cortex

Sensation from posterior tongue, pharynx, carotid body and carotid sinus →GVA→Inferior Ganglion
(non-synaptic)→ Jugular Foramen → Nuclei of Solitary Tract of Medulla→Thalamus→
To Hypothalamus if sent from carotid body/sinus
To Sensory cortex from other areas

Glossopharyngeal (IX) SVA   Effector→ Thalamus→Cortex

Taste posterior tongue→ SVA→ Inferior Ganglion (non-synaptic)→ Jugular Foramen→ Nuclei of Solitary
Tract of Medulla→Thalamus→ Gustatory Cortex (Insula)

Glossopharyngeal (IX) GSA   Effector→Thalamus→Cortex

Sensation from external ear→ GSA→Superior Ganglion (non-synaptic)→ Jugular Foramen→ Spinal
Trigeminal Nucleus of Medulla→Thalamus→Sensory Cortex

Vagus (X) GSA   Effector→Thalamus→Cortex

Sensation from Auricle, External Acoustic Meatus→ GSA→Superior Ganglion (non-synaptic)→ Jugular Foramen→Spinal Trigeminal Nucleus of Medulla → Thalamus→ Sensory Cortex

Vagus (X) GVA   Effector→Thalamus→ Cortex

Sensation from Thoracic/Abdominal viscera→GVA→ Inferior Ganglion→ Jugular Foramen→ Nuclei of Solitary Tract of Medulla→Thalamus→ Sensory Cortex

Vagus (X) SVA   Effector→Thalamus→Cortex

Taste Epiglottis and palate→ SVA→ Inferior Ganglion→Jugular Foramen→ Nuclei of Solitary Tract of Medulla→ Thalamus→Gustatory Cortex

Accessory and Hypoglossal   NONE

<u>PHEW!!!</u>

EFFERENT

EFFERENT

EFFERENT

EFFERENT

AFFERENT

AFFERENT

AFFERENT

AFFERENT

Notes/Diagrams/Dreams/Grocery List

Notes/Diagrams/Dreams/Grocery List

CHAPTER 6

Eye-Ear-Nose-Throat

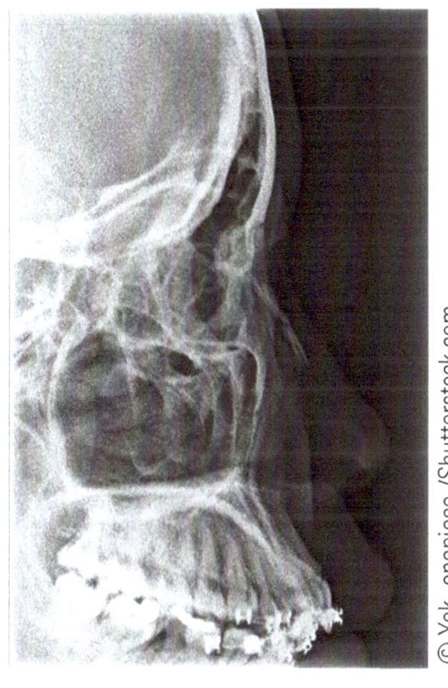

Eye
Ear
Nose
Throat

Eye, Ear, Nose, Throat

Eye Anatomy
Orbit

1. Superior Wall
   a. Frontal bone
   b. Sphenoid bone (Lesser wing)
2. Lateral Wall
   a. Zygomatic bone
   b. Sphenoid bone (Greater wing)
3. Inferior Wall
   a. Maxilla bone
   b. Zygomatic bone
   c. Palantine bone
4. Medial Wall
   a. Lacrimal bone
   b. Ethmoid bone
5. Foramen
   a. Optic canal
   b. Superior/Inferior Orbital Fissure

Palpebrae (eyelid)

a. 4 layers
  - i. Skin   *Thinnest skin of the body*
  - ii. Muscle
    1. <u>Orbicularis Oculi:</u>   **FACIAL NERVE**
       a. Closes the eyelid
    2. <u>Levator Palpebrae</u>   **OCULOMOTOR NERVE**
       a. Raise upper lid
  - iii. Connective tissue
    1. Contain <u>Tarsal Glands</u> (sebaceous)   *helps lids from sticking together*
  - iv. <u>Conjunctiva</u>
    1. Muscous membrane that lines inner surface of eyelids
    2. Covers the anterior eye surface (except cornea)

<u>Lacrimal Apparatus</u>   Protects eye through keeping eye moist and providing antimicrobes

1. Lacrimal Glands
   a. Tear secretion
   b. Facial N
2. Lacrimal Ducts
   a. Excretion from glands
3. Lacrimal Caruncle
   a. Sebaceous and sweat glands
4. Lacrimal Canaliculi
   a. Tear collection to bring to Lacrimal sacs
5. Lacrimal Sac
   a. Send tears to Nasolacrimal duct upon blinking
6. Nasolacrimal Duct
   a. Send tears to nasal cavity

<u>Eyeball Anatomy</u> (*function to be discussed later*)

3 Layers

Sclera	*outer white layer of eye continuous with transparent **cornea***
Choroid	*middle vascular layer*
Retina	*innermost neural layer*

<u>Eye Tunics of Eyeball</u>

<u>Outer Tunic/Fibrous Layer</u>   <u>Sclera/Cornea</u>

Sclera

- Continuous with Dura Mater
- Coat of the eyeball
- Provides attachment for extrinsic/intrinsic muscles
- Protection
- Optic Nerve
- Limbus

Cornea

- Transparent anterior aspect of the sclera
- Avascular; receives nutrients from lacrimal fluid and aqueous humor
- Innervated by Opthlamic aspect of Trigeminal N
- Very sensitive to touch so to elicit blinking or tears

Sclera Venus Sinus
- Removes aqueous humor from anterior chamber to venous system

Middle Tunic　　　　　　Choroid

Choroid
- Continuous with arachnoid/pia
- Between sclera and retina
- Vascular layer
- Vascular supply to retinal photoreceptors

Ciliary Body
- Anterior extension of Choroid
- Muscular and Vascular
- Connects the choroid with the Iris
- Contains Ciliary Processes and Ciliary Muscles

Suspensory Ligaments
- Ciliary body to Lens

When ciliary muscles are contracted → suspensory ligaments relax → lens thickens → near vision

When ciliary muscles are relaxed → suspensory ligaments taut → lens thins → far vision

Iris
- On the anterior surface of the lens
- This contractile diaphragm with central hole, the Pupil (transmission of light)
- Contains circular muscle fibers (contrict, dilate)

Pupil
- Transmission of light
- Size controlled by parasympathetically/sympathetically controlled muscles

Sphincter Pupillae:

   Decreases the Diameter
   Oculomotor GVE; Edinger-Westphal
   Parasympathetic drive

Dilator Pupillae:

   Increases the Diameter
   Sympathetic drive

Aqueous Humor
- Provide nutrients, immunoglobulins, maintain pressure
- Anterior and Posterior Chambers

Vitreous Humor
- Within the vitreous chamber (posterior cavity)
- Maintains shape of the eye

<u>Inner Tunic</u>	<u>Retina</u>
Outermost	Retinal Pigment Epithelium

- Single layer of cells to reinforce light absorption
- Reduce scattering of light
- The anterior aspect is non-visual

Innermost   Neural Retina (photo-R):

- Photoreception and impulse transmission

<u>Macula Lutea</u>

- Central retina
- Special photo-R cones

<u>Fovea Centralis</u>

- Depression in the Center of Macula
- Cones only
- Most acute vision

<u>Optic disc</u>

- Fibers from retina leave to optic nerve
- Lacks photo-r (blindspot);
- NO RODS/CONES

<u>Central Artery</u>

- Supplies retina except rods and cones
- A branch of the Opthalmic A
- Occlusion leads to blindness

Rods and Cones receive blood supply by a choroid capillary system

Light → Cornea → Aqueous humor → Pupil → Lens → Vitreous humor →Retina

<u>Notes/Diagrams</u>

Muscles

Superior rectus	Elevate Intorsion	**OCULOMOTOR (GSE)** *Oculomotor Nucleus*	**Ophthalmic A**
Inferior rectus	Depress Extorsion	**OCULOMOTOR (GSE)** ***Oculomotor Nucleus***	**Ophthalmic A**
Medial rectus	Adducts	**OCULOMOTOR (GSE)** ***Oculomotor Nucleus***	**Ophthalmic A**
Lateral rectus	Abducts	**ABDUCENS (GSE)** ***Oculomotor Nucleus***	**Ophthalmic A**
Superior Oblique	Intorsion of Upper pole	**TROCHLEAR (GSE)** ***Trochlear Nucleus***	**Ophthalmic A**
Inferior Oblique	Extorts Upper pole	**OCULOMOTOR (GSE)** ***Oculomotor Nucleus***	**Ophthalmic A**
Levator Palpebrae Superioris	Elevate upper eye lid	**OCULOMOTOR (GSE)** ***Oculomotor Nucleus***	**Ophthalmic A**
Orbicularis Oculi	Close eyelid	**FACIAL N**	**Facial and Superficial Temporal A**

Function and Vision to be discussed in detail later

Ear Anatomy

- The Stapes is the smallest bone of the body (grain of rice)
- Inner ear reaches full adult size when fetus is 20–22 wks old
- Hearing loss is the number one disability in the world
- SSA system; Special Sensory Afferent
- Detects sound frequencies from 20 Hz to 20,000 Hz (normal conversation ~300–3000)

Outer ear	
Helix **Antihelix** **Tragus** **Antitragus**	
Auricle	*Elastic Cartilage; Brings sound waves into EAM*
External Auditory Meatus	*Cartilage/Bone; Ceruminous Glands;*
Auditory Canal	*Transmit waves*
Tympanic Membrane	*Take in sound waves (vibration)*
Middle ear **Pharyngeotympanic Tube**	
Ossicles	**Malleus Anvil Stapes**
Stapedius Muscle CN VII *Tensor Tympani Muscle CN V3*	

Inner ear **(Bony Labyrinth)**	**1. Vestibule** **2. Semicircular Canals (3)** **3. Cochlea** • Receives blood supply from labyrinthine artery (Circle of Willis) • Contain Perilymph which suspends Membranous Labyrinth
Inner ear **(Membranous Labyrinth)**	• Sits within the bony labyrinth • Contains Endolymph **Scala Vestibuli of Cochlea** • Upper tube of Cochlea • ~ 2 turns to apex (**Helicotrema**) to communicate with the lower tube (Scala Tympani) • Contains Perilymph • Transmits waves towards the Scala Tympani and Round Window **Scala Tympanic of Cochlea** • Lower tube of Cochlea • Contains Perilymph • Terminates in the base at Round Window **Scala Media of the Cochlea (Cochlear Duct)** • Contains Endolymph • Between the two tubes (scala V and scala T) • **Vestibular Membrane:** Thin roof of Cochlear Duct • **Basilar Membrane:** Ligamentus floor of Cochlear Duct • Contains the Organ of Corti (Receptive Unit for hearing) **Organ of Corti** • Contains hair cells • **Tectorial Membrane:** Gelatinous roof • Rests on and supported by the Basilar Membrane • Analyzes frequencies
Cochlear Nerve **Vestibular Nerve**	

Nose Anatomy

External Nose

- Continous with the Frontal Bone, Nasal Bones, and Maxillary Bone
- Inferior external portion is cartilaginous

Nasal bones	
Lateral process of Septal Nasal Cartilage (paired)	
Septal Cartilage (middle; between lateral processes)	
Major Alar Cartilage (paired; Lateral and Medial Crus)	
Septal Nasal Cartilage	
Anterior Nasal Spine of Maxilla	
Alar Fatty Tissue (as comes to maxilla)	

Nasal Cavity

- Air enters:

Nares (Nostrils)	
Nasal Vestibule	*Vascular epithelium (hair cells);*
Roof	**Nasal, Frontal, and Sphenoid Bones** **Ethmoid Bone (Cribiform Plate for Olfactory N)**
Floor	**Maxilla and Palantine Bones**
Medial Wall (Septum)	**Ethmoid Bone (Perpendicular Plate)** **Vomer** **Septal cartilage**
Lateral Wall	**Ethmoid Bone (Superior/Middle Concha)** **Inferior Concha** **Nasal, Maxilla, Lacrimal, Sphenoid, Palantine Bones**

*The respiratory region is the lower nasal cavity; warms and moistens ai

Of Clinical Significance

Throat Anatomy

Larynx

- Voice/Respiratory Tract
- Level with C3-C6 above the Trachea
- Cartilaginous/Ligamentous/Membranous/Muscular

- Cartilages
  - Single Epiglottis   Thyroid   Cricoid

    Epiglottis

    Thyroid   _Laryngeal Prominence; Adam's Apple_

    Cricoid   _Level C6; end pharynx and Larynx_

  - Paired Arytenoid   Corniculate   Cuneiform

- Ligaments
  - Thyrohyoid membrane _Location for Laryngeal N_
  - Cricothyroid membrane

- External and Internal Muscles
  - External Elevators of the Larynx
    - Thyrohyoid
    - Stylohyoid
    - Mylohyoid
    - Digastrics
    - Stylopharyngeus/Palatopharyngeus

  - External Depressors of the Larynx
    - Omohyoid
    - Sternohyoid
    - Sternothyroid

  - Internal Laryngeal Muscles
    - Modulate sounds upon phonation
    - "vocal muscles"
    - Cricothyroid muscle
    - Cricoarytenoid muscles   _Lateral and Posterior_
    - Transverse Arytenoid muscle
    - Oblique Arytenoid muscle

False Vocal Cords (Vestibular Folds)
True Vocal Fold (Vocal Folds)

- Contain:
  - Vocalis Muscle
  - Vocal Ligament
  - Rima Glottidis

Spaces of the Larynx

- Vestibule
- Ventricles
- Rima Glottidis

<u>Nerves as supplied by the Vagus N</u>

- Superior Laryngeal N   *Through the Thyrohyoid membrane; above vocal folds*
  Provides sensation to mucosa above vocal folds of larynx
- Recurrent Laryngeal N   *Provides mucosa below vocal folds*
  *Muscle so larynx *except cricothyroid muscle*

<u>Arterial Supply</u>

- Superior Laryngeal A   *From Superior Thyroid A from External Carotid A*
- Inferior Laryngeal A   *From Inferior Thyroid A from Thyrocervical Trunk (Subclavian)*

<u>Venous Drainage</u>

Laryngeal V drain into Superior/Inferior Thyroid Veins

Notes/Diagrams

Pharynx

Pharynx

- Fibromuscular tube connecting nasal/oral cavities with larynx/esophagus
- 3 regions of Pharynx
  - **Nasopharynx**
  - **Oropharynx**
  - **Laryngopharynx**   *epiglottis to inferior cricoid cartilage*
- 3 Tonsils
  - **Pharyngeal Tonsil**   *Roof of Nasopharynx*
  - **Palantine Tonsil**
  - **Linguil Tonsil**
- Pharyngeal Muscles

Muscle	Function	Nerve	Artery
Circular Muscles			
Superior Pharyngeal Constrictor	Constrict pharynx when swallow	Vagus N through Pharyngeal Plexus	Ascending pharyngeal A *From External Carotid*
Middle Pharyngeal Constrictor	Contrict pharynx when swallow	Vagus N through Pharyngeal Plexus	Ascending pharyngeal A *From External Carotid*
Inferior Pharyngeal Constrictor	Contrict pharynx when swallow	Vagus N through Pharyngeal Plexus	Ascending pharyngeal A *From External Carotid*
Longitudinal Muscles			
Stylopharyngeus *Vertical muscles in between the Superior and Middle Constrictor*	Elevate pharynx and larynx when swallowing and talking	Glossopharyngeal N	Ascending pharyngeal A
Salpingopharyngeus	Elevate pharynx and larynx when swallowing and talking	Vagus N through Pharyngeal Plexus	Ascending pharyngeal A
Palatopharyngeus	Elevate pharynx and larynx; close nasopharynx	Vagus N through Pharyngeal Plexus	Ascending pharyngeal A

Thyroid Gland

Lateral Lobe
Isthmus
Parathyroid Glands

Putting it All Together

Functional Anatomy

Vision

Photoreceptors

- Chain of 3 neurons that project visual impulses via the **optic nerve** and the **lateral geniculate nucleus of Thalamus to the Visual Cortex occipital lobe**
- **Rods and cones:**
  - First order Neurons
  - Respond to light stimulation
  - Respond to changes in membrane potential NOT action potentials
  - *Glutamine is neurotransmitter (stimulatory)*

Rods 95% of photoreceptors 100 million	Contain **Rhodospin (visual purple); allow rods to capture more light** Sensitive to low – intensity light Night vision (scotopic vision) Can respond to only up to starlight levels of light intensity Produce colorless vision Provide general outlines Response of Rods is slow If lost; night blindness and loss of peripheral vision Fovea centralis of Macula Lutea LACKS RODS
Cones Most of our vision	Contain **Iodopsin** Low sensitivity to light; day vision (photopic) Provide sharp images Color vision Faster response to light than rods Require moonlight or greater light to function Less cones converge than rods thus the brain is able to better pinpoint the stimulation more accurately Fovea centralis of Macula Lutea dense in cones If lost; decrease in visual acuity (legally blind)

Several Rods Converge vs single cones have single fibers

Scotopic:	See by starlight	RODS
Mesoptic:	See in moonlight	RODS AND CONES
Photopic:	Better than moonlight	CONES

Bipolar Neurons:

- Second order neurons
- Relay stimulus from Rods and Cones to the **Ganglion cells**
- *Glutamate NT (stimulatory)*

Ganglion Neurons:

- Third order neurons forming the Optic Nerve
- Na VGC dense to activate a action potential
- Project to hypothalamus, superior colliculus (E-W nucleus), tectospinal nuclei, and **lateral geniculate nucleus** of thalamus
- *Glutamate (NT)*

Interneurons
Horizontal Cells
- Interconnect photoreceptors and bipolar cells
- Inhibit adjacent photoreceptors to help differentiate color
- *GABA-NT (inhibitor)*

Amacrine Cells
- Small cells with no axons
- Receive inputs from bipolar cells and project inhibitory signals to the ganglion cells (GABA)

Divisions of the Retina
Nasal Visual Field: (BINOCULAR): Both eyes
Left and Right Temporal Fields (MONOCULAR): One eye
Before we start: Remember **THE RETINAL IMAGE IS INVERTED AND REVERSED**
By reversing the image:
- Are able to visualize objects much larger than our eye (condense by crossing)
- Without this reversal, we would have a very limited view of our world (drinking straw)

Vertical	
1. Temporal Hemiretina	Receives image input from the nasal visual field Ganglion cells that project to the IPSILATERAL Lateral Geniculate Nucleus
2. Nasal Hemiretina	Receives image input from the temporal visual field Ganglion cells that project to the CONTRALATERAL Lateral Geniculate Nucleus
3. Upper Retinal Quadrant	Receive image from the lower visual fields Ganglion cells project via the LGN to the upper Calcarine Fissure
4. Lower Retinal Quadrant	Receive image from the upper visual fields Ganglion cells project via the LGN to the lower Calcarine Fissure

Visual Pathway
Photoreceptors → Bipolar Neurons → Retinal Ganglion Cells → Optic N → Optic Chiasm → Optic tract → LBN → Geniculocalcarine Tract → Primary Visual Cortex of Occipital Lobe

1) Ganglion Cells Project
- Nasal Hemiretina to Contralateral LGN
- Temporal Hemiretina to Ipsilateral LGN

2) Optic Nerve (CN II)
- Blood supply from the central retinal A
- Incapable of regeneration

3) Optic Chiasm
- Decussating fibers of the 2 nasal hemiretinae
- Non Decussating fibers of the 2 temporal hemiretinae
- Blood supply from Circle fo Willis; anterior cerebral and internal carotid

4) Optic Tract
- From Chiasm to LGN
- Contains Pupillary reflex fibers
- Blood supply from the Circle of Willis; Posterior Communicating Artery

5) Lateral Geniculate Nucleus of Thalamus
- Thalamic relay station for vision
- Has several layers receiving specific messages from specific hemiretinae
- Projects via the <u>visual radiation (geniculocalcarine tract)</u>
- Blood supply from Circle of Willis; posterior cerebral artery

6) Geniculocalcarine Tract (Optic Radiation)
- From the LGN to the Visual Cortex
- Blood supply via the Circle of Willis; Middle cerebral A, Posterior Cerebral A

7) Visual Cortex
- Banks of the Calcarine Sulcus
- Receives messages from the LGN (ipsilateral)
- Blood supply from Calcarine A (a branch of posterior cerebral A) and Middle cerebral A

Pupillary Light Reflex

Afferent input via the Optic N
Efferent output via the GVE of the Oculomotor N
Results in pupil constriction when light is on eye
- Light → Ganglion Cells→
- Optic N→
- Optic Chiasm→
- Ipsilateral Optic Tract→
- Edinger-Westphal Nucleus→
- Synapse with Ipsi/Contra Preganglionic Fiber of Oculomotor N→
- Ciliary Ganglion→
- Post Ganglionic Fiber to Sphincter Pupillae Muscle

Deficits After Lesions in Visual Pathways

Deficits After Lesions in Visual Pathways

Auditory Pathway

1. Air sound waves enter and reach the Tympanic Membrane
2. Tympanic membrane vibrates
3. Cause the Ossicle to vibrate
4. Stapes sends vibration to the Oval Window
5. The vibration at the oval window creates waves within the fluid of the scala tympani
6. Leads to vibration in the Basilar membrane
7. Hairs extending to the Tectorial membrane bend
8. Action potentials are generated and conducted through the spiral ganglion (SSA of cochlear N); leaves the temporal bone via the internal acoustic meatus
9. SSA of Cochlear N arrive at Dorsal and Ventral Cochlear Nuclei of the medullapontine junction

Olfaction

Olfactory Receptor Neuron Cell Bodies
   a. Located in the nasal mucosa (Olfactory Epithelium) at the roof of the nasal cavity
   b. Olfactory Cilia of these cells extend into the nasal cavity through the Mucous Layer of the Olfactory Epithelium

1. Stimulus via and Odorant
2. Odorant molecules arrive at the Cilia membrane
3. Bind to a binding protein
4. Stimulates the Opening of Na VGC and thus depolarization along the Olfactory Receptor Neurons
5. The axons of the receptor neurons will pass through the Foramina in the Cribiform Plate of Ethmoid
6. Fibers then enter the ipsilateral Olfactory Bulb in the Anterior Cranial Fossa
7. Fibers synapse with Mitral Cells in the Olfactory Bulb
8. The Mitral Cells axons will form the Olfactory Tract
   a. Lateral Olfactory Tract
      i. Largest bundle of axons exit olfactory bulb in this tract
      ii. Axons project to the
         1. Primary Olfactory Cortex (Pyriform Cortex)→ Prefrontal Cortex
         2. Amygdala→ Prefrontal Cortex
         3. Entorhinal Cortex (hippocampus, uncus)→ Prefrontal Cortex
   b. Medial Olfactory Tract
      i. Project to the hippocampus, hypothalamus, and brainstem

***NOTE: Unlike the other senses…no synapse occurs in the THALAMUS prior to cortex

*Sustentacular cells
- Supporting cells
- Help detoxify chemicals that come in contact with the olfactory epithelium

Taste

- Stimulated by chemical molecules
- Most sensitive areas of the tongue are
  - Tip of tongue for sweetness
  - Back of tongue for bitterness
  - Sides of tongue for saltiness and sourness

Gustatory Receptor Neurons
- Receptor cells are located in the taste buds
  - Taste buds are located in various Papillae which protrude from the tongue
  - Each taste bud has a pore at its tip so stimuli can enter
  - Taste receptor cells live for about 10 days and then have to be replaced
  - Special Afferent Neurons contact make contact with the gustatory receptor cells
  - Cell bodies for these afferents are located in the ganglion of
    - VII-Geniculate Ganglion→Solitary Nucleus
    - IX- Inferior Ganglion→ Solitary Nucleus
    - X- Solitary Nucleus

1. Salivary Fluids with chemical substances enter taste buds through the pore
2. Soak in the Microvilli which are at the head of the Taste Receptor Cell within the Pore
3. Specific receptors and their channels will open and close depending on the type of stimulus (Salty, sour.etc)
4. The end result is Depolarization of the Gustatory Afferent Neurons (Special Sensory; SVA/SSA)
   a. Anterior Taste Buds are innervated by the Facial Nerve (sweet, sour, salty)
   b. Posterior Taste Buds are innervated by the Glossopharyngeal N (sour and bitter)
   c. Epiglottis and Palate walls are innervated by the Vagus N
   d. Each of these will project into the SOLITARY NUCLEUS
5. Axons from solitary nucleus ascend via the SOLITARIO-THALAMIC TRACT
6. Terminate in the Ventral Posteromedial Nucleus of the Thalamus
7. The VPM then send their projections to the Gustatory Cortex

Swallowing (Deglutition)

Putting It All Together

Putting It All Together

<u>Putting It All Together</u>

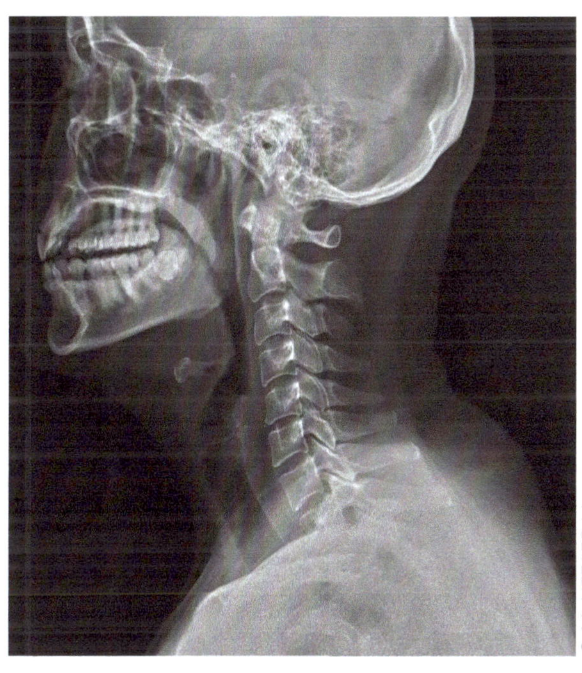

© M_LIZZARD/shutterstock.com

Neck, Face, Mastication

NECK

2 Major Triangles of the Neck

1. Posterior Triangle
   a. Boundaries
      i. Anterior   **Sternocliedomastoid**
      ii. Posterior   **Trapezius**
      iii. Inferior   Clavicle
      iv. Floor   Splenius Capitus, Levator Scapulae
            Anterior, Middle, Posterior Scalene Muscles
            Omohyoid
   b. Contains
      i. Accessory N
      ii. External jugular V
      iii. Occipital A, Subclavian A, Transverse Cervical A, Suprascapular A, Dorsal Scapular A
      iv. Occipital N, Dorsal Scapular, Suprascapular, Long Thoracic N, Brachial Plexus Roots, Trunks

    c. Triangles within the Triangle
       i. **Supraclavicular Triangle**
         1. Omohyoid
         2. Sternocliedomastoid
         3. Clavicle
         4. *N to subclavius crosses*
         5. *Lymph Nodes*
         6. *Pulse of the Subclavian A*
      ii. **Occipital Triangle**
         1. Trapezius
         2. Sternocleidomastoid
         3. Omohyoid

2. <u>Anterior Triangle</u>
    a. Boundaries
       i. Posterior   **Sternocliedomastoid**
      ii. Superior   Mandible
      iii. Anterior   Midline of neck
    b. Divided into additional triangles
       i. <u>Submandibular Triangle</u>
         1. Glandular
         2. Formed by: Mandible → Digastric Muscles
         3. Houses:
            a. Submandibular Gland
            b. Submandiblar Lymph Nodes
            c. Facial A
            d. **Stylohyoid Muscle**
            e. **Mylohyoid Muscle**
            f. **Hypoglossus Muscles**
      ii. <u>Submental Triangle</u>
        □ Lymph and Veins
        □ Boundaries
          • **Anterior Digastric Muscle**
          • Hyoid
          • Midline
        □ Floor
          • **Mylohyoid Muscle**
          • Genohyoid (deep)
        □ Contents
          • Submental Lymph Nodes

iii. <u>Carotid Triangle</u>
- ▫ Vascular Area
- ▫ Boundaries
  - **Posterior Digastric**
  - **Sternocleidomastoid**
  - **Omohyoid**
- ▫ Contents:
  - Internal Jugular V
  - Common Carotid/Carotid Sinus
  - Vagus N
  - Hypoglossal N
  - Ansa Cervicalis
  - Cervical Lymph Nodes
  - Spinal Accessory N

iv. Muscular triangle
- ▫ Boundaries
  - **Omohyoid**
  - Sternocleidomasotid
  - Midline
- ▫ Contents
  - **Sternohyoid**
  - Thyrohyoid
  - **Sternothyroid**

<u>Drawings</u>

<u>Drawings</u>

<u>Drawings</u>

Cervical Plexus
- Ventral Rami of C1-C4
- Innervates most neck muscles
- Cranial Nerves will also innervate choice neck muscles
- Posterior to Jugular V

C1	With Hypoglossal N to Geniohyoid Thyrohyoid	*Supraclavicular* C3, C4	*Sensory*
Ansa Cervicalis C1-C3 (loop)	Infrahyoid Muscles	Phrenic C3-C5	Motor to diaphragm
Lesser Occipital N C2	*Sensory to neck*		
Great Auricular N C2, C3	*Sensory parotid/ear*		
Transverse Cervical C2, C3	*Sensory anterior neck*		

Arterial From External Carotid A

Superior thyroid (heading to thyroid)	Thyroid Gland, Larynx, Infrahyoid Muscles
Ascending Pharyngeal (heading to digastric)	Pharyngeal region, middle ear, meninges, prevertebral muscles
Linguil (Heading to hypoglossal muscle)	Tongue
Facial (Heading under mandible)	Face
Occipital (Heading towards mastoid process)	Sternocliedomastoid
Posterior Auricular (Heading towards external meatus)	Posterior ear
Maxillary (heads behind upper Mandible)	Maxillary region
Superficial temporal (heads to temporal region)	Face, Temporalis, Scalp

Veins to Internal Jugular V
1. Sigmoid Sinus
2. Occipital (from posterior)
3. Pharyngeal (from pharynx (anterior))
4. Facial (from mandible)
5. Linguil (from under mandible)
6. Superior Thyroid

<u>Notes/Diagrams</u>

Structures at the Root of the Neck

Most of previous Modules

Musculature
Anterior Scalene
Middle Scalene
Posterior Scalene
Levator Scapulae
Posterior Omohyoid

Arteries
Brachiocephalic A

Common Carotid A   to   Internal Carotid A   External Carotid A

Subclavian A

      Vertebral A

      Thyrocervical Trunk

            Inferior Thyroid A

            Ascending Cervical A

            Transverse Cervical A

            Suprascapular A

            Internal Thoracic A

            Dorsal Scapular A

Veins
Subclavian V
Internal/External Jugular V
Brachiocephalic V

Nerves
Vagus N
      Recurrent Laryngeal N
      Superior Laryngeal N

Phrenic N

<u>Drawings:</u>

<u>Pre-Vertebral Neck</u>

<u>Muscles</u>
Longus Colli
Longus Capitis
Rectus Capitis Anterior
Rectus Capitis Lateralis

<u>Nerves within Pre-vert. Neck</u>
Facial
Glossopharyngeal
Vagus
Spinal Accessory
Hypoglossal
Ansa Cervical
Cervical Ganglia

<u>Vessels within Pre-vert Neck</u>
Carotid A
Vertebral A
Internal Jugular V

<u>Notes:</u>

<u>Doodles</u>

Summary of Muscles

M	O	I	F	N	A
Platysma	Fascia of Pect. Major and Deltoid	Inferior Mandible Facial Skin	Frown Tense the skin of the neck	Facial N *Cervical Branch* **SVE/BE**	Facial A *From External carotid*
Sternocliedomastoid	Manubrium of Sternum Clavicle	Mastoid Process Superior Nuchal Lin	Bilateral Ext. Ipsilateral Lat. Flex Contralateral rotation	Spinal Accessory N Ventral Rami C2-C3	Transverse Cervical *Thyrocervical Trunk of Subclavian A*
Trapezius	Superior Nuchal Line Nuchal Ligament Spines C7-T12	Lateral Clav. Acromion Spine of Scapula	ST Elevation ST Retraction ST Depression Neck: Lateral Flex	Spinal Accessory N	Suprascapular Transverse Cervical *Thyrocervical Trunk of Subclavian A*
Posterior Scalene	Cervical Transverse	Rib 2	Neck Ipsilateral lateral flexion Elevate rib	Ventral Rami C6-C8	Vertebral A *Subclavian A*
Middle Scalene	Cervical Transverse	Rib 1	Neck Ipsilateral lateral flexion Elevate rib	Ventral Rami C3-C8	Vertebral A *Subclavian A*
Anterior Scalene	Cervical Transverse	Rib 1	Neck Ipsilateral Lateral Flexion Elevate Rib	Ventral Rami C5-C7	Vertebral A *Subclavian A*
Suprahypid					
Mylohyoid	Mandible	Hyoid Bone	Elevate Hyoid Elevate floor of mouth Elevate tongue (speech/swallow)	Mandibular Branch of Trigeminal (V) *SVE/BE; Motor Nucleus of V*	Facial A *External Carotid A*
Geniohyoid (deep)	Mandible	Hyoid Bone	Pulls hyoid up and out; Shortens floor of mouth Widens pharynx	Ventral Rami C1 *GSE*	Facial A Linguil A *External Carotid*
Stylohyoid	Styloid Process of Temporal	Hyoid	Elevate and Retract Hyoid	Facial N *SVE/BE; Facial Nuclei*	Facial A *External Carotid*
Digastric Anterior Posterior	Mandible Mastoid Process to Hyoid	Hyoid	Depress mandible Elevate and Stabilize Hyoid	Mandibular Branch of CN V *SVE/BE; Motor Nucleus of V* Facial N *SVE/BE; Facial Nuclei*	Facial A *External Carotid*
Infrahyoid					
Sternohyoid	Sternum and Clavicle	Hyoid	Depress Hyoid	Ansa Cervicalis	Superior Thyroid A *External carotid*
Omohyoid	Scapula	Hyoid	Depress Hyoid Stabilize Hyoid	Ansa Cervicalis	Superior Thryoid A *External carotid*
Sternothyroid	Posterior Sternum to Thyroid Cartilage	Thyroid Cartilage	Depress Hyoid and Larnyx	Ansa Cervicalis	Superior Thryoid A *External carotid*

Thryrohyoid	Thyroid Cartilage	Hyoid	Depress Hyoid Elevate Larynx	Ventral Rami C1	Superior Thryoid A *External carotid*
Prevertebral Muscles	Mixed O to I		F	N	A
Longus Colli	Anterior Cervical Vertebrae Bodies and Transverse Processes		Neck Flexion Neck Contralateral Rotation	Ventral Rami C2-C6	Vertebral A *Subclavian A*
Longus Capitus	Anterior Cervical vertebrae to Anterior Occipital Bone		Neck Flexion	Ventral Rami C1-C3	Vertebral A *Subclavian A*
Rectus Capitis Anterior	Anterior Atlas to Anterior Occipital Bone		Neck Flexion	Ventral Rami C1-C2	Vertebral A *Subclavian A*
Rectus Capitis Lateralis	Atlas to Lateral Occipital Bone		Neck Flexion Head Stabilization	Ventral Rami C1-C2	Vertebral A *Subclavian A*
Sub-Occipital Muscles					
Rectus Capitus Posterior Major			Extension of Head Rotate same side	Suboccipital N fom Dorsal Rami C1	Vertebral A Occipital A
Rectus Capitus Posterior Minor			Extension of Head	Suboccipital N fom Dorsal Rami C1	Vertebral A Occipital A
Superior Oblique Capitus			Extension of Head Lateral Flexion	Suboccipital N fom Dorsal Rami C1	Vertebral A Occipital A
Inferior Oblique Capitus			Rotation	Suboccipital N fom Dorsal Rami C1	Vertebral A Occipital A

A MOMENT TO BREATHE!!

FACE

In General

Facial Expression Muscles

- Motor Innervation by the **Facial Nerve (SVE/BE)**
  - Facial Motor Nuclei in Pons → Through Internal Acoustic Meatus → Geniculate Ganglion → Stylomastoid Foramen → Muscles of Facial Expression
  - **Branches of the Facial N to ID in Face**
    - Exit Stylomastoid Foramen
    - Under Parotid Gland
    - **Temporal Branches**
    - **Zygomatic Branches**
    - **Buccal Branches**
    - **Mandibular Branch**
    - **Cervical Branch**
- Sensory innervations by the **Trigeminal N (3 divisions; GSA) and Cervical Plexus**
  - **All receive sensory information from the region for which they are located**
  - **Opthlamic Division V1**
    - Through the Superior Orbital Fissure
  - **Maxillary Division V2**
    - Through the Foramen Rotundum
  - **Mandibular Division V3**
    - Through Foramen Ovale
- Receive blood supply from the External Carotid A and Internal Carotid A
  - **Facial A**   *External carotid A*
  - **Superior Temporal A**   *External carotid A*
  - **Opthalmic A**   *Internal carotid A*
- Venous drainage from Facial V, Mandibular V, Opthalmic V which drains to Internal Jugular V

Summary of Facial Expression Muscles

M	Landmarks	F
Frontalis	Forehead Attach to aponeurosis	Elevate eyebrow and forehead Wrinkle forehead
Corrugator	Medial superior orbit to Skin of medial eyebrow	Pull eyebrow down and medial
Orbicularis Oculi	Surrounds eye	Close eyelid
Procerus	Between Eyes	Wrinkles bridge of nose Anger expression
Nasalis	Paired on side of nose	Flare nostrils
Orbicularis Oris	Surround mouth	Compress lips against teeth Protrude lips
Levator Labii Superioris	Superior lip to skin of nose	Raise upper lip Flare Nose
Levator Anguli Oris	Superior lip to maxilla	Smile
Zygomaticus Major/Minor	Oris to Zygomatic	Mouth upward and back
Risorius	Lateral Oris to Mandible	Mouth laterally
Depressor Labii Inferioris Depressor Anguli Oris	Lower Oris ro Mandible	Depress lower lip-draw mouth laterally Depress angle of mouth
Mentalis	Lower Oris to Chin	Raise and Protrude lower lip Wrinkle chin

Buccinator	Mandible to Angle of Mouth	Compresses cheek against teeth; mastication Suck Blow
Platysma	Pectoral region to lower mouth	Tense skin of neck

MASTICATION

In general

Temporomandibular Joint

- Condylar process of the mandible to the Mandibular fossa of the temporal bone
- Synovial modified hinge joint
  - Has a Articular Disc
- Ligaments
  - Stylomandibular Ligament
  - Sphenomandibular Ligament

Of Clinical Significance

Salivary Glands

Submandibular gland   Sublingual gland   Parotid gland

Muscles of Mastication

- Mandibular Division of the Trigeminal N
- Maxillary A and Superficial Temporal A of the External Caratid A
    - **i. Temporalis Muscle**   *Elevate Mandible; Posterior aspect Retrudes Mandible*
    - **ii. Masseter Muscle**   *Elevate and Protrude Mandible; Deep fibers Retrude Mand.*
    - **iii. Lateral Pterygoid**   *Side to Side Movements; Protrude Mandible; Grind*
    - **iv. Medial Pterygoid**   *Side to Side Movements; Elevate, Protrude Mandible, Grind*

Putting It All Together

Putting It All Together

Thorax

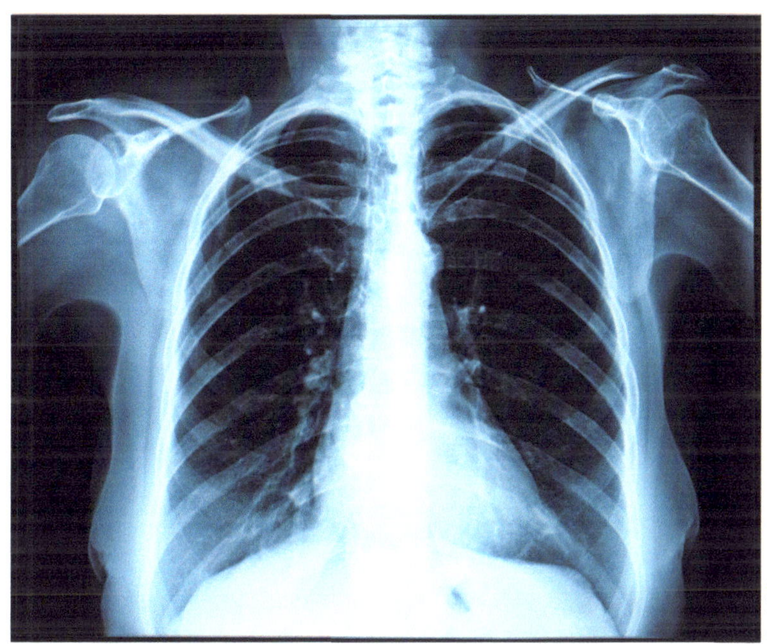

Thorax
Breast
Mediastinum

Thorax

Bones of the Thorax

12 pairs of ribs

True 1–7	direct to sternum
False 8–10	indirect to sternum
Floating 11, 12 end in posterior musculature	

Head, neck and body of rib
Tubercle and angle of rub

Costal cartilage

12 thoracic vertebrae with costal facets (superior, inferior, and transverse)

Intervertebral discs

Sternum
Jugular Notch
Manubrium

Sternal Angle
Body
Infrasternal Angle
Xiphoid Process

Intercostal Spaces (muscle, vessels, and nerve)

<u>Notes:</u>

Surface Anatomy

<u>*Key Landmarks:*</u>
Jugular Notch
Sternal angle (of Louis)
Nipple
Xiphiod porcess

<u>*Lines:*</u>
1. Median line (midsternal line)
   a. *Longitudinally down sternum*
2. Mid-clavicular line
   a. *Longitudinally through mid-clavicle; parallel to median*
3. Anterior axillary line
   a. *Longitudinally at anterior axillary fold;*
   b. *Border of pectoralis major*
4. Mid-axillary line
   a. *Mid-axilla*
5. Posterior axillary line
       Borders by latissimus dorsi
6. Scapular line   *Run vertically through inferior angles of scapula*
7. Mid-verterbal line

Arteries of the Thoracic Wall

Derived from:
1. **Thoracic Aorta** →
   a. <u>Posterior Intercostal Arteries</u>
   b. <u>Subcostal Artery</u>
2. **Subclavian Artery** →
   a. <u>Internal Thoracic Artery</u> → <u>Anterior Intercostal Arteries</u>

<u>Notes/Diagram:</u>

Veins of the Thoracic Wall

- Accompany the intercostals arteries and nerves
- Most superior in the bundle within intercostal space
- 11 posterior intercostals veins (bilateral)
- 1 subcostal vein (bilateral)

1. **Anterior intercostals veins** → Internal Thoracic Vein
2. **Posterior intercostals veins** → Azygous Vein (right side) and Hemiazygous Vein (left)

Azygous Vein:
- *Runs up the RIGHT side of the vertebrae*
- *Receives blood from right posterior walls of the thorax and abdomen to deliver to superior vena cava*
- *No symmetry with the vein on the left side*

Hemiazygous Vein:
- *Runs up the LEFT side of the vertebrae*
- *Receives blood from the left posterior walls and abdomen*
- *At ~9th thoracic vertebrae it will run across the vertebrae to the right (behind aorta) and end in azygous vein*

Nerves of Thoracic Wall

- 12 pairs of thoracic spinal nerve → anterior/posterior rami
- Anterior rami T1-T11 form <u>intercostals nerves</u>
- <u>Rami Communicantes</u> help the anterior rami to *communicate* with the <u>sympathetic trunk</u>

Muscles of the Thoracic Wall

- Review functions, O, I, N, Arterial Blood Supply for
  - Latissimus Dorsi
  - Serratus Posterior Sup/Inf
  - Serratus Anterior
  - Pectoralis Major/Minor
  - Erector Spinae (Iliocostalis, Longissimus, Spinalis)
  - Scalenes

Muscle	O	I	F	N	A
External Intercostal *Always closest to the vertebrae*	Inferior border of rib above	Superior border of rib below	Elevate rib cage Inspiration	Intercostal N	Anterior/Posterior ntercostal A
Internal Intercostals *Always closest to the sternum*	Costal cartilage of rib below	Costal cartilage of rib above	Depress rib cage Forced expiration	Intercostal N	Anterior/Posterior Intercostal A
Innermost Intercostal	Internal aspect of inferior rib	Internal aspect of superior rib	Assist inspiration and expiration Elevates and depresses rib	Intercostal N	Posterior Intercostal
Subcostalis *Seen in vertebral view* **	Inner angles of rib	Superior border of rib below	Depress ribs (aid)	Intercostal N	Posterior Intercostal Musculophrenic
Transverse Thoracis *Seen in posterior rib cage*	Posterior sternum and xiphoid process	Inner costal carti-lage of True	Depress costal cartilages; expiration	Intercostal N	Anterior Intercostals Internal Thoracic A

Breast Tissue

Anatomy checklist:

Nipple
Areola
Suspensory Ligaments (Cooper's Ligaments)
- *attach the mammary glands to the dermis*
- *help support the mammary gland lobules*

Axillary Tail
- *inferior lateral edge of pectoralis major to the axillary region*

Retromammary Space
- *Between the breast and the deep pectoral fascia*
- *Contains fat to allow movement*

<u>Vasculature of the Breast Tissue</u>

Subclavian A → Internal Thoracic A →
   <u>Medial Mammary A</u>
   <u>Anterior Intercostal A</u>

Subclavian A/Axillary A → Lateral Thoracic A → <u>Lateral Mammary A</u>

Thoracoacromial Trunk → <u>Pectoral A</u>

Lymphatic Drainage of Breast

Important due to its role in the metastasis of cancer cells
Lymph will pass from the nipple, areola, and the mammary gland lobules to the Areolar Lymphatic Plexus

1. Areolar Plexus
    a. MEDIAL Quadrants of the breast drain into
        i. Parasternal Lymph Nodes (bilateral) to
        ii. Bronchomediastinal Lymphatic Trunk to
        iii. Subclavian Lymphatic Trunk to
        iv. Lymphatic Duct to
        v. Brachiocephalic Vein
    b. INFERIOR Quadrants of the breast drain into abdominal lymph system
    c. LATERAL Quadrants of the breast drain into
        i. Axillary Lymph Nodes (pectoral, humeral, subscapular, central, and apical)
            1. Pectoral   *lateral border of pectoralis minor*
            2. Humeral   *Pectoralis major as it approaches the humerus*
            3. Central   *At Glenoid fossa area of scapula*
            4. Apical   *medial border of the pectoralis minor*
        ii. From Axillary Nodes to
        iii. Supra and Infra clavicular nodes to
        iv. Subclavian Lymphatic Trunk to
        v. Lymphatic trunk to
        vi. Brachiocephalic Vein
        vii. **75% of lymph

<u>Drawings</u>

THORACIC CAVITY

1. **Pulmonary Cavities** contain lungs and pleurae
2. **Mediastinum** contains heart, great vessels, trachea, esophagus, lymph nodes

Pleura

Each lung is enclosed in a *Pleural Sac*

Parietal Pleura	Lines the pulmonary cavities Adheres to the thoracic wall, mediastinum, and diaphragm Consists of a: Cervical pleura: Forming over the apex of the lung; extending mid trachea Costal pleura: covers the internal surfaces of the sternum, ribs, intercostals muscles, and lateral thoracic vertebrae Mediastinal pleura: Central/medial aspect of thoracic cavity Diaphragmatic pleura: Covers the superior surface of diaphragm on each side of mediastinum
Visceral Pleura	Covers the lungs Adherent to all its surfaces including within fissures Cannot be dissected from the lungs

Lung

Right Lung

Contains 3 lobes
   Superior
   Middle
   Inferior

Fissures
   Horizantal Fissure: *Superior and Middle of Rt Lobe*
   Oblique Fissure: *Separates the Inferior Lobe of Rt lobe*

Hilum
   Bronchus
   Rt Pulmonary A
   Rt Pulmonary V
   Lymph

Left Lung

Contains 2 lobes
   Superior
   Inferior

Fissure
   Oblique Fissure: *Superior and Infeiror of Left Lobe*

Hilum
   Bronchus
   Left Pulmonary A
   Left Pulmonary V
   Lymph

   Lingula: *Lower medial projection of the superior lobe (middle lobe)*
   Cardiac Notch: *Just above lingual*

Bronchi

Trachea will divide (bifurcate) at the **Bifurcation of the Trachea (Carina)** at the sternal angle

Tracheal Bifurcation (carina) → **Rt and Left Primary Bronchi**

Right is
   wider and shorter
   more vertical
   passes directly to the hilum of the right lung

Left is
   passes inferiolateral, inferior to the aortic arch, and anterior to the esophagus to reach hilum of left lung

Will then branch into **Secondary (lobar) bronchi**   3 on right   2 on the left

Secondary Bronchi become a **Tertiary Bronchiole** → Conducting Bronchiole → Respiratory Bronchioles → Aveoli

Diaphragm

- Aperatures:

Aperature	Level	Passing through
Caval Opening	*T8-T9*	*Inferior vena cava*
Esophageal Hiatus	*T10*	Esophagus Vagal trunks
Aortic Hiatus	*T12*	*Abdominal aorta* *Thoracic duct* *Splanchic nerves*

- Vasculature of Diaphragm

Superior Phrenic A	*From the Thoracic Aorta* *Pass to the posterior surface of the diaphragm* *Will merge with the musculophrenic and pericardiophrenic A*
Inferior Phrenic A	*Variable in origin* *Superior to celiac trunk* *Right Phrenic: Passes behind the inferior vena cava* *Left Phrenic: Passes behind the esophagus*
Pericardiophrenic A	*Off Internal thoracic* *Long and Slender; moves with phrenic nerve*
Musculophrenic A	*Off internal thoracic* *Behind costal cartilage of false ribs*

- Innervation
  - Phrenic nerve
    - Motor, sensory, autonomic

Diaphragm will contract during inspiration moving the diaphragm down thus increasing the thoracic cavity size, decreasing intra-thoracic pressure

When relaxed there is a recoil lending to the diaphragm moving up thus decreasing thoracic cavity size once again- Expiration

MEDIASTINUM

- *Found in between the pulmonary cavities*

Superior Mediastinum	Superior boundary is the first rib Inferior boundary is the sternal angle to intervertebral disc T4-5 Contains: SVC, Brachiocephalic veins, aortic arch, thoracic duct, trachea, esophagus, vagus nerve, left recurrent laryngeal nerve, phrenic nerve, thymus
Anterior Mediastinum	Anterior to the pericardium Posterior to sternum and transverse thoracic Contains lymph nodes, fat, and connective tissue
Middle Mediastinum	Between right and left pulmonary cavities Contains heart, pericardium, phrenic nerve, great vessels, arch of azygos vein, main bronchi
Posterior Mediastinum	Posterior to pericardium Contains esophagus, thoracic aorta, azygos and hemiazygos, thoracic duct, vagus nerves, sympathetic trunks, splanchnic nerves

Pericardium

- Fibroserous sac that encloses the heart and roots of the great vessels
- Posterior to the sternum and 2-6 costal cartilage (T5-T8)

Fibrous Pericardium	External layer Strong, dense fibrous tissue Protects heart against sudden overfilling
Parietal Pericardium	Internal surface of the fibrous layer Serous membrane Overlies the visceral layer
Visceral Pericardium	Forms the epicardium which is the outer layer of the heart wall
Pericardial Cavity	Space between the parietal and visceral layers of serous pericardium Thin film of serous fluid that allows the heart to contract free of friction

- Vasculature of the Pericardium
  - **Pericardiophrenic** A (a branch of Internal Thoracic A)
- Innervation to the Pericardium
  - **Phrenic Nerve**
    - Sensory information of pain
    - Referred pain to the ipsilateral superior shoulder
    - Fibrous and Parietal Pericardium
  - **Vagus N**
    - Parasympathetic fibers
    - Visceral Pericardium
  - **Sympathetic Trunks**
    - Visceral Pericardium

HEART

Layers of the Chambers

Epicardium	*Outermost, visceral serous pericardium*
Myocardium	*Cardiac Muscle*
Endocardium	*Lines the chambers of the heart and valves*

Functions of the Anatomy

ATRIAL STRUCTURES	
Superior Vena Cava	Bring upper body deoxygenated blood Opens to the RA at the 3rd costal cartilage
Inferior Vena Cava	Bring lower body deoxygenated blood Opens to the RA at the 5th costal cartilage level
Auricles	Muscular pouch projecting from atria Increase the capacity of atria
Pectinate Muscles	Peel the auricle back Stretch to improve capacity of the atria
Tricuspid Valve	Valve between the Right Atrium and Right Ventricle Opens when pressure of the right atrium is greater than the right ventricle
Bicuspid Valve	Valve between the Left Atrium and Left Ventricle Opens when pressure of the left atrium is greater than the left ventricle
Coronary Sinus	Collecting blood from the outer coronary sinus; venous blood to right atrim
Fossa Ovalis	Closed upon birth; allowed bypass of lungs before birth as the lungs are then underdeveloped
VENTRICLE STRUCTURES	
Chordae Tendinae	Think of cords of a parachute Papillary muscles to the valvular cusps (tricuspid=3 cusps)
Papillary Muscles	Attached the to the ventricular wall Contract → tighten the chordate tendinae → opening of the valve 3 papillary muscles (ant, posterior, and septal)
Trabeculae Carnae	Irregular muscle projections Much like papillary muscles; pull on the chordate tendinae for opening valve
Interventricular Septum	Muscular Wall between the ventricles Houses the Bundle Branches
Pulmonary Semilunar Valve	Valve between the Right Ventricle and the Pulmonary Trunk
Aortic Semilunar Valve	Valve between the Left Ventricle and the Aorta
Ligamentum Arteriosum	Connects pulmonary trunk and aortic arch; *Fixes the aorta at deceleration; risk of ruptured aorta* *Recurrent laryngeal nerve of the left vagus n will loop around the aortic arch posterior to the Lig. Art.*

*Aortic Arch Pulmonary Trunk Pulmonary Arteries Pulmonary Veins

ARTERIES of the Heart

<u>RIGHT CORONARY ARTERY</u>:
Right ascending aorta → Right Coronary Artery (RCA) → through Coronary Sulcus →
Serving
- Right atrium through small atrial branches
- SA/AV nodes
- Posterior Interventricular Septum

<u>RIGHT MARGINAL ARTERY</u>:
RCA → Passes to inferior aspect and apex →
Serving:
- Right ventricle
- Apex of heart

<u>POSTERIOR INTERVENTRICULAR ARTERY</u>:
RCA (67%) → to posterior apex of heart
Serving:
- Right and left ventricles
- Posterior Interventricular septum

<u>LEFT CORONARY ARTERY</u>:
Left ascending aorta → LCA →
Serving:
- Much of LA and LV
- AV bundles

<u>ANTERIOR INTERVENTRICULAR ARTERY</u>
LCA → through anterior interventricular groove to apex
Serving:
- Rt and L ventricles
- Anterior interventricular septum

<u>CIRCUMFLEX ARTERY</u> *(runs on the lower border of the left auricle)*
LCA → Left in AV sulcus to the posterior aspect of the heart
Serving:
- Left atrium and left ventricle
Will form anastomoses with the RCA posteriorly
<u>LEFT MARGINAL ARTERY</u>
Circumflex artery → Left border of heart →
Serving:
- Left ventricle

Venous Drainage

Middle Cardiac, Great Cardiac, and Small Cardiac to Coronary Sinus to Rt. Atrium

Drawing Space

Drawing Space

Putting It All Together

Putting It All Together

Abdomen

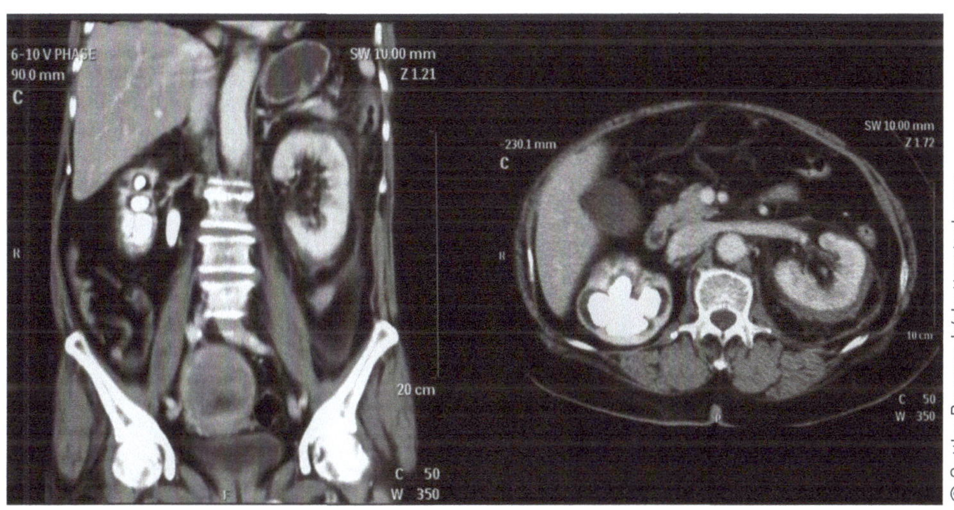

Abdomen
Muscles
Blood Supply
Innervation
GI Viscera

Abdomen

- Skeletal muscle
- Abdominal cavity
- Viscera

Quadrants of the Anterior Abdominal Wall

Defined by 2 Planes:

1. Transumbilical Plane:
   a. Passes horizontally through the umbilicus
2. Median Plane:
   a. Passes vertically along Median Line

RIGHT UPPER QUADRANT	Liver (rt lobe) Gallbladder Pylorus of stomach Duodenum Pancreas (head) Right adrenal medulla, kidney, hepatic flexure Ascending Colon (superior) Transverse Colon (right)
LEFT UPPER QUADRANT	Liver (left lobe) Spleen Stomach Jejunum, proximal Ileum Pancreas (body, tail) Left kidney, left adrenal medulla, left splenic flexure Transverse colon(left) Descending colon (superior)
RIGHT LOWER QUADRANT	Cecum Appendix Ileum Ascending colon (inferior) Rt ovary, ureter, spermatic cord, uterus (if pregnant) Urinary bladder when full
LEFT LOWER QUADRANT	Sigmoid colon Descending colon (inferior) Left ovary, left ureter, spermatic cord Uterus (if enlarged) Urinary bladder (if very full)

9 Regions of Abdominal Cavity

- Used clinically to locate abdominal organ and pain sites
- Divided by 4 planes
  - **Horizantal Planes:**
    - Subcostal Plane
      - Pass horizontally through the inferior border of 10th costal cartilage (bilateral)
    - Transtubecular Plane
      - Pass horizontally through the iliac tubercles (level of L5)
  - **Vertical Planes**
    - Mid-Clavicular Plane (rt and left)
      - Pass vertically through mid clavicles to mid-inguinal

Rt Hypochondriac Region	Pain in region may be due to lesions in biliary system; right lobe of liver, gall bladder, hepatic flexure of colon
Epigastric Region	Pain in region may indicate lesion to lower esophagus, stomach, duodenum, left liver lobe, biliary system, pancreas, transverse colon
Left Hypochondriac Region	Pain in region may be due to lesion to the Fundus of stomach, spleen, tail of pancreas, splenic flexure of colon
Right Lateral Lumber Region	Pain in region may be due to Right kidney, Right adrenal gland, Right ureter, Ascending colon
Umbilical Region	Pain due to lesions to abdominal aorta, inferior vena cava, lower body of stomach, head and body of pancreas, small intestine, mesentery,
Left Lumbar Region	Pain due to lesions to left kidney, left adrenal gland, left ureter, descending colon
Right Iliac (inguinal) Region	Pain due to lesions to Lower ascending colon, Cecum, Appendix, Right ovary
Hypogastric Region	Pain due to lesions in Uterus, Urinary bladder, Sigmoid Colon, Rectum
Left Ilac (inguinal) Region	Pain due to lesions to Sigmoid colon, Lower descending colon, left ovary

Anteriolateral Abdomen

Skin

Superficial Fascia   *Fatty layer*

Investing Layer   *Covers muscle layers*

Abdominal Muscles

Endoabdominal Fascia

Extraperitoneal Fat Layer

Peritoneum

Of Clinical Significance

Abdominal Muscles

Nerves

ROOT	NERVE	LANDMARK	SUPPLIES
T12	Subcostal N	Runs just under the 12th RIB; over the quadratus lumborum	External Oblique Internal Oblique Transverse Abdom Rectus Abdom *Has cutaneous branches serving lateral hip*
L1	Iliohypogastric N	Out the lateral aspect of Psoas Major; runs over Qaudratus Lumborum; Pierces Transverse Abdominus just above iliac crest to move into external oblique	Internal Oblique External Oblique Transverse Abdom. *Has cutaneous branches serving lateral butt and pubis*
L1	Ilioinguinal N	Through the Lateral Psoas Major; Over the Quadratus Lumborum; and travels along the Iliac crest to Inguinal Canal	External Oblique Internal Oblique Transverse Abdom
Others in Plexus			
L2-L3	Lateral Femoral Cutaneous		*Anterior and Lateral skin of the femur*
L1-L2	Genitofemoral		*Skin of external genitalia*
L2-L4	Femoral		*Anterior thigh muscles* *Psoas, Iliacus*
L2-L4	Obturator		*Medial thigh muscles*

<u>Drawings</u>

<u>Drawings</u>

Arterial Supply

- ○ Subclavian A → Internal Thoracic Artery →
  - ■ **Musculophrenic A**
  - ■ **Superior Epigastric A**
- ○ Aorta
  - ■ **Posterior Intercostal A**
  - ■ **Subcostal A**
  - ■ **Lumbar A**
- ○ External Iliac A
  - ■ **Inferior Epigastric A**
  - ■ **Deep Circumflex Iliac A**

- • Veins are Similar to Arteries   *Drains to Subclavian V/Axillary V and Femoral V*

Muscles of Abdominal Region

Muscle	O	I	F	N	Blood
Psoas Major Psoas Minor	Lumbar V	Lesser Trochanter of Femur	Flexion of vertebral column	Femoral N	Lumbar A
Iliacus	Iliac fossa	Lesser Trochanter of Femur	Flexion of vertebral column	Femoral N	Iliolumbar A
Quadratus Lumborum	Iliac Crest	Lower rib Lumbar transverse process	Flex vertebral column	T12-L3	Iliolumbar A
External Oblique	Lateral ribs 5-12	Linea Alba Pubic Tubercle Iliac Crest	Compress wall Support Viscera Flexion Rotation	Intercostal Subcostal Iliohypogastric Ilio-inguinal	Superior and Inferior Epigastric A
Internal Oblique	Ribs 10-12 Linea Alba	Thoracolumbar fascia Iliac crest Inguinal Ligament	Compress wall Support viscera Flexion Rotation	Intercostal Subcostal Iliohypogastric Ilioinguinal	Superior and Inferior Epigastrics Deep Circumflex Iliac A
Rectus Abdominis	Xiphoid Process Costal Cartilage 5-7	Pubic crest	Flexion Compress wall	Intercostal Subcostal	Superior and Inferior Epigastric
Transverse Abdominis	Costal Cartilage lower ribs Iliac Crest Inguinal Ligament	Linea Alba Pubic Crest	Compress wall Support viscera	Interccostal Subcostal Iliohypgastric Ilioinguinal	Deep circumflex iliac A Inferior Epigastric A

Notes:

Inguinal Region

- ASIS to pubic tubercle
- Structures exit and enter the abdominal cavity
- Site of herniations

Inguinal Canal:

Superficial Inguinal Ring	Medial opening to the External Oblique Aponeurosis just lateral to the Pubic Tubercle
Deep Inguinal Ring	Opening in the Transversalis Fascia Lateral to Inferior Epigastric A/V
Inguinal Canal	Transverse abdominal aponeurosis
Anterior Wall	External and Internal Oblique muscle aponeuroses
Posterior Wall	Transversalis Fascia
Floor	Inguinal Ligament Lacunar Ligament

*The Transversalis Fascia → Retroinguinal Space (Extraperitoneal Tissue) → Peritoneum

Of Clinical Significance

Arterial Supply of Abdomen

Supply primarily derived from:
Abdominal Aorta
1. Celiac Trunk
2. Superior Mesenteric A
3. Inferior Mesenteric A

Aortic Hiatus

R/L Inferior Phrenic A	*Moves superiorly; Just above Celiac Trunk*	
Superior Suprarenal A	*Branches from Inferior Phrenic A to Suprarenal*	
Celiac Trunk		
1) Left Gastric A	*Moves towards the Lesser curvature of Stomach*	
Esophageal A	*Branches from L Gastric to Esophagus*	
2) Common Hepatic A	*Right sided Branch*	
Gastroduodenal A	*First branch of CHA which dives under Duodenum*	
a) Right gastroepiploic A	*To Rt aspect of Greater Curvature*	
b) Superior pancreatico-duo A	*Follows duodenum and Pancreas*	
Right Gastric A	*Second branch of CHA to Lesser Curvature*	
Hepatic A	*Continuation of CHA*	
a) Cystic A	*To GallBladder*	
b) R/L Hepatic A	*To Liver*	
3) Splenic A	*A Left Branch dives under stomach to Spleen*	
Short Gastric Arteries	*Fundus*	
Left Gastroepiploic A	*Left Greater Curvature*	
Pancreatic A (Great)		
Middle Suprarenal Arteries	*Branches just under Celiac Trunk to Mid-Suprarenal*	
Renal Arteries	*To Kidneys*	
Inferior Suprarenal A	*To inferior aspect of suprarenal*	
Superior Mesenteric A		
Inferior Pancreaticoduodenal A	*To Pancreas and Duodenum*	
Jejunal and Ileal Branches	*Jejunum/Ileum*	
Middle Colic A	*Ascending Colon, Transverse Colon, Ileum,*	
Ileocolic A	*Cecum, Appendix*	
Appendicular		
Ant/Post Cecal		
Ascending Colic		
Right Colic		
Lumbar Arteries		
Gonadal Arteries		
Inferior Mesenteric A	*Transverse Colon, Sigmoid Colon, Proximal Rectum*	
Left Colic		
Sigmoid		
Superior rectal		
Common Iliac		
Median Sacral A		

Drawings:

Venous Drainage from Abdomen

- Hepatic Portal System   *drains GI, pancreas, gallbladder, spleen into liver sinusoid system to vena cava*

1. Inferior Mesenteric V, Splenic V, Gastric V, Superior Mesenteric V drain into **Hepatic Portal V**
2. Empties blood into the **venous sinusoids of the liver** where hepatocytes act to detoxify blood
3. Detoxified blood empites from sinusoids to the **Central Vein**
4. Central V empties to the **Hepatic veins**
5. Hepatic Veins empty into **Inferior Vena Cava**
    a. **Also emptying into the Inferior Vena Cava**
        i. **Common Iliac V**
        ii. **Renal V**
        iii. **Gonadal V**
        iv. **Inferiror Phrenic V**
        v. **Hepatic V**

Portal-Systemic Anastomosis allow rerouting of venous return to heart if others become occluded
- Esophageal V to Azygos Vein (systemic) or Left Gastric V (portal)
- Rectal V to Inferior Vena Cava (systemic) or Inferior Mesenteric V (portal)
- Paraumbilical V to Epigastric V (systemic)

Nerve Innervation to Abdomen

Sympathetic Nervous System in Region (Sympathetic Chain – Paraverterbral)

- Ascending and desending preganglionic sympathetic GVE an GVA fiber
- Convey preganglionic sympathetic fibers to abdominopelvic cavity
- They pass through the paraverterbal ganglia of sympathetic rami **without** synapsing
- Then enter the **splanchnic nerves** which send to prevertebral ganglion of abdominal cavity

Ganglions in descending order

Celiac
Aorticorenal
Superior Mesentaric
Inferior Mesenteric

1. **Splanchnic Nerves :** Carry GVE/GVA to the viscera SPLANCH = ORGAN

Thoracic Splanchnic Nerves *Prevertebral Ganglion*	Preganglionic sympathetic GVE fibers with cell bodies in the lateral horn of spinal cord	**Greater S N T5-T9**
		Synapses with postganglionic fiber in the ***CELIAC GANGLION to supply the Celiac plexus:***
		Stomach, Liver, Gallbladder, Bile Ducts, Pancreas, Gonads
	Preganglionic sympathetic GVA fibers with cell bodies in dorsal root ganglia	*Directly synapses (with not post ganglionic fiber) to the Adrenal Medulla*
		Lesser S N T10-T11
		Synapses with postganglionic fiber in the ***AORTICORENAL GANGLION*** *to supply the Kidneys and Ureters*
		Synapses with the postganglionic fiber in the ***SUPERIOR MESENTERIC GANGLION*** *to supply the Intestines*
		Least S N T 12
		Synapses with postganglionic fiber within RENAL PLEXUS (no ganglion) to supply Kidneys and Uerters
		The least may also synapse in one of the mesenteric ganglion to supply the intestines.
Lumbar Splanchnic Nerve L1-L3	Preganglionic sympathetic fibers orginate in the abdominal sympathetic trunk (L)	*Synapses with the* ***SUPERIOR MESENTERIC GANGLION*** *to supply the Intestines*
		Synapses with the ***INFERIOR MESENTERIC GANGLION*** *to supply the Intestine*
		Synapses to the ***HYPOGASTRIC PLEXUS*** *(no ganglion) to* ***PELVIC PLEXUS*** *to supply sigmoid colon, rectum, bladder, prostate, external genitalia*

*Sympathetic Innervation to the Digestive Tract will

- DECREASE peristalsis and GI blood flow so other areas can receive blood as needed;
- Contracts the internal anal sphincter to prevent defacation;
- Promotes glycogenolysis in liver;
- Slows urine production and contraction of internal sphincter of bladder to prevent urination

Parasympathetic influence on Abdomen

<u>Anterior and Posterior Vagal Trunks</u>	Continuation of right and left vagus nerve Preganglionic parasympathetic and GVA fibers	To **ABDOMINAL AORTIC PLEXUS** Supplying parasympathetic innervations to the smooth muscle of GI to the beginning of the descending colon
<u>Abdominal Autonomic Plexus</u>	Contain both sym and parasym Surround the abdominal aorta and its branches Interconnects the celiac, sup/inf mesenteric plexi	Celiac Plexus *spleen, liver, upper GI* Sup Mes Plexus *Upper GI* Renal Plexus *Kidney* Inf Mes Plexus *Lower GI*
<u>**Pelvic Splanchnic N**</u> S3-S4	Parasympathetic Derived from S2-S4 ventral rami Convey Preganglionic parasympathetic fibers to **Hypogastric (Pelvic) Plexus**	Supplies the descending colon Sigmoid, and rectum Pelvic Viscera

<u>Drawings</u>

<u>Drawings</u>

Abdominal Viscera

Peritoneum

- **Serous membrane**
- **Parietal peritoneum**:
  - Lines the abdominal and pelvic walls and the inferior surface of the diaphragm
  - Innvervated by:
    - Phrenic
    - Lower intercostals
    - Subcostal
    - Ilihypogastric and ilioinguinal
  - Sensitive to pressure, pain, heat, and cold; pain is localized
- **Visceral peritoneum**:
  - Covers the viscera
  - Innervated by:
    - Visceral nerves;
  - Insensitive to touch, heat, cold; sensitive to stretch
  - Pain is not localized (referred pain)

Reflections of the Peritoneum

- Serve as pathways for neurovascular structures
- Omentum:
  - Fold of peritoneum which extends from stomach to adjacent abdominal organs
    - **Lesser Omentum**
      - Extends from the liver to stomach
      - *Hepatoduodenal and Hepatogastric Ligaments*
      - Transmits Left and Right Gastric A/V
      - Contains Portal and Biliary A/V/ Duct
    - **Greater Omentum**
      - Aprons from stomach and intestines
      - *Gastrophrenic, Gastrosplenic, and Gastrocolic Ligament*
      - Transmits Rt and Left gastroepiploic A/V
- **Mesenteries**
  - Mesentary of small intestine
    - Fan like fold
    - Transmits superior mesenteric a/v and intestinal a/v
    - Transmits lymphatics
  - Transverse Mesocolon
    - Connects transverse colon to posterior abdominal wall
    - Transmits colic vessels
  - Sigmoid Mesocolon
    - Sigmoid colon to pelvic wall
    - Transmits sigmoid vessels
  - Mesoappendix
    - Appendix to the mesentery of the ileum
    - Transmist appendicular vessels

Peritoneal Cavity

- A space between parietal and visceral peritoneum
- Contains fluid which lubricates allowing movement of viscera
- Fluid also contains leukocytes and antibodies
- No organs in the peritoneal cavity

- In males this is a completely closed sac
- In females it is opened via the uterine tubes, uterus, and vagina

Intraperitoneal:
- Almost completely covered by visceral peritoneum forming a closed sac (hand in balloon)

Partially retroperitoneal/Retroperitoneal:
- Outside of the cavity
- Partially covered by the peritoneum (usually only one surface)
  - Kidney only has peritoneum on anterior surface

Intraperitoneal	Both Partial Retroperitoneal	Retroperitoneal
Stomach	Duodenum	Rectum
Jejunum	Ascending colon	Adrenal Gland (suprarenal)
Mesentary	Descending colon	Kidneys
Ileum	Liver	Ureters
Cecum	Pancreas	Seminal Vescicles
Transverse Colon	Gall Bladder	Aorta
Appendix	Urinary Bladder	IVC
Sigmoid Colon		
Spleen		
Uterus/Ovary		

Of Clinical Significance

GI Viscera

<u>Esophagus (Abdominal Aspect)</u>
- Muscular tube extending from diaphragm to cardiac orifice of the stomach
- <u>Esophageal sphincter</u>
  - Circular layer of smooth muscle at the distal aspect of esophagus
  - Contraction prevents gastric contents from moving back into the esophagus
- Innervation
  - Parasympathetic
    - Vagal
      - Right and left vagus nerves branch to form the **esophageal plexus**
      - LARP; Left vagus to the anterior surface; Right vagus to the posterior surface
  - Sympathetic
    - Thoracic
      - Greater splanchic nerves
        - Preganglionic in lateral horn
        - Synapse with postganglionic in the celiac prevertebral ganglion
        - To esophageal plexus
- Vasculature
  - Abdominal Aorta → Celiac Trunk → Left Gastric A → **ESOPHAGEAL BRANCHES**
  - Adbominal Aorta (or celiac trunk) → **LEFT INFERIOR PHRENIC A**
  - Venous Drainage
    - To the portal venous system via the left gastric vein

<u>Stomach</u>
- Intraperitoneal (completely covered by peritoneum)
- Left hypochondriac and epigastric regions
- **Curvatures**
  - Lesser: Shorter superior border   *Lesser Omentum*
  - Greater: Longer inferior border   *Greater Omentum*
- **Notches**
  - Angular notch:
    - At pyloric end of lesser curvature; where body of stomach junctions with the pylorus
  - Cardiac notch:
    - At the Cardiac aspect as enter the fundus
- **Openings**
  - Cardiac orifice:
    - Opening between esophagus and stomach
  - Pyloric orifice
    - Opening between stomach and duodenum

- **Regions**
  - Cardia
  - Fundus
    - 5th rib level; inferior to apex of heart
    - Contains **rugae** folds increasing surface area of digestive ability
    - Expands to make room for more food
  - Body
    - Contains rugae for digestion
    - Most of digestion of proteins and starches
  - Pylorus
    - Pyloric atrum: the first wide aspect leads into the
    - Pyloric canal: the narrow aspect leads into the
    - Pylorus: region of the **pyloric sphincter;** circular layer of smooth muscle
    - To the Pyloric orifice → Duodenum

1. The sphincter is **constricted by sympathetic stimulation** and **relaxed by parasympathetic stimulation**
2. Stomach produces **HCL** destroying bacteria and **Pepsin** which is a protein digesting enzyme;
   a. both within the fundus and the body
3. Stomach produces the hormone **Gastrin** which stimulates release of HCL and aids in grinding motion of stomach;
   a. occurs in the Pyloric antrum
   b. Parasympathetic fibers of vagus nerve stimulate gastrin secretion
- **Innervation**
  - **Parasympathetic:** *(dorsal motor nucleus of vagus nerve)*
    - Left vagus → Anterior Vagal Trunk → Celiac Plexus
    - Right vagus → Posterior Vagal Trunk → Celiac Plexus
  - **Sympathetic**
    - Greater Splanchnic N → Celiac Ganglion → Stomach
- **Vasculature**
  - **Arteries**

Artery	From	Course references
Left Gastric A	**Celiac Trunk** *From abdominal aorta*	Rides along the lesser curvature Will anatomose with right gastric at lesser curvature
Right Gastric A	**Common Hepatic Artery** *From celiac trunk*	Will anatomose with the left gastric along the lesser curvature
Right Gastroepiploic A	**Gastroduodenal A** *From common hepatic A*	Descends posteriorly to the gastroduodenal junction to the greater curvature
Left Gastroepiploic A	**Splenic A (hilum of spleen)** *From celiac trunk*	From posterior aspect will descend to the greater curvature
Short Gastric A	***Splenic A***	To the Fundus of the Stomach

- **Veins**
  - Same as arteries emptying into **Hepatic Portal Vein**

Small Intestine

- Pyloric orifice → ileocecal junction
- Digestion and absorption (water, electrolytes, Ca, and iron)
- Duodenum, jejuneum, ileum
- **Duodenum;** *(epigastric region)*
  - *Superior:*
    - *From the pylorus; mobile, free section*

- ○ *Descending:*
  - ■ *Junction of foregut and midgut*
  - ■ *Area where the* **common bile duct** *and* **main pancreatic ducts open (hepatopancreatic ampulla)**
  - ■ *Fat digestion*
- ○ *Inferior (Transverse):*
  - ■ *Longest aspect*
  - ■ *Crosses the inferior vena cava and aorta*
  - ■ *Will then ascend into the Ascending aspect*
- ○ *Ascending:*
  - ■ *Terminates at the duodenojejunal junction where it joins the jejunum*
  - ■ *Held in a fixed position by the* **Suspensory Ligament of Treitz (attached with diaphragm)**
    - □ *Contraction widens the angles of the duodenojejunal junction increasing motility*
    - □ *Malrotation of gut syndrome in children*
- • **Jejunum**
  - ○ Left hypochondriac Region
  - ○ Held in place by the posterior mesentery
  - ○ Fat and Protein digestion
  - ○ Thicker walled and larger diameter than the Ileum
  - ○ Greater vascularity vs Ileum
  - ○ Has **Plica Circulares (circular folds);**
    - ■ Tall and closely packed folds
    - ■ Slows passageway of food and increases surface area
- • **Ileum**
  - ○ Longest part of the SI
  - ○ Most nutrients are absorbed
  - ○ Umbilical and Right Iliac Regions
  - ○ Joins the Large intestine at an **ileocecal fold**
    - ■ Sphincter less valve
    - ■ Passes into the Cecum of LI
  - ○ **Peyer's Patches**
    - ■ Lymph tissue
  - ○ Less Plica and absent in the most distal part
- • **Innervation**
  - ○ **Parasympathetic**
    - ■ Dorsal Motor Nucleus of Vagus → Posterior Vagal trunk (Right Vagal) → synapse with postganglionic fiber in Celiac Plexus/Superior Mesenteric Plexus
    - ■ Increases secretions and motility; restore post sympathetic
  - ○ **Sympathetic**
    - ■ Thoracic → Through Paravertebral → Lesser Splanchnic N → Aorticorenal Ganglion → Pregang continues to the Superior Mesentaric Ganglion → Postganglionic fiber to Small Intestine
    - ■ *some resources also site the Celiac Ganglion as being target of Lesser Splanchnic also*
    - ■ Reduce secretion and motility
    - ■ Vasoconstriction thus reducing digestion and allowing blood to be available where needed
  - ○ The SI is insensitive to most pain stimuli..however, it is sensitive to stretch (distention)

- **Vasculature**

Artery	From	Reference	Supplies
Superior Pancreaticoduodenal A *Anterior and Posterior*	*From* Gastroduodenal A *From Hepatic* A *From Celiac Trunk*	*From behind the gastroduo-denal junction*	Superior duodenum Descending duodenum
Inferior Pancreaticoduondenal A *Anterior and Posterior*	*From* Superior Mesenteric A *From the Abdominal Aorta*	*Just inferior to stomach* *Head of Pancreas*	Inferior duodenum Ascending duodenum
Arterial Arcades to Vasa Recta	Superior Mesentaric	*Jejunum:* *6-8 and LONG* *Ileum:* *4-6 and SHORT*	Jejunum and Ileum
Veins	Superior Mesentary Vein will empty into the Hepatic Portal Vein		

Large Intestine

- From Ileocecal junction to the anus
- Convert contents from the ileum into semisolid feces by absorbing water, salts, and electrolytes
- Stores and lubricates the feces with mucous
- **Tanae Coli**
  - ○ 3 narrow bands of longitudinal muscle coat
  - ○ Produce the **Haustra**
- **Epiploic Appendages**
  - ○ Peritoneum covered sacs of fat along the teniae
- **Cecum**
  - ○ Pouch of the LI
  - ○ Right Iliac Region
  - ○ Surrounded by peritoneum
  - ○ Ileocecal junction to the Ascending Colon
  - ○ No mesentery so can be lifted freely
  - ○ **Appendix**
    - Intestinal diverticulum (outpouch)
    - Triangular mesentery to cecum (mesoappendix)
    - Lymph tissue with the smooth muscular wall
    - McBurney's Point
      - □ Spinoumbilical line
  - ○ **Innervation**
    - **Parasympathetic:** Dorsal Motor N of Vagus N → Preganglionic to synapse in Superior Mesenteric Plexus → Postganglionic to Cecum and Appendix
    - **Sympathetic:** Lower Thoracic (T10-12) →
      - □ Lesser Splanchnic → Aorticorenal Ganglion with no synapse → Superior Mesenteric Ganglion → Cecum and Appendix
      - □ Least Splanchnic → Superior Mesenteric Ganglion → Cecum and Appendix
  - ○ **Vasculature**

Ileocolic A	*From* Superior Mesenteric *A* *From Aorta*	Last branch of Superior Mesenteric	Ileum Cecum Ascending colon
Appendicular A	*From* Ileocolic A	Passes between the layers of mesoappendix	Appendix

- **Colon**
  - **Ascending Colon**
    - Cecum to Right Lobe of Liver
    - **Right Colic Flexure (hepatic flexure)** to the Transverse Colon
    - Narrower than cecum
    - Covered by peritoneum anteriorly and bilaterally
    - Separated from the anterolateral abdominal wall by great omentum
  - **Transverse Colon**
    - Largest aspect
    - Most mobile aspect
    - Right colic flexure to the **Left Colic Flexure (Splenic Flexure)**
    - Transverse mesocolon (Mesentery)
  - **Descending Colon**
    - Splenic Flexure to Sigmoid Colon
    - Covered by peritoneum anteriorly and bilaterally
  - **Sigmoid Colon**
    - S-shaped
    - Descending colon to Rectum
    - Iliac fossa to 3rd sacral (joins rectum)
    - Sigmoid Mesentery allowing movement
- **Innervation of Colon**
  - **Parasympathetic**
    - Ascending and Transverse via Dorsal Motor N of Vagus → Vagal Trunks → Superior Mesenteric Plexus
    - Left Transverse, Descending to Sigmoid → Lateral Horn → synapse in pelvic plexus
  - **Sympathetic**
    - Ascending and Transverse via Least Splanchnic → Sup Mes Ganglion
    - Lumbar segments → Lumbar Splanchnic N → Inferior Mesenteric Ganglion

Vasculature

Superior Mesenteric	**From** Aorta (L1)	Runs in mesentery to ileocecal junction	Branches to provide supply
Inferior Mesenteric	**From** Aorta (L3)	Descends to the left of the aorta	Descending colon
Middle and Right Colic A	**From** Superior Mesenteric		Right: Ascending Middle: Transverse
Left Colic A	**From** Inferior Mesenteric		Descending colon
Sigmoid A	**From** Inferior Mesenteric		Descending and Sigmoid
Superior Rectal A	**From** Terminal Inferior Mesenteric A		Proximal Rectum

Notes:

Liver

- Largest visceral organ
- Largest gland of the body
- Production and secretion of bile used to emulsify fats
- Detoxifies blood that comes from digestive region
- Storage of glycogen (glycogenolysis occurs)
- Storage of triglycerides
- Production of blood coagulants (fibrinogen and prothrombin)
- Production of anticoagulants (heparin)
- Production of bile pigments from the breakdown of hemoglobin (bilirubin)
- Reservoir for blood and platelets
- Storage of certain vitamins, Fe, etc
- Manufactures RBC in fetus
- Located in Right hypochondriac and Epigastric Regions
- 4 lobes to identify

RIGHT LOBE	*Big; With Gall Bladder and IVC*
LEFT LOBE	*Small*
QUADRATE LOBE	*Within Left Lobe* *Along Gall Bladder* *Receives Blood from the Left hepatic A* *Drains Bile into Left Hepatic Duct*
CAUDATE LOBE	*Along IVC* *Receives blood from Right and Left hepatic A* *Drains bile into Right and Left Hepatic Ducts*
LOBES	*Contain:* *Portal Triad:* 　1. *Hepatic Artery*: Branch of celiac trunk providing blood to hepatocytes Empty into sinusoids to gct to hepatocytes 　2. *Hepatic Portal Vein*: GI blood to sinusoiods to hepatocytes to be detoxified; clean blood is then emptied into central vein to Hepatic Vein to vena cava 　3. *Biliary Duct*: Hepatocytes produce Bile → Bile Caniculi → Biliary Duct → Right and Left Hepatic Ducts → Common Bile Duct → Gallbladder or Duodenum
Falciform Ligament	In between Right and Left Lobes Attaches liver to the anterior wall of abdomen
Round Ligament	Between Left lobe and Quadrate Lobe

<u>Notes, Diagrams, Doodles</u>

Gall Bladder

- Rt 9th costal cartilage and Lateral border of Rectus Abdominis
- Inferior liver between Right and Quadrate Lobes
- Receives bile from **Cystic Duct**, Stores Bile, and Releases bile upon fat stimulation via **Common Bile Duct**
- Stimulated to contract by hormone **cholecystokinin** parasympathetically when food arrives in duodenum
- Receives blood from the **Cystic A** from the Right Hepatic A
- Gallstones
  - Choleliths formed from solidification of bile filtrates esp cholesterol crystals (bile pigments with calcium)
  - Bile crystallizes and forms sand → gravel → stones
  - Can become lodged in fundus where they may ulcerate throught the wall into the transverse colon or duodenum
  - Can be lodged in bile ducts where they obstruct bile flow to the duodenum → jaundice
  - Can be lodged at the **hepaticopancreatic ampulla** where they block bilary and pancreatic ducts; may lead to pancreatitis
  - 4-F's
    - Fat, Fertile, Females > 40 yoa
  - Cholecystitis
    - Inflammation of the GB due to obstruction of the cystic ducts by gallstones
    - Right hypochondriac and epigastric pain radiating to back or right shoulder

Pancreas

- Right and Left Hypochondriac, Epigastric, and Left Hypochondriac
- Head of the pancreas is wrapped in the C-shape formed by aspects of duodenum
  - **Hepaticopancreatic ampulla** location
- Exocrine function:
  - Produces digestive enzymes that help digestion
- Endocrine function:
  - Secretion of insulin and glucagon
- Ducts
  - Main Pancreatic Duct
    - Begins in tail of pancreas → to the head
    - Carries pancreatic digestive juice containing digestive enzymes
    - Joins the common bile duct to from hepatopancreatic ampulla (vater) to the duodenum
  - Accessory Pancreatic Duct
    - Begins at the head
    - Empties just above the major ampulla

**Blood supply to the Pancreas:

1. Celiac Trunk → Splenic A → **Greater Splenic A and Dorsal Splenic A** *surround body and tail*
2. Celiac Trunk → Common Hepatic A → Gastroduodenal A → Anterior/Superior Pancreaticoduodenal A *over the head*
3. Aorta → Superior Mes A → Posterior/Inferior Pancreaticoduodenal A *inferior head*

Spleen

- Vascular lymphatic organ
- Against diaphragm in left hypochondriac region (ribs 9-11)
- Filters blood (removes old RBC and platelets)
- Acts as blood reservoir, storing RBC and platelets

- Immune response
- Produces mature lymphocytes, macrophages, and antibodies
- Blood supply via Celiac trunk to Splenic A
- Integrity supplied by **Gastrosplenic Ligament and Splenorenal Ligament**

Notes/Diagrams

Drawings

Drawings

Drawings

Posterior Abdomen Viscera

Kidney/Urinary Tract

**Superiorly on the kidney is the Suprarenal Glands with no function with the kidney

Kidneys → Ureters → Urinary Bladder → Urethra

Remove excess water, salts, protein metabolic waste to urine
Reclaim what we need
Fascia and Fat

The Kidney is enclosed within a renal fascia which houses fat; therefore there is a direct layer of fat surrounding the kidney's capsule

The kidney moves upon change of position and respiration (3cm) and the fat assists this mobility

Kidneys
- Extends from T12 to L3 in the standing position
- The right kidney is lower (rib 12) than the left (Rib 11 and 12) due to the large right lobe of the liver
- The Hilus is on the medial border of the kidney
  - Allows entrance of the ureter, renal artery and vein, and renal nerves
- Nephron and surrounding Peritubular Capillaries
  - Renal Corpuscle contains the Glomerulus which will act to filter incoming blood (from renal artery)
  - Filtered blood (filtrate) will →
    - Proximal Tubule →
      - reabsorbs 85% vitamins, amino acids, proteins, glucose, from the filtrate
      - *water will follow these items out of the filtrate*
  - Filtrate continues to →
    - Loop of Henle →
      - Reabsorption of primarily water from the filtrate
  - Filtrate continues to →
    - Distal Tubule →
      - Sodium reabsorption from filtrate with some water
  - Filtrate continues:
    - Convoluted Tubule →
      - Here is the last adjustment ; primarily items are secreted back into filtrate at this point as need
    - Collecting Duct →
    - **Papillae of Medullary Pyramids**
    - **Minor Calyx**
    - **Major Calyx**
    - **Renal Pelvis**
    - **Ureters**

**1300 ml of blood flows through the kidney per minute; 125 ml is filtered in tubular system per minute; about .7 ml of that filtered blood is excreted as urine; Rest may be reabsorbed as nutrients, water, etc; can return to circulation via the renal vein

- Has 2 regions
  - **Medulla**
    - Innermost aspect consisting of the renal pyramids; and aspects of the nephron; Loop of henle, Aspects of distal tubule, aspects of collecting duct; renal papilla which fits into minor calyx
  - **Cortex**
    - Outer aspect consisting of the nephron; Renal corpuscle, proximal tubule, aspects of distal tubule, aspects of the collecting duct

- Nerve innervations:
  - PARASYMPATHETIC VAGUS
  - SYMPATHETIC
    - Lesser Splanchnic N

Ureter

- Muscular tube beginning at renal pelvis extending to urinary bladder
- May be obstructed by kidney stones as it joins the renal pelvis or where it enters the urinary bladder as these are the narrow aspects
- Receives blood supply from the *aorta, renal artery, gonadal artery, common iliac a, internal iliac artery, umbilical a, superior and inferior vesical a, and middle rectal arteries*
- Innervated by the LUMBAR SYMPATHETIC and PELVIC PARASYMPATHETIC SPHLANCHIC NERVES

Bladder

Innervated

Sympathetic: Lumbar and Sacral via Hypogastric plexus
Parasympathetic: Pelvic Plexus (pelvic splanchnic)

Notes/Diagrams

Putting It All Together

Putting It All Together

CHAPTER 10

Pelvis-Perineum

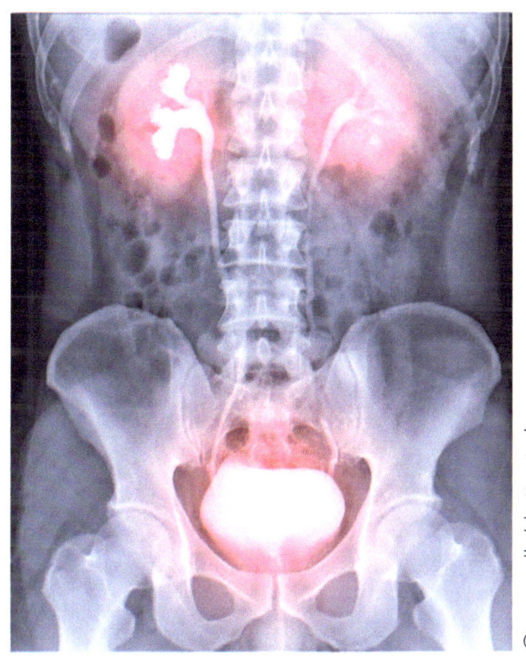

Pelvis and Perineum

PELVIS

Surface Anatomy Review

Umbilicus	Level of L3-L4
Iliac Crest	Level of L4 Bifurcation of Abd Aorta to Common Iliac Arteries
ASIS	Inguinal Ligament
Inguinal Ligament	External Oblique Aponeurosis Separation of Lower abdomen from Lower Limb
Pubic Tubercle	Inguinal Ligament
PSIS	Dimpling

*Review bony landmarks of the Pelvis as needed

**Review Joints as needed

- Lumbrosacral Joint
  - Cartilaginous
- Sacroiliac Joint
  - Synovial plane
- Sacrococcygeal Joint
  - Cartilaginous
- Pubic Symphysis
  - Cartilaginous

Pelvic Inlet

- Superior aperature of the pelvic cavity
- Boundary between pelvic cavity and abdominal cavity
- Bounded by the **Linea Terminalis**
  - Pubic symphysis → Pectineal Line → Arcuate Line (of Ilium) → Anterior Ala (sacrum) → Sacral Promontory

Pelvic Outlet

- Inferior aperature of pelvic cavity
  - Inferior pubic symphysis → Inferior Rami (pubis) → Ischial Tuberosity → Sacrotuberous Ligament → Tip of posterior Coccyx

Male vs Female

	Male	Female
Bones	Large, Thick, Heavy	Small, Thin, Light
Pelvic inlet	Heart shaped	Oval shaped
Pelvic outlet	Small	Large; Ischial tuberosities are more lateral
Pelvic cavity	Narrow and Deep	Wide and Shallow
Sacrum	Long and Narrow	Short and Wide
Obturator Foramen	Round	Oval or Triangular
Pubic arch	Narrow	Wide

Pelvic Diaphragm

- Forms the **pelvic floor**
- Supports all pelvic viscera
- When contracts it will raise the pelvic floor
- Flexes the **anorectal canal** during defecation and assists with urination (voluntary)
- Helps direct the fetal head towards the birth canal

Arteries of Pelvic Diaphragm

- Branches of the **Internal Iliac**
  - **Pudendal** → TO Perineum and external genitalia
  - **Obturator** → TO medial thigh
- Major branches of the Internal Iliac

Artery	Origin	Reference	Supplies
Posterior Division of Internal Iliac A			
Iliolumbar	Posterior Div of I.I.		N/A
Lateral Sacral A	Posterior Div of I.I.		N/A
Superior Gluteal A	Posterior Div of I.I.	Through greater sciatic notch superior to the piriformis	Piriformis GMax, Med, Min TFL

Anterior Division of Internal Iliac A			
Obturator A	Anterior Div. of I.I.	Lateral pelvic wall and exits through obturator foramen	Medial pelvic floor muscles, medial thigh
Umbilical A	Anterior Div. of I.I.	Anterior abdominal wall Gives rise to *superior vesical A*	*Will be cut at child birth and become the Superior Vesical A*
Superior Vesical A *In male and female*	Anterior Div. of I.I. from umbilical A	Multiple branches stem from SVA	Urinary bladder Pelvic aspect of ureter Ductus deferens Seminal vesicals
Inferior Vesical A *In Males*	Anterior Div. of I.I.	Runs to inferior bladder	Inferior Urinary bladder Prostate Seminal vesicals Ductus deferens
Uterine A	Anterior Div. of I.I.	Lateral to uterus Anastomosis with ovarian artery	Uterus Vagina Uterine tubes Ligaments of the uterus
Vaginal A *Female version of Inferior Vesical A*	From the Uterine A	Lateral to ureter and descends along lateral vagina	Inferior urinary bladder Vagina Cervix Distal Ureter
Middle Rectal/Inferior Rectal	Anterior Div of I.I.	Inferior in pelvis to lower rectum	Inferior rectum Seminal vesicals
Pudendal	Anterior Div. of I.I.	Exits pelvis through the greater sciatic notch; enters perineum through the lesser sciatic notch	Perineum
Inferior Rectal	From Pudendal A		Anal canal

<u>Drawings</u>

Nerves

1. **Pudendal N**
   a. Structures of the perineum
   b. Sensory innervations to the genitalia
   c. Motor innervations to the perineal muscles, sphincter urethra, external anal sphincter
2. **Nerves to Levator Ani**
   a. Levator ani and coccygeus muscles
3. Sympathetic
   a. Lumbar Splanchnic to Hypogastric Plexus
4. Parasympathetic
   a. Through Pelvic plexus

Muscles of the Diaphragm

Levator Ani: Puborectalis, Pubococcygeus, Iliococcygeus. Ischiococcygeus (coccygeus)

- Together raise the pelvic floor
- Maintain continence; when relaxed allow defecation and urination
- Contraction occurs when the thoracic diaphragm and abdominal wall muscles compress the abdominal and pelvic contents thus resisting intra-abdominal pressure (if not the contents would enter the pelvic outlet; occurs when coughing, sneezing, vomiting, lifting heavy items, etc

M	O (anterior attachment)	I (posterior attachment)	F	N	A
Levator Ani (3)					
Puborectalis	Body of Pubis	Anal Canal Rectum U-shaped sling	Maintain the anorectal angle; maintains fecal continence **(when relaxes, the angle will increase assisting defecation)**	Nerve to Levator Ani (S4 ventral rami)	Pudendal A *From internal iliac A* Inferior Rectal A Inferior Gluteal
Pubococcygeus	Body of Pubis	Coccyx Perineal Body	Support pelvic viscera; **resist intra-abdominal pressure;** Controls urine flow; Reduces chance of incontinence; **proper head positioning of baby's head at childbirth**	Nerve to Levator Ani (S4)	Pudendal A *From internal iliac A* Inferior Rectal A Inferior Gluteal
Iliococcygeus	Obturator Fascia	Coccyx Perineal Body	Support pelvic viscera; **resist intra-abdominal pressure**	Nerve to Levator Ani (S4)	Pudendal A *From internal iliac A* Inferior Rectal A Inferior Gluteal
Ischiococcygeus (coccygeus)	Ishial Spine	Lower Sacrum Coccyx Perineal Body	Hold pelvic viscera; **resist intra-abdominal pressure**	S4, S5	Obturator A

Notes:

PERINEUM

- Diamond shaped space
- Bounded by (same as pelvic outlet)
  - Pubic symphysis
  - Ischiopubis rami
  - Ischial tuberosities
  - Sacrotuberous ligaments
  - Tip of coccyx
- Floor is skin and fascia
- Roof is pelvic diaphragm
- Two subsections
  - Urogenital triangle (anteriorly)
  - Anal triangle (posteriorly)
  - Divided by a line joining the ischial tuberosities

Urogenital Triangle

1. Superficial Perineal Space
   a. Lies between the **Inferior Fascia of the Urogenital Diaphragm (Perineal Membrane)** and **Superficial Perineal Fascia (Colle's)**
   b. Contains (only ones discussed in lab manual are provided)
      i. **Bulbospongiosus muscle and fascia**
      ii. **Ischiocavernous muscle and fascia**
      iii. **Bulb of penis/vestibule bulbs**
      iv. **Crus of penis and clitoris**
      v. **Pudendal A/V**
      vi. **Perineal N**
      vii. **Central Tendon (perineal body)** *Fibromuscular mass in the center of perineum between anal canal and penis bulb/vagina*
      viii. **Greater Vestibular (Bartholin's Glands) (in females)** *lubricates vagina*

<u>Notes/Diagrams</u>

2. Muscles of Superficial Perineal Space

M	O Posterior	I Anterior	F	N	A
Ischiocavernosus	Inner ischial tuberosities, Ischiopubis ramus	Crus of clitoris or Crus penis (base of corpus cavernosum) i	Maintain clitoral and penile erection *Compresses the crus and deep veins thus decreasing venous return*	Perineal branch of Pudendal N	Pudendal A
Bulbospongiosus	Perineal body	Males: Fascia of corpus cavernosum Females: Dorsal aspect of clitoris	Support perineal body; Males: Compress penis bulb this impeding venous return so to maintain erection Works with the ischiocavernosus to constrict the cavernosum to expel final urine and final semen in ejaculation Females: Compress vestibular bulb to maintain erection Constricts vaginal orifice (sphincter)	Perineal branch of Pudendal N	Pudendal A
Superficial Transverse Perineal Muscle	Ischial tuberosities	Bilateral Perineal Body (central tendon)	Stabilizes perineal body	Perineal branch of Pudendal N	Pudendal A

3. Deep Perineal Space
    a. Between the superior and inferior fascia of urogenital diaphragm
    b. Contains
        i. Deep transverse perineal muscle
        ii. Sphincter urethra
        iii. Urogenital triangle
        iv. Bulbourethral Glands (males) *helps lubricate the urethra for passage of sperm*

<u>Notes</u>

Anal Triangle
- Only discussed in general
- Contains
  - Obturator Internus _lateral rotation_ _nerve to obturator internus_
  - Sphincter Ani Externus _closes the anus_ _inferior rectal nerve_
  - Levator Ani Muscles _Pelvic floor muscles_ _ventral rami S4_

<u>Notes:</u>

Female Perineum (external genitalia)

- Mons pubis   *superior apex of labia majora*
- Labia Majora   *outermost folds; sebaceous glands; sweat glands*
- Labia Minora   *innermost folds; surround the vestibule of vagina*
- Clitoris   (prepuce of clitoris) *superior apex of the clitoris; erectile tissue*

                    Clitoris is just inferior to prepuce

                    Contains corpora cavernosum, crus, and glans of clitoris

- Vaginal vestibule   *space between the labia minora;*
                    *Contain **external urethral orifice, vaginal orifice, and greater vestibular glands** (Bartholin's Gland)*

Innervated by: Blood Supply:

- Ilio-inguinal N   Pudendal A
- Iliohypogastric N   labial branches   *labias, vestibule*
- Pudendal N   clitoral branches *clitoris, sup vagina*

Female Pelvic Viscera and Ligaments

Uterus	Thick muscular organ Superior to the Bladder Opens into the vagina via the cervix Hormone and ANS responsive to stretch cervix and allow fetus passage Uterine and Vaginal A from Internal Iliac Least Splanchnic and Lumbar Splanchnic (Sym) Pelvic Plexus (Para)
Round Ligament of Uterus	Cord-like structure Posterior uterus extending laterally towards deep inguinal canal Passes through inguinal canal to insert at labia majora fascia Continuous with **Ligament of Ovary**
Broad Ligament of Uterus	Folding of peritoneum over uterus and uterine tubes Drape-like descending from uterine tubes to the cervix and lateral pelvic wall Uterine A and Vein within
Uterine tubes (Fallopian)	Extend laterally from uterus towards ovaries Open over each ovary at the **Infundibulum** (finger-like **Fimbrae**) which directs the released ovum into uterine tubes towards uterus Fertilization will take place in the **ampulla of the uterine tubes** The narrowest aspect of tube is the **isthmus** Ovarian plexus (nerve) Uterine (from internal iliac) and Ovarian A (abdominal aorta)
Ovary	**Ligament of Ovary: Ovary to Uterus** **Suspensory Ligament of Ovary: Lateral ovary to lateral pelvic wall** **Broad ligament will also support the ovary** Innervated by ovarian plexus (para/sym) Ovarian A
Vagina	Passage from Cervix of Uterus to Vaginal Orifice

Male Perineum (external genitalia)

- Penis
  - **Root of Penis**
    - Attachment point of penis to the pelvis via the urogenital diaphragm
    - Consists of 2 crus and a bulb
    - Covered by the ischiocavernosus and bulbospongiosus
  - **Body of Penis**
    - Erectile tissue
      - **Corpora Cavernosa (2)**
        - Expand and become firm at erection
        - Blood supplied by the **dorsal artery of the penis,** a branch of the pudendal A
      - **Corpus Spongiosum (1)**
        - Spongy tissue that surrounds the urethra
        - Allows the urethra to remain opened during erection
        - Becomes the **Glans Penis**
  - **Glans Penis**
    - Glans is covered by the **prepuce** (foreskin)
    - **External urethral orifice**
  - **Raphe**
    - Fused labia
    - Scrotal and penile
  - Innvervated by the pudendal nerve
    - Parasympathetic via pelvic plexus
    - Sympathetic via lumbar (pelvic) splanchnic nerves
  - Blood supply
    - Pudendal A
      - Artery of bulb of penis
      - Dorsal A of penis
      - Deep arteries of penis
- Scrotum
  - Derived from the anterolateral abdominal wall
  - Provides support to the testes
  - Organs of the scrotum

Testes	Homologous to the ovary in females Reproductive and endocrine Produce sperm (spermatogenesis) Produce testosterone`
Tunica Albuginea of Testes	Fibrous layer surrounding the testicles *Also found in the penis surrounding the cavernosa* *Also found in female surrounding the ovaries*
Tunica Vaginalis of Testes	Serous covering of the testis (peritoneum) **Hydrocele:** Accumulation of serous fluid Commonly mistaken for tumor

Spermatic Cord	Begins at the deep inguinal ring and lateral to inferior epigastric vessels Passes through the inguinal canal Contents include: Testicular Artery (aorta) Artery to ductus deferens Cremasteric A Genital branch of genitofemoral nerve Cremaster muscles (elevating scrotum) Ductus deferens Lymphatic vessels Sympathetic testicular nerves Pampiform plexus (veins draining scrotum and testis)
Epididymis	Coiled tube Testicle to Ductus Deferens Storage and maturation of spermatozoa (2–3 months) Release sperm into the ducts deferens
Ductus deferens (Vas)	Carrying away vessel Transport sperm from epididymis to urethra during ejaculation Smooth muscle walls contract propelling sperm fwd (peristalsis)

Notes/Diagrams

Male Pelvic Viscera

Seminal Vesicals	Between the rectum and bladder Produce thick seminal fluid which will pass into the **Ejaculatory Ducts** during ejaculation Hypogastric plexus (sym) Pelvic plexus (para) Inferior visceral A Middle rectal A
Prostrate Gland	Most superior aspect of urethra will pass through 5 Lobes: 4 are glandular and produce secretions added to ejaculate 1 is muscular anterior to the urethra Surrounded by thin capsule

Urethra	• Prostatic Urethra
	  – Widest and most dilatable
	  – Involuntary continence
	  – Transmission of semen into urethra
	• Membranous:
	  – Shortest, narrowest
	  – Least dilatable
	  – Voluntary continence
	• Spongy Urethra
	  – Surrounded by corpus spongiosum
	  – To External urethral orifice

Notes/Diagrams/ DOOOOODLE

Putting It All Together

Putting It All Together

YOU DID IT!!!

CONGRATULATIONS ON A JOB WELL DONE!!

WHITE COATS ARE COMING!!!!

(You're Gonna Miss This!!!)